Samuel Laing

Problems of the Future

and Essays

Elibron Classics
www.elibron.com

Elibron Classics series.

ISBN 978-1-4021-3566-8 (paperback)
ISBN 978-1-4212-9144-4 (hardcover)

This Elibron Classics Replica Edition is an unabridged facsimile of the edition published in 1893 by Chapman & Hall, London.

PROBLEMS OF THE FUTURE
AND ESSAYS

PROBLEMS OF THE FUTURE AND ESSAYS.

1500	copies	printed	September 1889.
1500	,,	reprinted	November 1889.
1000	,,	,,	December 1889.
1000	,,	,,	March 1890.
1000	,,	,,	September 1890.
1000	,,	,,	March 1891.
1000	,,	,,	January 1892.
1000	,,	,,	September 1892.
1000	,,	,,	February 1893.
1000	,,	,,	November 1893.

PROBLEMS OF THE FUTURE

AND ESSAYS

BY

S. LAING

AUTHOR OF "MODERN SCIENCE AND MODERN THOUGHT,"
"A MODERN ZOROASTRIAN," ETC.

ELEVENTH THOUSAND.

LONDON: CHAPMAN AND HALL, Ld.

1893

Richard Clay & Sons, Limited,
London & Bungay.

CONTENTS.

CHAPTER I.

SOLAR HEAT.

CHAPTER II.

WHAT THE UNIVERSE IS MADE OF.

CHAPTER III.

CLIMATE.

CHAPTER IV.

THE GLACIAL PERIOD.

CHAPTER V.

TERTIARY MAN.

CHAPTER VIII.

THE RELIGION OF THE FUTURE. AGNOSTICISM AND CHRISTIANITY.

PART I.

CHAPTER VIII. (*continued*).

PART II.

CHAPTER VIII. (*continued*).

PART III.

CHAPTER IX.

THE HISTORICAL ELEMENT IN THE GOSPELS.

CHAPTER X.

SCEPTICISM AND PESSIMISM.

CHAPTER XI.

CREEDS OF GREAT POETS.

CHAPTER XII.

ARMED EUROPE.

CHAPTER XIII.

TAXATION AND FINANCE.

CHAPTER XIV.

POPULATION AND FOOD.

PROBLEMS OF THE FUTURE
AND ESSAYS.

INTRODUCTION.

"Men, my brothers, men the workers, ever reaping something new,
That which they have done but earnest of the things that they shall
do."
TENNYSON'S *Locksley Hall.*

THE traveller in the Alps, after struggling up through dense fir woods, in which his view is limited to a few yards, emerges on grassy slopes, where swelling ridges and rocky peaks appear to bound the horizon. Weary and scant of breath, he thinks if he can surmount these his labour will be ended, and a free view enjoyed, with nothing but the vault of heaven above him. But no! when these heights are scaled, he sees before him ridge behind ridge of loftier summits, and in the background of all, the glittering peaks of Jungfraus and Matterhorns, standing out white and seemingly inaccessible, against the deep blue sky.

But if he is a practical mountaineer he knows that, grim as are the glaciers and precipices which girdle their icy fortresses, they are not invincible to human effort; and as the foot of man has stood on some of the loftiest summits, he feels assured that it will stand on those which remain unscaled.

So it is with modern science. For centuries it had to grope its purblind way through dense jungles of superstitious ignorance, where misty shapes of theological and metaphysical speculation obscured the real facts of the universe, or were mistaken for them. At length, and comparatively quite recently, the human intellect emerged into the light of day, and gaining the first heights, began to acquire accurate ideas of the true laws and constitution of the universe. The progress, once begun, went on at an accelerated rate, until in the last half century it has carried with it in an impetuous torrent old creeds and cherished convictions, like so much drift-wood floating on the surface of Lake Erie, when caught by the current which hurries it down the Falls of Niagara.

So irresistible and so wide-spread has been the advance of science, that at first sight we are perhaps disposed to overrate it, and to fancy, like Alexander, that no more worlds remain to conquer, or that, at most, a few unimportant territories are still unannexed. But the true man of science knows differently. He sees ridge still rising behind ridge, and at every step wider horizons opening, with distant peaks that still baffle the boldest climber.

But he no longer gazes at them with aimless wonder, or if he fails to understand them, invents a high-sounding phrase to disguise his ignorance. His faith is firm in the laws of Nature, and he feels assured that whatever lies within their domain is discoverable, and will, sooner or later, and probably sooner rather than later, be discovered.

In former works I have attempted to give some popular view of what modern science has actually accomplished in the domains of Space, Time, Matter, Energy Life, Human Origins, and other cognate subjects. In this, I will endeavour to point out some of the "Problems of the Future," which have been raised but not solved, and are pressing for solution.

In both cases I address myself to what may be called the semi-scientific reader. The advanced student of science will find little which he does not already know. Those who are ignorant of the first elements of science, and, like Gallio, care for none of these things, will scarcely understand or feel an interest in the questions treated of. But there is a large, and I believe rapidly increasing class, who have already acquired some elementary ideas about science and who desire to know more. Curiosity and culture are in effect convertible terms: the wish to know is the first condition of knowing. To many who are in this stage of culture, but who have neither the time nor faculty for following up closely the ever-widening circle of advanced thought, it may be interesting to get some general and popular idea of some of the unsolved problems which have been raised by modern science, and are occupying the thoughts of the men who lead its van.

In selecting a few among the many questions which have been thus raised, I have been guided by this principle. In the course of nature I must have left this earth before they have been solved. If the option were given me of paying it a short visit fifty or a hundred years hence, what are the questions which I should ask with the most eager curiosity, and to which I should expect to get a satisfactory reply?

They are partly scientific questions, respecting the age of the earth, the constitution of the sun and solar system; the ultimate nature of matter and energy, the beginnings of life, the origin and antiquity of man; partly religious, social, and political questions which are looming on the horizon and engaging the attention of thinking men.

I do not pretend to have exhausted the list, but I hope I may have done something to give definiteness and precision to the ideas of some of the educated public who are not specialists upon various questions which are now pressing forward and waiting for solution.

CHAPTER I.

SOLAR HEAT.

Difference between Astronomers and Geologists—The former say twenty, the latter two hundred millions of years—Argument of Astronomers—Amount of Heat received from Sun—How Supply kept up—Meteorites—Gravity—Method of Calculation—Result: Supply of Heat cannot have lasted more than ten to fifteen millions of years—Case of Geologists—Progress of the Science—Theological—Theologic-Scientific—Scientific—Uniformity of Conditions—Proved by Fossil Remains—By Temperature and Atmosphere—Assuming uniformity, time required—Instances—Solent River—Eocene Lake—Lake of Geneva—Coal Measures—Geology based on Facts—Mathematical Conclusions on Theory—If Heat comes from Gravity, where does Gravity come from—Gravity really unknown—Different Theories as to Solar Heat—Lockyer and Crookes—Sun-spots—Magnetic Storms—Conservation of Energy.

One of the most interesting and perplexing scientific problems of the day is that raised by the conflict between physicists and geologists as to the duration of solar heat.

Leading mathematicians, such as Sir W. Thomson and Helmholtz, assign twenty, or more probably, ten millions of years as the outside possible past duration of a supply of heat from the sun, sufficient to maintain the earth under conditions enabling it to support life. Lyell, and a majority of the best geologists, consider that one hundred to two hundred millions of years are required to account for the undoubted facts of geology since life began. Each side support their case by arguments, which, taken by themselves, seem conclusive. And yet the gap between the two is so wide that it cannot be bridged over by mutual concessions, and it is evident that there must be some fundamental error in the assumed data on one side or the other.

The mathematicians base their argument on the supply of solar heat. They say the present amount of heat radiated by the sun is a measurable quantity; the principle of the conservation of energy shows that this heat cannot be self-supplied, but must be a transformation of pre-existing energy; the only sufficient energy we know of is that of the mechanical force generated by the contraction of the sun as it cools. This, again, is a measurable quantity, and the outside amount of mechanical power generated by contraction of the sun's mass to its present volume by gravity, would not supply the present amount of heat for more than twenty millions, or more probably for more than ten or fifteen millions of years.

This forms a chain of reasoning, every link of which seems to be solidly welded. Let us examine each link in detail. The amount of solar heat received at the earth's surface has been carefully measured by Herschell, Pouillet, and other eminent observers, the principle being to intercept a beam of sunshine of known dimensions, and make it give up its heat to a known mass of water or other substance, measuring accurately the rise of temperature produced in a given time. The result is this: the heat, measured by Calories, or units of heat sufficient to raise the temperature of one kilogramme of water one degree Centigrade, received per minute by one square metre exposed perpendicularly to the sun's rays at the upper surface of the atmosphere, ranges from Pouillet's estimate of 17·6 to that of Forbes' 28·2 Calories, the difference arising mainly from the different allowance made for absorption by the atmosphere, and the highest estimate being proved by Langley's observations at a high altitude to be the most reliable.

From this it is easy to calculate the amount of heat received by the earth from the sun in a given time. Herschell puts it in this striking way. The amount of heat received on the earth's surface, with the sun in the zenith, would melt an inch thickness of ice in two hours and thirteen minutes. But, if it be assumed that the sun radiates heat equally in all directions, the earth intercepts only an almost infinitesimally small amount of this heat. In fact, only the proportion which the earth's surface bears to the surface of a sphere whose centre is in the sun, and its radius the distance of the earth from the sun, or about ninety-three millions of miles. This proportion is $\frac{1}{2,200,000,000}$. But even this minute fraction is sufficient to melt yearly, at the earth's equator, a layer of ice of more than one hundred and ten feet thick. So, as Sir W. Thomson puts it, if the sun were a mass of solid coal, and produced its heat by combustion, it would burn out in less than six thousand years. Of course this calculation depends on the assumption that the sun radiates heat equally in all directions into space. It is difficult to conceive how this can be otherwise, for, as far as we know, all heated bodies at the earth's surface do so, and all impulses which cause waves in an elastic medium, such as we know to be the case with heat and light, propagate these waves in all directions.

Assuming therefore that the sun gives out this enormous amount of heat, where does it come from, and how is the supply kept up, uniformly or nearly so, for millions of years? The law of the conservation of energy says, in effect, that something cannot be made out of nothing, and that all special forms of energy, such as heat, light, electricity, and mechanical power, are convertible into one another, and are simply transformations of one original fund of energy. If so, the sun's heat must be kept up by energy transformed into heat from some other form. It cannot be from combustion, which is a chemical action, for we have seen that a sun of solid coal would be burned out in six thousand years. It must be from mechanical force, which we know as a fact to be convertible into heat in a definite and ascertained proportion.

Now what are the sources of mechanical power known in the case of the sun? Two—the impact of aerolites, and the shrinkage of the sun as it contracts, which latter resolves itself into an effect of gravity.

Both are real causes. Aerolites fall on the earth and generate heat, the smaller ones, or shooting stars, being set on fire and burnt up by the friction of the atmosphere; the larger ones reaching the earth in masses of stone, singularly like those ejected from deep-seated volcanoes, and with their surfaces glazed by intense heat. If such meteors fall on the earth, it is reasonable to suppose that vastly more must fall on the sun, with its vastly greater surface and attracting power. And it is to be noted that comparatively small masses might generate large amounts of heat, for the amount of mechanical force, and therefore of heat, generated by arrested motion, increases with the square of the velocity. A body weighing 8·339 kilogrammes falling from a height which gave it a velocity of one metre per second, would generate one calory of heat, or enough to raise the temperature of one kilogramme of water by 1° Centigrade. But the same body moving with the velocity of a cannon-ball, or 500 metres per second, would generate two hundred and fifty thousand times as much heat; and if moving with a velocity of 700,000 metres per second, which is about the velocity with which a body would fall into the sun from the distance of the earth, the heat produced would be nearly two million times as great.

Sir W. Thomson has calculated that a quantity of matter equal to about one-hundredth of the mass of the earth falling annually with this velocity on the sun's surface, would maintain its present radiation indefinitely. It is clear therefore that if this amount of meteoric matter really falls on the sun its heat might be maintained. But many objections have been raised to such a supposition.

To explain the sun's heat we must have a cause that is not only sufficient to generate its total amount, but also one which generates it uniformly. If the sun were a target kept at an intense white heat by showers of meteoric small shot peppering into it, how is it that this stream of small shot is incessant and uniform?

Only small portions of the total meteoric mass revolving round the sun can be captured by it gradually, as their orbits are contracted. An extra supply, as some solid body or enormous comet with its attendant meteoric train falling into the sun, would raise its temperature above, while a deficient supply would lower it below the average, and a comparatively slight variation in the sun's temperature would destroy existing conditions of life on the earth.

Another objection to the meteoric theory is, that it would require such a large mass of meteoric matter revolving in space as might be expected to exercise a perceptible effect on the motions of the planets, both by the law of gravity and by the retardation due to a resisting medium. And this is specially true of the orbits of comets which approach the sun very closely. As meteors do not fall from a state of rest straight into the sun, but revolve round it with planetary velocities, they can only fall into it by being drawn inwards in gradually contracting spirals, until they reach a point where they impinge on the sun or its atmosphere. Hence a vastly greater amount of meteoric matter must be revolving round the sun in the space near it, than can be captured and generate heat in any single year. But several comets are known to have almost grazed the sun's atmosphere, and emerged from it to continue to describe their elliptic orbits and return true to time, as predicted by calculations based on the known laws of gravity acting on them from the sun and planets alone, in a non-resisting medium.

Consider what this means. Comets are bodies of such immense volume and extreme rarity that one of them got entangled among Jupiter's satellites and thrown out of its course, without affecting in the slightest perceptible degree the motions of those satellites. How could such comets, rushing closely round the sun with enormous velocities, avoid showing perturbations, if they encountered any considerable mass of meteoric matter?

The theory of meteorites, to which reference will be made in a future chapter, meets many of these difficulties, and strengthens the case for a meteoric origin of a large part of solar heat, but it hardly accounts for the uniformity of the supply, and is hardly yet so generally accepted as to supersede the older theory that the main source of the sun's heat is to be sought in the transformation of the mechanical energy of gravity, as its volume contracts.

Assuming this theory, the principle on which the supply of solar heat is calculated is the following. We know the amount of heat given out by each square metre of the sun's surface, and we know the height from which a given weight must fall to generate this heat when its motion is arrested. We know also that this heat will be the same whether the motion is suddenly or gradually arrested. Now in this case the given weight is that of a long narrow cone of matter, whose base is one square metre at the sun's surface, and its apex a point at the sun's centre. Knowing the sun's diameter and mean density, it is easy to calculate the weight of such a cone if we suppose it to be solid. Its weight is equivalent to that of 244,000,000 tons of solar heaviness at the sun's surface. To reduce this to terrestrial tons, and their equivalent in horse-power, we must allow for the difference of weight or gravity, at the respective surfaces of the sun and earth.

Reduced to terrestrial figures, in which one horse-power is 270 metre-tons per hour (*i.e.* a ton lifted 270 metres in an hour), the horse-power at the sun's surface is 10 metre-tons. But the radiation from each square metre of the solar surface in heat per hour is equivalent to 78,000 horse-power in energy, or to that of 780,000 metre-tons. An easy calculation shows that to supply energy at this rate for a year, our supposed cone of 244,000,000 tons must fall one metre in 313 hours, or about 35 metres in a year. Refined mathematical calculations are requisite to show how this result is effected, if we suppose, as is probable, that the mass of matter forming the sun, instead of being solid, existed first in the nebulous or gaseous state, and gradually contracted into a fluid mass in which convection currents are constantly carrying down surface layers which have become cooler by radiation, and replacing them by ascending currents from the hotter and denser interior. These calculations have been made by mathematicians of undoubted competence, with the result that the dynamical equivalent of the heat radiated from the sun in a given time is practically the same as if it were solid.

This result shows that if the sun has contracted to its present size, from a volume extending far beyond the orbit of the remotest planet, Neptune, it has furnished about eighteen million times as much heat as it now supplies in a year; and that with its present dimensions it must contract at the rate of 35 metres per year, or one per cent. of its radius in 200,000 years.

Allowing for the increasing density of the sun as shrinkage proceeds, the problem works out that if the sun's radiation of heat has been uniform for the last fifteen millions of years, the solar radius must then have been four times greater than it is now; and that if the present supply were maintained by shrinkage alone, for the next twenty millions of years, the sun must have shrunk to half its present size. But these figures must be greatly reduced by several considerations. They are based on Herschell's and Pouillet's figures for the total activity of solar radiation, but Forbes and Langley have shown that the allowance made for absorption of solar heat by the earth's atmosphere was insufficient, and that the real amount of heat radiated by the sun is greater than was supposed by Pouillet in the ratio of 1·7 to 1. This diminishes the past and future periods of solar radiation in the same proportion, reducing the past period from fifteen to nine millions of years, and the future from twenty millions to twelve. Moreover, when the sun's surface was four times larger, it must have given out more heat than at present, and more than existing conditions of life in geological times could support. If, therefore, the sun's shrinkage from gravity has been the sole or principal source of its supply of heat, it is difficult to see how life and the existing order of things on the earth can have lasted for more than ten millions of years at the outside.

So far the mathematicians seem to have it all their own way, and, as often happens when the plaintiff's case only has been heard, it seems to be conclusive. But what say the defendants—the geologists? They also base their case on an undoubted principle, and on undeniable facts. The principle is that of the uniformity of existing causes; the facts, those of actual experiment and observation.

Geology, in the pre-Lyellite days, passed through two stages, the theological and the theologico-scientific. The theological, which prevailed universally until the present century, was based on the belief that the book of Genesis, instead of being a sort of poetical prelude to a collection of ancient writings of religious and moral import, was a strictly literal and scientific narration of what actually took place, every word of which was imparted by a Divine revelation, which it was impious to explain away or to dispute. Geology was therefore confined very much to searching for facts in Nature confirming this narrative. Thus when fossil-shells were observed on mountain-tops, they were adduced as incontrovertible proofs of Noah's deluge; and even a sceptical and encyclopædic mind like that of Voltaire could only attempt to palliate this proof by suggesting that the shells were dropped from pilgrims' hats while crossing the Alps on their way to Rome. The period when such a ridiculous suggestion could be made by an accomplished scholar seems thousands of years from us, and yet it occurred in the last century. The naïve and infantile narrative of the Noachian deluge is now taken no more seriously than are the little wooden arks, with their contents of pigmy animals, which with other toys amuse the nursery.

The next stage was what may be called the theologico-scientific, when the facts and laws of Nature began to be recognized; but the old dogmatic faith was still so prevalent, that these facts and laws were viewed through a theological medium, and attempts were made to reconcile the Bible and science, by distorting the conclusions of science, and giving the statements of Genesis a general and allegorical, rather than a literal meaning. This was the era when days were expanded into periods, universal deluges contracted into local floods, and when miraculous catastrophes and creations were invoked *ad libitum*, to bring geological and zoological facts into some sort of possible accordance with the non-natural versions of plain words into which Scriptural texts were evaporated. This school included, in its time, some eminent men, such as Buckland and Hugh Miller, and it still lingers on the outskirts of science, as may be seen by Mr. Gladstone's essay on the *Proem to Genesis*. But with all the leaders of science it is quite extinct, and the prevailing tone of thought has become Darwinian, as universally as a century ago it was theological. Differences may exist as to the details of Darwin's theory, and the extent of its application in some of the more recondite causes of variation, but no one of any authority in science doubts that evolution, under fixed laws, is the key to the secrets of the universe, and that one original impress, and not perpetual miracle, or secondary interference, has been the real course of Nature.

In geology this conviction has been embodied in what is known as Lyell's Law of Uniformity. If any one wants to get a clear idea of what this means, let him go to the British Museum and look at a slab of sandstone from the Silurian formation. He will see precisely what he may see to-day on the sands of Southend or Margate. Ripple marks of a gently flowing or ebbing tide, worm castings, or even little pits showing where rain-drops had fallen on the wet sand, and these pits higher on one side than the other, showing the size of the drops, the force of the wind, and the direction from which it was blowing. The inference is irresistible that at this immensely remote period the winds blew, the rain fell, the tides ebbed and flowed, sand-banks were formed, and worms or sand-eels burrowed in them, as they do at the present day. Or look at a piece of chalk through a microscope, and you will find it mainly composed of the microscopic shells of a minute form of animal life, the

Globigerina, which, gradually falling to the bottom of a deep ocean like the finest dust, have accumulated strata more than a thousand feet in thickness. Precisely the same thing is going on in the Atlantic to-day, where deep-sea dredgings bring up a Globigerina ooze, which affords a safe bed for the submarine telegraph. Or take another instance. A shell called the Lingula, about the size of a small mussel, is found abundantly in the Silurian, and even in the earlier Cambrian formations; and another shell, the Terebratula, in the Devonian. Both are found living at the present day, not only of the same genus, but identically of the same species. It is evident that no great change can have taken place in the conditions of oceanic life since these mollusks lived and flourished in Silurian and Devonian seas.

Nor can the condition of the atmosphere have greatly changed since the time of the air-breathing Silurian scorpion, whose fossil remains show him to be scarcely distinguishable from the present scorpion.

In fact, the atmosphere affords one of the most conclusive proofs of the uninterrupted maintenance of existing conditions during an enormous period. When we say enormous time, the term is used with reference to any recent or historical standard as applicable to the period when geology practically commences; that is, with the first dawn of life disclosed by fossils in the Cambrian era, or beyond that with formations like the Laurentian, which can be clearly proved to be sedimentary and metamorphic. But no geologist ventures to extend this doctrine of uniformity beyond the date when fossils appear, or to deny that though the laws of Nature are the same, the conditions must have been totally different in the earlier stages of the planet, when it was cooling and condensing into its present form. Nor could he deny that, even within this comparatively recent period, there may have been changes of existing conditions, as we know indeed from the alternations between the Glacial period and those of higher and more uniform temperature. But his position is that such changes have been of the same order, and owing to similar causes as those which now prevail; and that when a known cause, given a sufficient time, will produce an effect, it is unphilosophical to assume miracles, catastrophes, or a totally different order of things, in order to reduce the time to some procrustean standard of theoretical prepossession.

To Sir C. Lyell belongs the credit of having established this doctrine of uniformity on an unassailable basis, and made it the fundamental axiom of geological science. By an exhaustive survey of the whole field of geology, from the earliest formations in which life appears, down to the present day, he has shown conclusively that while causes identical with, or of the same order as, existing causes, will, if given sufficient time, account for all the facts hitherto observed, there is not a single fact which proves the occurrence of a totally different order of causes. This, of course, applies only to the geological record commencing with the commencement of organic life on the earth, and not to the earlier astronomical period when the planet was condensing from nebulous matter, and slowly cooling and contracting. Nor does it imply absolute uniformity with existing conditions, for changes in climate, temperature, distribution of sea and land, and otherwise, have doubtless occurred from the slow operation of existing causes. But it excludes all fanciful theories of cataclysms, annihilating each successive era with its life, and introducing a new one;

earthquakes throwing up mountain chains at a shock; deluges sweeping over the face of the earth, and so forth, in which even eminent geologists used to indulge thirty or forty years ago. While no competent geologist of the present day would like to affirm positively that there may not have been, in past ages, explosions more violent than that of Krakatoa, lava streams more extensive than that of Skaptar-Jokul, and earthquakes more powerful than that which uplifted five or six hundred miles of the Pacific coast of South America six or seven feet, it may be doubtful if he could point out a single instance since the Silurian epoch where such was demonstrably the case.

Assuming the principle of uniformity, the time requisite to explain the facts of geology becomes a matter for approximate calculation. Not readily in years or centuries, for our historical measuring-yard does not extend beyond seven thousand years, when we find a dense population and high civilization already existing in Egypt; but in periods of which we can form some approximate idea.

To understand the full force of the evidence, it is necessary to study carefully the works of Lyell, Croll, Geikie, and other authorities on geology; but some idea of the sort of periods which are required for gauging Time back to the commencement of life may be arrived at from a few instances.

The tests of geological time are derived mainly from two sources—denudation and deposition. The present rate of denudation of a continent is known with considerable accuracy, from careful measurements of the quantity of solid matter carried down by rivers. The Mississippi affords the best test, both because the measurements have been made with the greatest accuracy, and because the conditions of the vast area drained by it and its tributary rivers afford a better average of the rate of continental denudation, including as it does a great variety of climates and geological formations, and being singularly free from exceptional influences. The rate thus deduced is one foot from the general surface of the basin in six thousand years. Now the measured thickness of the known sedimentary strata is about 177,000 feet. The proportion of sea to land is three to one, and the bulk of the deposition of the waste of land must have been laid down within a comparatively narrow margin of the sea nearest to land. On these data Wallace calculates that the time required to deposit this 177,000 feet would be 28,000,000 years, taking the rate of denudation at one foot in 3000 years, or 56,000,000 years, taking the rate deduced from the Mississippi. But it must have been much more than this, for the stratified rocks are to a great extent composed of the *débris* of older strata, which have been deposited, upheaved, and again denuded. Most of the known stratified rocks must have been in this way denuded and deposited many times over. Nor is there any good reason for supposing that the rate of denudation was materially greater in former, than in recent geological eras. On the contrary, the recent Glacial period, by grinding down solid rock into loose materials, and, as the ice and snow melted, causing more torrential inundations of rivers, must have tended to accelerate denudation.

Another proof of the enormous amount of solid rock which has been removed by denudation, is afforded by the faults or cracks in the earth's crust, which have in many cases displaced strata by thousands of feet, all traces of which displacement have been subsequently planed down to one uniform surface. Thus the great fault which separates the Silurians of the south of Scotland from the Devonian and Carboniferous region to the north of it, is estimated by the Geological Survey at 15,000 feet. A mountain mass of this height, terminating in a steep cliff at the fault, must have existed to the south of it, composed mainly of the Devonian strata which now stop abruptly at the north edge of the fault. At present there is no inequality of the surface at the fault, and therefore 15,000 feet or nearly three miles of rock must have been removed by denudation. And what is most important, the time in which this denudation was effected is fixed as having occurred in the interval between the Devonian and Carboniferous periods, for while no trace of the former formation is found south of the fault, the limestones and coal-measures of the latter lie directly on the Silurian rocks. At the rate of denudation deduced from the Mississippi observations of one foot in 6000 years, the removal of those three miles of rock would have required 90,000,000 years for the interval between two of the geological formations.

Croll, in his recent work on Stellar Evolution, gives a number of similar instances, one in the Appalachian Mountains, in which the vertical displacement is not less than 20,000 feet, bringing the upper Devonian strata on one side opposite to the lowest Cambrian on the other. Of course we cannot assume these enormous intervals of time to have actually occurred, but they are quite sufficient to show the absolute impossibility of reconciling geological facts with any estimate of the duration of solar heat derived from the theory of contraction by gravitation.

Take another instance from a more recent period. There is a dried-up Eocene lake in North America, which once occupied an extensive area in the States of Wyoming and Nebraska, formed by streams running down from the Wahsatch, Uintah, and other mountain ranges, which are Eastern outliers of the great backbone of the continent—the Rocky Mountains. It was gradually silted up by a deposit of more than 5000 feet, or a mile thick of clays and sands, a portion of which has since been carved by the rain and weather into the singular formation of isolated castle-like bluffs and pyramids, known as the "bad lands." It is full of remains of Eocene animals, often of huge size and of a peculiar type. How long must it have taken to silt up a lake larger than Lake Superior, with tranquil deposits of fine mud and sand? The nearest approximation towards such a calculation is afforded by the silting up of the Lake of Geneva. Swiss geologists have calculated from the rate of advance of the delta in historical times, that it may have taken 90,000 or 100,000 years since the silting process began, which could only be after the first Rhone glacier, which once extended to the Juras, had shrunk back to the head of the lake. This calculation may be right or wrong, but certainly a vastly longer time must have been required to silt up a vastly larger lake to a depth of 5000 feet. And if anything, one would expect the process of silting up to have been slower, for in the Eocene period there were no glaciers, or melting snow-fields, to accelerate the denudation which must have gone on *pari passu* with the deposit. If we consider the geological evidence more in detail, we find it all pointing to the same conclusion of immense antiquity.

Thus, if we take the coal-measures which form only a part of one formation—the Carboniferous. Each seam of coal consists of the consolidated debris of a forest. With every seam there is an under-clay in which the trees and ferns grow; and a roof of shale or sandstone deposited on it when this floor was submerged. The bulk of the coal is frequently composed of the microscopic spores of the ferns and club-mosses which formed the principal vegetation of these forests. The time required is therefore that for the accumulation of vegetable matter, consisting mainly of fine spore-dust, to a depth sufficient, under great compression, to give the seam of solid coal. In Nova Scotia, and other localities, the coal-measures have a thickness of 12,000 feet, made up of seam upon seam of coal, each with its under-clay and roof, implying a separate growth, submergence, and elevation.

Sir J. Dawson and Professor Huxley, who have studied the subject minutely, calculate that the time represented by the coal-measures alone would be six millions of years. In other words, the time required for this one subordinate member of one geological formation, would be half the total time assigned by Thomson and Helmholtz for the total possible past duration of the present supply of solar heat.

Those who fully consider and appreciate any one of these instances will not be astonished to hear that Sir C. Lyell, after carefully going over and summing up the various lines of evidence afforded by the 100,000 feet of stratified and fossiliferous formations above the

Cambrian, came to the conclusion that two hundred millions of years was the probable, and one hundred millions the *minimum* possible duration of the existing order of things that would explain the facts. And all subsequent discoveries, and the best geological opinions, go to confirm this estimate. Thus, when Lyell made his estimate, the great Laurentian system of gneissic and other rocks which underlie the Cambrian was scarcely known, or assumed to be a primitive portion of the earth's crust of Plutonic origin. But it is now clearly proved to be bedded, and therefore an aqueous deposit from the denudation of older rocks, though the minor signs of stratification have disappeared, owing to metamorphism under heat and pressure. This at once adds 30,000 feet to the known thickness of deposited strata. It is not positively known to have contained life, for with the doubtful exception of the Eozoon Canadiense, the fossils, if any, have disappeared during this process of metamorphism; but it contains indirect evidence of life on the most extensive scale. Thus, great quantities of graphite or plumbago are found in it, and as ordinary coal can be traced first into anthracite and then into graphite, the inference is strong that the Laurentian graphite must, like coal, have originated from masses of vegetable matter. It contains also great beds of limestone, similar to those which, in later formations, are known to have originated from the remains of corals and other hard parts of marine animals, which derived their skeletons from calcareous matter dissolved in sea-water. Large beds of iron ore are also found, which, in later formations, owe their origin to the solution of peroxide of iron and its deoxidation by organic agency. There is thus, therefore, evidence of the existence of life on a vast scale in this lowest of all formations, which of itself adds more than a fourth to the thickness of the whole of the previously known deposited strata of the earth's crust, and therefore to the time presumably required for their deposit.

And yet, as we have seen, mathematicians affirm with equal confidence that Lyell's figures must be divided by at least ten, or probably by twenty, to arrive at the ten millions of years which is their estimate of the time for which the sun has given out its present life-sustaining amount of light and heat, and this short period has to provide not only for geological time, but for the far larger time during which the earth was passing through its earlier stages, and condensing from a gaseous vapour.

It is evident that there must be some fundamental error on one side or the other, which some day will be detected, for the laws of nature are uniform, and there cannot be one code for astronomers and another for geologists. I am inclined to think that the error will be found in some of the assumptions of the physicists. The data of geology seem more certain and more capable of verification by an appeal to facts. Thus, the rate at which rocks waste away, and lakes silt up; the amount of solid matter carried down by rivers, and the number of feet or inches per square mile thus denuded in a given time, are all matters of approximate and tolerably accurate observation and calculation. But of the nature and constitution of the sun we really know very little, and are only beginning to get some glimpses of them during the past ten or twenty years by the aid of the spectroscope. The sun, as we see it, is not fluid, for if it were its rotation must make it protuberant at the equator, which it is not. It is not solid, for if it were its equatorial region could not rotate, as it does, more rapidly than that nearer the pole. We know its apparent volume and its mean density; but we do not know how this density is distributed. The conditions of matter under such extreme temperature and pressure are quite conjectural. For aught we know to the contrary, the sun may have a nucleus much smaller and much heavier than we are in the habit of assuming.

Above all, what makes me distrust these mathematical calculations respecting the sun's heat is, that they do not really solve the problem, but only remove it one step further back. Heat, they say, can be nothing but transformed mechanical power; but where does the mechanical power come from? From gravity. And where does the gravity come from? They cannot tell. It is the old Hindoo cosmogony over again. The world rests on an elephant; the elephant on a tortoise. But what does the tortoise rest on?

We are accustomed to speak of gravity as the one well-known and established fact of the universe. And so it is as regards the various motions which result from it, and the fact of its being an attribute of all matter from atoms to stars. But of its real essence and *modus operandi* we know nothing; less even than in the case of some of the other forms of energy into which it can be transformed. In the case of light, for instance, we know that it is caused by waves or vibrations of an exceedingly elastic and imponderable medium or ether diffused through space. We can measure and count these vibrations, and know the velocity with which the light-wave travels, and trace its effects from impact on the eye, through the retina and optic nerve up to the cells of the brain.

But in the case of gravity we know none of these things, and cannot even form a conception of how one mass of matter can act upon another, without connection and apparently without requiring time for the transmission of the impulse. Is it a pulling or a pushing force? We do not even know this, and are not one whit advanced beyond the saying of Newton that he could not conceive how one body could act on another without some physical connection between them.

It seems to me that Sir W. Thomson starts from the assumption that gravity is the one fundamental form of energy from which all other forms, such as light and heat, are derived by transformation. But what a mere drop in the ocean is the energy of gravity compared with the atomic and molecular energies, which now in a latent and now in an active form build up the universe of matter? How incalculably small must the gravity of the sun be, compared with the sum of the energies of the atoms of which its mass is composed.

If it were permissible to hazard a conjecture where there is no proof, it would be that gravity may turn out to be one, and that by no means the most important, manifestation of the primitive fund of energy, which underlies the atoms of which all matter is composed.

Various ingenious attempts have been made to explain the cause of gravity, as that of strain or stress of some intervening medium, or space-filling, incompressible fluid; or by Le Sage's theory of infinite impacts of ultramundane corpuscles, partially screened in the direction in which gravity acts by the bodies which attract one another. But Clark Maxwell and other accomplished mathematicians have shown fatal objections to all these theories, and Tait in his *Properties of Matter* sums up the latest results almost in the identical words used by Newton in his letter to Bentley—" In fact, the cause of gravitation remains undiscovered."

Again, who can tell what is the constitution of the infinite space through which our solar system and the universe of visible stars are travelling, with a velocity which has been estimated in some cases as high as two hundred or even four hundred miles per second?

These facts of the proper motions of the stars, and especially of what are known as the " runaway stars," seem conclusive against the assumption that gravity is the sole and primitive form of energy, from which all other forms, such as heat and light, are derived by transformation. These star-motions are apparently in straight lines, in a variety of directions, and the velocities are such that it is impossible to account for them by any conceivable action of the force of gravity. Professor Newcomb has shown by mathematical calculation that the gravitation of the whole universe, assuming it to contain 100,000,000 of stars, each on the average five times larger than the sun, would require to be sixty-four times greater than it really is, to have given one star (1830 Groombridge) the velocity of 200 miles per second which it actually possesses, or to be able to arrest its flight through space. Of course this applies with greater force to a star like Arcturus, moving with a velocity of 400 miles per second. The amount of energy of a star like this, whose volume has been computed to be eleven times greater than that of the sun, moving with a velocity of 400 miles per second, must be enormously greater than any energy exerted by it in the form of gravitation, and if its motion were arrested, the heat engendered must be in an even larger proportion, seeing that it depends on the square of the velocity, than any heat which could be supplied by its gradual contraction, on the theory applied by Thomson and Helmholtz to solar heat.

After all, what do we really know of the contents of space except this, that it contains a vast number of stars which are suns like ours, scattered at enormous distances from one another, and also innumerable meteorites? And also this, that the phenomena of light and heat prove the existence of waves of known dimensions vibrating with known velocities, and transmitted at a known rate; which waves compel us to assume a medium or ether with certain calculable qualities. But these qualities are so extraordinary that it may almost be doubted whether such an ether has a real material existence, and is anything more than a sort of mathematical entity. Its elasticity must be a million million times that of air, which, as we know, is equal to a pressure of about 15lbs. to the square inch; the number of its oscillations must be at least 700,000,000,000,000 in one second of time; and it must be destitute of any perceptible amount of the ordinary qualities of matter, for it exerts no gravitating or retarding force, even on the attenuated matter of comets moving through it with immense velocities.

Beyond this we can only conjecture that space may contain a number of larger meteors or dark suns, rushing through it in all directions, and possibly in the state of dissociated atoms the elements of substances such as carbon and oxygen, which are locked up in the earth's crust through the medium of life and vegetation, in vastly greater quantities than could be afforded by any conceivable supply derived from the atmosphere. And it may be conjectured also that variations of temperature may exist in different regions of space, helping to account for the secular variations of temperature at the earth's surface, such as are shown by the Glacial period or periods.

Even if we confine ourselves to the sun itself, leaving these cosmic speculations to be discussed in a subsequent chapter, we find the greatest uncertainty prevailing as to the conditions under which it exerts and generates heat. Thus, Professor Young says, "The sun's mass, dimensions, and motions are, as a whole, pretty well determined and understood; but when we come to questions relating to its constitution, the cause and nature of the appearances presented upon its surface, the periodicity of its spots, its temperature, and the maintenance of its heat, the extent of its atmosphere, and the nature of the corona, we find the most radical differences of opinion."

Take the case of the spots. These were originally attributed by Herschell to cyclones in the sun's atmosphere, showing us glimpses, as through a funnel, of a cool and dark solid body below; by others they have been thought to be splashes caused by the downfall of large masses of meteoric matter; by some to be volcanic eruptions throwing up vast scoriæ; and finally, as the most probable solution, to be great whirlwinds, or cyclonic convection currents, by which the cooler gases of the sun's atmosphere are sucked down and replaced by hotter gases from the interior. But none of these theories give an explanation of the observed fact that these sun-spots have a regular *maximum* and *minimum* period of about eleven years. Nor do they give the slightest clue to the other remarkable fact that the outburst of large sun-spots often produces an apparently instantaneous effect on the earth's magnetism; causing electric telegraphs to write with a tongue of fire, magnets to oscillate violently, the Aurora Borealis to appear, and otherwise indicating what is known as a magnetic storm.

It is pretty clearly established that the spots are cooler than the sun's general surface, but not sufficiently so as to affect its general temperature, or the course of the seasons upon the earth; but the far more inexplicable effect upon terrestrial magnetism is attested by too many observations to be at all doubtful.

This opens up a new and quite unexplained field of speculation as to the sun's electric energy. The physicists, who treat the attractive form of gravity as the sole cause of the sun's energy, and convert it all into heat, take no account of the energy which manifests itself as a repulsive force, and takes the form of electricity. And yet electricity is one of the transformable manifestations of energy as much as heat or mechanical power, and the phenomena of comets' tails are sufficient to show that, under certain conditions, the sun can exercise an enormous repulsive force. The question also may be raised whether, after all, it is certain that heat is radiated out in all directions, so that out of 1,000,000 units of the life-giving energy of the sun, 999,999 are absolutely wasted in space, and one only is utilized. Electricity, so far as we know, cannot exist without two opposite poles, implying reciprocal action. Do the sun-spots, which affect the earth's magnetism, radiate out an equal amount of magnetic energy in all directions into space? If not, how can we be sure that heat, into and out of which electricity and magnetism can be transformed, does so?

As Professor Young observes, "perhaps we assume with a little too much confidence that in free space radiation does take place equally in all directions," and he asks "whether the constitution of things may not be such that radiation and transfer of energy can take place only between ponderable masses; and that too, without the expenditure of energy upon the transmitting agent (if such exist) along the line of transmission, even in *transitu.* If this were the case, then the sun would send out its energy only to planets, meteors and sister-stars, wasting none in empty space; and so its loss of heat would be enormously diminished, and the time-scale of the planetary system would be correspondingly extended."

The same difficulty applies in the case of gravity. We only know it as an attractive force reciprocally exerted between two bodies in the proportion of their masses and inverse squares of distances. Is it radiated out in all directions into empty space, where it meets with no reciprocally attracting body? This affects not only the permanent maintenance of the supply of gravity, but goes even deeper to the fundamental axiom of all modern conceptions, whether scientific or philosophical, of the universe, viz., the "Conservation of Energy." You cannot make something out of nothing; you cannot *create* energy or matter, but only *transform* them. Good; but how about that which is one of the principal manifestations of energy in the universe—that of gravity? You can catch limited portions of it, transform them into mechanical power, and then backwards and forwards as you like into heat, light, chemical action, electricity and magnetism, neither losing nor gaining a particle of the original energy by any of these transformations. A water-wheel may turn a dynamo, which generates electricity that may be stored in accumulators, and turn a wheel a hundred miles off; and, if you could eliminate waste and friction, the second wheel would give out exactly what the weight of the falling water put into the first one. But whence came the gravity which made the waterfall and the wheel turn? Was it itself a transformation of heat or electricity? If not, what was it, and how came it there? If Thomson and Helmholtz assume an infinite fund of energy in the form of gravity to account for heat, why shall they not as well assume an infinite fund of heat to account for gravity? And if heat is dissipated by use until it is exhausted, or reduced to one stationary average of temperature, and worlds and suns die, why should gravity be gifted with perpetual youth, and escape the general law of birth, maturity and death?

These are problems which the present cannot answer. Possibly the future may, but in the meantime we shall do well to keep a firm footing on solid earth, and rely on conclusions based on ascertained facts and undoubted deductions from them, rather than on abstract and doubtful theories, even if they are presented to us in the apparently accurate form of mathematical calculation. Or, to bring this chapter to a practical result, we shall be more likely to arrive at just views respecting the constitution of the earth and its inhabitants by following Darwin and Lyell as our guides, than by accepting astronomical theories which would so reduce geological time as to negative the idea of uniformity of law and evolution, and introduce once more the chaos of catastrophes and supernatural interferences.

CHAPTER II.

WHAT THE UNIVERSE IS MADE OF.

Shooting Stars: their number, velocity, size—Connection with Comets—Composition—Spectra—Meteorite Theory—Genesis of Stars and Nebulæ—Further stage of Theory—Impact Theory—Dark Suns in Space—Temperature of Visible Stars—Their proper Motions—New Stars—Variable Stars—Facts better explained by Impact Theory—Laplace's Theory—Based solely on Gravity—Not inconsistent but insufficient—Even Impact Theory not last step—Stony Masses made of Atoms—What are Atoms—Chemical Elements—Attempts to reduce them to one—Hydrogen—Helium—Mendelejoff's Law—Atoms Manufactured Articles—All of one Pattern—Vortex Theory—What behind Atoms—The Unknowable.

WHAT is the universe made of? Such is the question which has been asked in many ages and countries by earnest men looking up at the starry vault of heaven, and down into the recesses of their own minds. The latest reply of science is, that it is made of shooting stars. The idea may seem paradoxical to those whose only knowledge of shooting stars is derived from an occasional glimpse on a clear night, when they have seen something like a small rocket flash across the sky, apparently close to the earth, out of darkness into darkness, reminding them of some human life—

"Qui file, qui file et disparait"

And yet it is now presented to us by eminent authorities, and supported by a long array of serious scientific arguments.

What do we know as certain facts with regard to shooting stars?

1. They are vastly more numerous than any one has an idea of who has not watched them continuously for many nights. Astronomers who have kept a record for many years assure us that the average number seen by one observer at one place on a clear moonless night is fourteen per hour, which is shown by calculation to be equivalent to twenty millions daily for the whole earth. But the number of meteorites met with by the earth can only be the minutest fraction of those circulating in space. The orbits of those we see do not coincide with the ecliptic, but lie in planes inclined to it at all sorts of angles, and apparently having no relation to the plane in which the earth travels round the sun, or to the solar system. The chances are almost infinite against our minute speck of a planet encountering any single meteor, or stream of meteors, thus traversing space in all directions, and as we do encounter some seven thousand millions of these small bodies in the course of each year, their total number must be an almost infinite multiple of this large figure. Moreover, the sun, with its attendant system, is rushing through space with a velocity of some 20 miles per second, and therefore carrying us into new regions of the universe at the rate of some six hundred millions of miles per annum, and yet meteorites are met with everywhere. Granting, therefore, that each separate meteorite may be very small, not exceeding on the average a fraction of an ounce in weight, and that even in meteor streams they may be,

as some astronomers have calculated, 200 miles apart, the aggregate amount of this meteoric matter in space must be practically almost infinite.

2. They are not terrestrial phenomena moving in the lower atmosphere, but celestial bodies moving in orbits and with velocities comparable to those of planets and comets. Their velocities are seldom under ten miles a second or over fifty, and average about thirty, the velocity of the earth in its orbit round the sun being eighteen.

3. They are of various composition, comprising both a large majority of smaller particles which are set on fire by the resistance of the earth's atmosphere, and entirely burned up and resolved into vapour long before they reach its surface; and a few larger ones, known as meteors, which are only partially fused or glazed by heat, and reach the earth in the form of stony or metallic masses.

4. They are not uniformly distributed through space, but collect in meteoric swarms or streams, two at least of which revolve round the sun in closed rings which are intersected by the earth's orbit, causing the magnificent displays of shooting stars which are seen in August and November.

5. They are connected with comets, it having been demonstrated by Schiaparelli that the orbit of the comet of 1866 is identical with that of the August swarm of meteors known as the Perseids, and connections between comets and meteor streams have been found in at least three other cases. The fact is generally believed that comets are nothing but a condensation of meteorites rendered incandescent by the heat generated by their mutual collision when brought into close proximity.

6. Their composition, as inferred from that of the

larger meteors which reach the earth, is identical or nearly so with that of matter brought up from great depths by volcanic eruptions. In each case they consist of two classes: one, composed mainly of native iron alloyed with nickel, the other of stony matter consisting mainly of compounds of silicon and magnesium. Most meteorites consist of compounds of the two classes, in which the stony parts seem to have broken into fragments by violent collision, and become embedded in iron which has been fused by heat into a plastic or pasty condition.

At this point our positive knowledge of meteorites from direct observation ceases, and we have to be guided by the spectroscope in further researches. This marvellous instrument enables us, by analyzing the light transmitted to us by all luminous objects however composed and however distant, to ascertain their composition as accurately as if portions of them had been brought down to earth and could be analyzed in our laboratories. We can tell whether they are gaseous, liquid, or solid; whether they shine by intrinsic or reflected light; and by comparing the lines in their spectra with those of known terrestrial elements, whether they contain those elements, or are made up of matter in a state unknown to us. The first result of spectroscopic discoveries was to establish the fact that the sun, stars, nebulæ, comets, and meteorites, all show such an identity in their spectra with some one or more of those of terrestrial elements, as to leave no doubt that the composition of matter is uniform throughout the universe.

Further experiments, of which Mr. Norman Lockyer's paper, read to the Royal Society, affords the latest and most complete summary, carry this knowledge farther.

They show that spectra are not fixed and invariable, but change according to the conditions of heat, pressure and otherwise, affecting the bodies from which the spectra are given out. Thus the spectrum of a comet in perihelion, when its component parts are crowded together and intensely heated by the sun, is very different from that of the same comet when it is at a great distance from the sun, either in advancing towards it or receding from it. Thus the spectrum of the great comet of 1882 when nearest the sun exhibited many of the lines obtained in the laboratory from the vapours of sodium, iron, and magnesium at the temperature of the Bunsen burner. As it receded the lines gradually died out until a very few were left; and in the comet of 1886-7, when last seen, all had died out except one line of magnesium. Thus carbon also, which is such an important ingredient in organic life, appears and disappears in cometary spectra according to the conditions of pressure and temperature.

What Mr. Lockyer has done is to show that all the varied spectra and classes of spectra, given out by suns, stars, nebulæ, comets and shooting stars, can be reproduced from actual meteorites which have fallen to the earth, by experiments in the laboratory, with the exception only of those of stars which, like Sirius, are glowing at a transcendental temperature far exceeding that of our sun, and which cannot be approached by the electric arc in any form of intense heat which can be obtained in our present earth. Thus the "spectrum of the sun can be very fairly reproduced (in some parts almost line for line) by taking a composite photograph of the arc spectrum of several stony meteorites between iron meteoric poles."

We are now in a position to understand the

meteorite theory of the universe. Granted that the number of meteorites in space is practically infinite, and that they tend to coalesce into streams, their collisions supply an equally unlimited fund of heat upon which we can draw at pleasure. The amount of heat developed by each collision is the transformed energy of the mechanical force. This force, and consequently this heat, increases with the square of the velocity. Thus, if a tropical hurricane, moving at the rate of 100 miles an hour, uproots trees and levels houses, the same mass of air moving with the mean meteoric velocity of 33⅓ miles per second, would exert a force of one hundred and forty-four million times greater. We know from the explosion of dynamite that when a gas expands very much quicker than the air can get out of its way, the effect is as if the blow of a tremendous steam-hammer were inflicted on an unyielding anvil; and we can readily conceive, therefore, how meteorites are almost invariably burnt up and dissipated, even in the rare air of the upper atmosphere, and how their repeated collisions in space might generate any required amount of heat.

Suppose, therefore, in the beginning of things, space filled by an innumerable multitude of these little stony masses, composed of the one, or possibly two or three, primitive elements of matter, moving in all directions, with immense though different velocities, coalescing into streams and colliding, we have a basis out of which suns, stars, planets, satellites, nebulæ and comets might be formed. The looser aggregations, giving fewer collisions and less heat, form comets and nebulæ, and the clash of two mighty streams gives us suns like Sirius in a state of intense luminosity and temperature. As these cool and contract by radiating out their heat, they pass into the second stage of stars of which our sun is one, still glowing with heat and light, but cooled down to a point at which the primitive elements can combine and form secondary ones, which can be detected by the spectroscope, and identified with those with which we are familiar as chemical elements upon earth. As cooling proceeds, they pass from the white-hot into the red-hot stage, and finally into the cold and lifeless non-luminous stage of burnt-out suns. Not, however, necessarily to die, for in the chances of infinite time these dead and invisible masses may collide together, and at a blow regain their youth, and commence the cycle anew as suns of the first order.

There is grandeur in the idea which, to a certain extent, reproduces what the kinetic theory of gases teaches as to the clash of innumerable atoms darting about in all directions, producing the temperature and pressure of a gas in a confined space. Only here, instead of atoms—so small that one of them is of the size of a rifle bullet, compared to the earth—we have stony masses for atoms, stars and nebulæ for molecules, and instead of glass jars or bladders, the whole universe.

This, however, is only the first stage of the theory. What are these little stony bodies, and how did they come there? The only answer we can give is derived from the constitution of those larger meteor-stones which actually fall on the earth and can be examined. They have invariably the appearance of fragments torn from larger bodies by collisions or explosions, and there is no reason for doubting that what they appear to be they are.

This carries us back to the impact theory of which

a full account is given in the work recently published by Dr. Croll on Stellar Evolution. It supposes that for an almost infinite time, an almost infinite number of dark stars, or cold and non-luminous solid bodies of stellar magnitude, have been rushing about in an unlimited space in all directions, and with enormous velocities. Occasionally they collide, and, as mechanical principles show, generate an intense heat, more than sufficient to convert their whole mass into glowing gas, at a temperature which may possibly dissociate its atoms, with the exception of some fragments from the shattered surfaces which are thrown off into space by the sudden generation of explosive gas. That there really are such dark suns rushing through space appears certain from what we know respecting the constitution of the visible stars. We find them exhibiting all ranges of temperature, from the intense heat of the white stars like Sirius, to that of the duller red stars like Arcturus, our own sun occupying an intermediate position; while our moon affords an example of a dead world, which from its smaller size has cooled more rapidly. As the moon is, so must the red stars inevitably become in a sufficient number of millions of years, if the laws of nature continue uninterrupted. And their proper motions, rushing through space in different directions with velocities ranging up to 400 miles per second, must continue after they have become dark, as long as the first law of motion holds good, that bodies in motion cannot generate changes of motion of themselves, but must continue to move forwards in a straight line unless acted upon by some external force.

Among bodies thus rushing in different directions collisions must occasionally occur, and it is a matter of simple calculation that the mechanical force converted into heat by such collisions, is amply sufficient to produce any temperature that may be required to create new suns and nebulæ, and to account for all the phenomena which are actually observed.

Moreover, the existence of such dark bodies is established by direct observation. That fragmentary masses, weighing several cwts, come in from space and fall upon the earth is a *fact.* So also is it a *fact* that bright stars, some of them like the famous new star in Cassiopæa, brighter than stars of the first magnitude, suddenly blaze out and gradually disappear. The impact theory accounts for this, while the nebular theory, or any hypothesis based solely on the contraction of a mass of nebulous vapour under the law of gravity, entirely fails to do so. Again, the phenomena of variable stars can best be explained by assuming either that such stars pass periodically through dense streams of meteoric matter, increasing their light, or else that large dark bodies are periodically interposed between us and the stars, and thus diminish it. The constitution also of comets, and of many nebulæ, as disclosed by the spectroscope, is far better explained by the impact than by the nebular theory. In fact, it is inconsistent with the latter theory, which can give no account of comets, meteorites, or other phenomena, which imply small dissociated portions of matter, moving in streams or aggregating in nebulæ, and rushing with immense velocities in paths inclined to each other at different angles, and which have no relation to the rotating plane of the solar or any other system. Even within the limits of the planetary system there are many facts which are better explained by the theory of impact than by that of contraction. For instance the great differences in the inclination of the axes of rotation of many planets and satellites to the plane in which they revolve about the sun and their primaries. But after all there is no real inconsistency between the impact theory and that of Laplace. The former takes up the history of the universe at an earlier stage, and supplies a mass of gas or cosmic matter, at a higher temperature, and with that temperature longer maintained by repeated collisions and indraught of meteorites, than is assigned to it by the nebular hypothesis, but ultimately a great deal of this gas must resolve itself into such a medium as Laplace supposes, contracting and forming whirls under the operation of gravity. The triumphs of mathematical science deduced from Newton's law of gravity were so signal, that it is not surprising that it should have been assumed that gravity, and gravity alone, was the fundamental law which would explain everything. But, as often happens, increasing knowledge has rendered many things uncertain which appeared to be certain. Problems which seemed simple have become complex, and it has become apparent that the universe contains many forms of motion, and many manifestations of energy, which cannot be explained by the laws of gravity. For instance, the runaway stars, the world of meteorites, the proper motions of molecules and atoms, and the requisite duration of solar heat to account for the undoubted facts of geology. The law of gravity and the nebular theory were a great step towards reducing the phenomena of the universe to one great uniform law; but the theory of impact takes up the history at an earlier stage, and carries us one step further towards infinity and eternity. If the whole stellar universe is not, so to speak, the crop of a single season, but an indefinite succession of crops, stars being born and dying, dying and being renewed, without appearance of a beginning or an end, the vista of existence is vastly enlarged. But even this is not the last step towards the unknowable. Granted that these dark suns are facts, they are not ultimate facts. They are matter, and matter is made up of molecules, and molecules of atoms. Judging from the fragments which reach the earth, and the teachings of the spectroscope, meteoric matter is composed of a few atoms identical with those which are the most common elements of terrestrial chemistry. Hydrogen, nitrogen, sulphur, iron, nickel, calcium, silicon, and aluminium, are the principal, if not the sole constituents of meteoric stones, and are those the lines of one or more of which appear in the spectra of stars, nebulæ, meteors, and comets, according to their conditions of temperature and pressure. What then are these atoms? There are some seventy of them known to chemists as ultimate elements; that is to say, which are not further resolvable by any means available in our laboratories. But no one can suppose that this is really the ultimate fact, and that original matter really consists of seventy indivisible units, ranging in weight from the 1 of hydrogen to the 240 of uranium, and more than half of them consisting of exceedingly rare elements, which play no appreciable part in the construction of any form of matter. The mind refuses to accept the conclusion that such little mole-hills as yttrium, zirconium and gallium, only known as minute products of a few of the rarest minerals, really present unsurmountable obstacles to the science which has scaled Alps, measured light-waves, and weighed stars.

Accordingly, constant attempts are being made to

reduce atoms to one simple element, and to one comprehensive law. The problem is not yet solved; but it is being attacked on various sides, and almost every day brings us nearer towards a solution. Hydrogen first put in a claim to be the primitive element, as being the lightest, and it is remarkable that the weight of a very large proportion of the other elementary atoms is an exact multiple of that of the hydrogen atom. The spectral lines of hydrogen are also the last seen in those of the hottest stars, where all secondary combinations may be supposed to be dissociated. This hydrogen theory, which was first proposed by Prout, can hardly be said to be firmly established, as there are some important elements, such as chlorine and sodium, which do not correspond with the law of being simple multiples of hydrogen. Still the agreement is too close in a number of cases to be accidental, and the latest researches show that by halving the hydrogen atom, that is, supposing this atom to be composed of two-linked atoms, the deviations from the law may be reduced within limits which may be fairly attributable to errors in the delicate operations requisite for fixing atomic weights. Mr. Crookes suggests that helium, which is only known from a single line in the solar spectrum, and which is apparently lighter than hydrogen, may be this half-hydrogen-atom, and thus be the ultimate element out of which all other atoms are manufactured. For Herschell and Clark Maxwell both arrived at the conclusion that "atoms bear the impress of being manufactured articles."

It is, in fact, certain that some relation exists among them, for the Russian chemist Mendelejeff has shown that if the atomic weights of the known elements are arranged in a consecutive order, they show what is called a periodical law. That is, the other qualities of atoms, such as specific heat, affinity, atomicity, &c., rise with the weights up to a certain point, then fall, then rise again, and so describe a sort of zig-zag line like those we see of the readings of the barometer on a weather chart. Only this atomic zig-zag seems to follow a certain law, so that groups of elements which have similar qualities recur at nearly fixed intervals.

The meaning of this law is not yet clear, but it is so certain that it enabled Mendelejeff to predict the discovery of three new elements which have since been found, filling up gaps in the series which his law required.

The nearest approach to a mathematical explanation of this law is afforded by the discovery that if the cube roots of the atomic weights were used as ordinates instead of the weights themselves, which is equivalent to taking volumes instead of lines to represent the atomic weights, the zig-zag line resolves itself into a regular curve, which is identical with, or very closely resembles, the logarithmic curve well known to mathematicians.

What the effect of these laws may be is not yet fully known, but they all point towards the conclusion that the atoms which we call elementary are all really manufactured out of some one atom or sub-atom, which is the primary element of matter. Where are they manufactured? Crookes says on the outside of the universe, wherever that may be, and that they are destroyed or dissociated when they reach the position of the lowest potential energy, which is in the centres of the largest stars. This may or may not be true, but it shows the direction in which speculation is tending, and carries our conceptions of the possibilities of the universe far beyond the limits of the nebular hypothesis and the results of the law of gravity.

This also may be said of the atoms, whatever sort of manufactured articles they may be, they are manufactured to the same pattern, like the nuts and screws of a large locomotive or gun factory. The hydrogen-atom gives the same spectral lines, which means that it vibrates, and starts or absorbs ether-waves precisely in the same manner, whether it exists in Sirius, in the nebula of Orion, or in a jar of gas in a laboratory.

The problem of atoms is being attacked from another side. What, after all, are atoms, or the primary protyle or sub-atom, if we can succeed in tracing them back to one origin? The general idea is that of an almost infinitesimally small, but still finite, unit of matter, impenetrable, indivisible, and endowed with enormous energies, both of velocity and attractive and repulsive forces. Various other ideas have been started. Some have considered them as mere centres of force without parts or magnitude; others as condensed portions of a continuous matter; but all these theories are open to fatal objections, and the conception of atoms has pretty well settled down to that of small separate bodies floating, like buoys, in an ocean of ether, that is, of the still rarer, all-pervading medium which transmits light and heat. This accounts best for all the phenomena hitherto observed, and may be said to hold the field. The only serious competitor with it is the vortex theory of Helmholtz and Thomson, which assumes atoms to be revolving rings of a perfect fluid pervading space. The general idea is given by the rings of smoke which occasionally escape from the lips of smokers. These rings persist for a long time, glide before the knife so as to be indivisible, and when two of them collide they rebound and vibrate. In a word, they behave in many respects very like atoms, and refined mathematical calculations show that if we could suppose them formed and rotating, not in air, but in what is called a perfect fluid, incompressible, possessing inertia, and yet offering no resistance whatever to motion through it in any direction, such vortex-rings would be indeed indivisible and indestructible, and might well be what we call atoms. The theory is extremely ingenious, but it has hardly yet got beyond the stage of a mathematical speculation, like space of four dimensions, which has no relation to actual facts known by observation and experiment.

To begin with, there is absolutely no proof of such a medium as is required. It is already difficult enough to realize the conception of such a medium as ether, though its waves can be measured, and its existence is imperatively demanded by the phenomena of light and heat. But to suppose a second and equally all-pervading medium, of a different nature, and with properties still more inconceivable, is too much for the imagination, and would require to be supported by undoubted facts. Again, atoms have weight, and the supposed medium has no weight and offers no resistance. Why should a portion of it acquire weight by being made to rotate; and what conceivable cause could set it rotating? But the most fatal objection is, how could such rings continue to rotate unless some external or centripetal force counteracted the centrifugal tendency to fly off from the circumference in a straight line? Rotation implies centrifugal force, and matter which possesses inertia and obeys the first law of motion must inevitably fly off, unless acted on by some other force. In the case of the earth, the sun's attraction supplies the centripetal force; in the case of a wheel or bicycle the molecular cohesion of the solid parts; in the case of smoke-rings the resistance of the air. But what supplies it in the vortex-rings, rotating in a perfect fluid, which offers no resistance to any motion?

It will be seen that the problem of atoms, involving that of the ultimate constitution of matter, is fast advancing towards some definite solution; but it is not yet solved, and is a problem of the future. Seeing, however, the wonderful advances which have been made in the last half-century, and specially in the last few years, it is impossible to doubt that, as in the case of gravity, some future Newton will sum up in some comprehensive law all the scattered facts which point in the same direction towards the unity of the universe, and the persistence of evolution from the simplest to the most complex.

But even when this triumph of science has been attained, the question remains as insoluble as ever—Whence came this primeval matter and primeval energy?

I recollect as a boy looking up at the stars, and asking myself what does all this mean? Where did it come from, and what is beyond it? The only answer was a sort of painful ache, as of straining the eyes to see in the darkness. And now that, thanks to the discoveries of modern science, I can see so much beyond the visible stars, far off into the infinitely great, far down into the infinitely small, far back into infinite Time—at the end of all I am not one whit advanced beyond that feeling of boyhood. I gaze with straining eyes into the Unknowable, and gaze in vain. Others may see, or fancy they see, something behind the knowable phenomena of the universe, linked together by invariable laws. Some a personal God, others a design like human design, a living whole, ideas in a Universal Mind, illusion, Maya Nirvana, what not. For my own part, if I candidly confess the truth to myself, I can only say with Tennyson—

"Behold! I know not anything,"

and content myself with the only creed which seems to me certain, that of trying to do some little good in my generation, and leave the world a little better rather than a little worse for my individual unit of existence.

CHAPTER III.

CLIMATE.

Conflict between Geology and Astronomy—Geology asserts Uniformity of Climate until Recent Times—Astronomy asserts Inclination of Earth's Axis to be invariable, and therefore Climates necessary—Evidence for Warm and Uniform Climates—Greenland—Spitzbergen—Impossible under Existing Conditions—Heat, Light and Actinism—Invariability of Earth's Axis—Causes of Higher and more Uniform Temperature—Cooling of the Earth—More Heat from the Sun—Warmer Regions of Space—More Carbonic-dioxide—Would not explain Uniformity of Temperature—Excess of Oxygen—Modification of Species—Configuration of Sea and Land—Croll's Theory—Displacement of Earth's Axis—Inclination of Axis of Planets and Moon—Unsolved Problems of the Future.

Geology and astronomy are in conflict on other questions as well as that of the time during which a sufficient supply of solar heat has rendered the earth habitable. The conditions of that supply are as important as the total quantity, and these conditions depend mainly on climate. Geology seems to show that during the vast lapse of time embraced by fossil records from the Cambrian to the close of the Tertiary period, there were no well-marked zones of climate, and the conditions of life were uniform, or nearly so, throughout the whole earth. Astronomy, on the other hand, asserts that the vicissitudes of the seasons, with their corresponding zones of climate, must have existed from the beginning as they now are. Geology relies on undoubted facts. Coral formations, which require both a warm and an equable climate, and cannot live in a temperature below 66° Fahrenheit, were found by Captain Nares in Greenland, in latitude 81° 40′. Ammonites of the same genera and even of the same species are found alike in Melville's Island and in India; and Ichthyosauri have been met with in Greenland and Spitzbergen. Lyell, Dana, and all modern geologists agree that in primordial times there were "no zones of climate," "no marked difference between life in warm and cold latitudes;" "warm Arctic seas all the year round."

This continued until what is, geologically speaking, quite the other day, the close of the Tertiary period. In Spitzbergen, latitude 78° 56′, are found the remains of a luxuriant Miocene flora, comprising species like the common cypress, which now grow in the Southern United States and California. Magnolias and zamias are found in Miocene strata in Greenland in latitude 70°.

These species, it must be observed, require not only a warm but an equable climate. They would be killed by a single severe night's frost, and yet they grew and flourished where the winter night now lasts for four months, and where the thermometer has registered more than 100° below freezing-point. The difference between summer and winter temperature in high Arctic latitudes exceeds 100° Fahrenheit, and whatever may have been the initial temperature, this difference of heat, due to solar radiation, must have been added and subtracted every year, as long as the earth's axis of rotation preserved its present obliquity to the plane of the ecliptic in which the earth revolves round the sun. If the temperature of Spitzbergen was from any cause high enough to prevent the thermometer from falling below zero in winter, it must have risen in summer far above the extremest tropical temperature at which life and vegetation are possible.

Nor is it a question of temperature only, but of light and the actinic rays of the solar beam, which are equally essential for vegetation. A luxuriant forest vegetation, including such forms as the magnolia and cypress, could no more flourish under any conditions now known to us in Spitzbergen, than they could if shut up for four months in a dark cellar. And yet with the present obliquity of the axis, the sun must have been below the horizon in those latitudes from November till March.

At present, as we go north from the equator towards the Arctic circle, we find species changing to accommodate themselves to the change of environment. Palms are succeeded by oaks and beeches; these again by pines and birches, and these by dwarf willows and lichens, until all vegetation, except of the very humblest forms, dies out as we approach the pole. But in the geological records of earlier periods no such changes are discernible. The Miocene magnolia of Spitzbergen is not even a greatly modified magnolia, but of the same species as the magnolia of the present day. The Miocene cypress is the common cypress. If there were no such science as astronomy, geology would point to the conclusion that until after the Miocene period climate was uniform; there were no distinct zones or seasons, and therefore no obliquity of the earth's axis, or at any rate nothing like the present amount. With these conditions there would have been perpetual spring, and all we should require would be a higher average temperature for the whole earth. But to this conclusion astronomy opposes an inflexible *non possumus*. If there is one thing more certain than another, it is that mathematical calculations, based on Newton's law of gravity, explain all the movements of the solar system. They do so with a certainty that enables us to predict the places of the earth, moon, and planets years beforehand, with absolute accuracy. And if there is one thing more certain than another in these calculations, it is that no permanent change is possible in the inclination of the earth's axis. The earth now spins, in twenty-four hours, round an axis inclined at an angle of $66\frac{1}{2}°$ to the plane on which it revolves round the sun in a year. It must always have so spun, for there is no cause known to science by which, when this rotation was once established, the inclination of the axis could have been permanently altered. The plane of the equator shifts its position slowly on that of the ecliptic, owing to various minor actions of the force of gravity, the principal one being the precession of the equinoxes, due to the protuberant matter at the earth's equator; and thus in 22,000 years, it makes a complete circuit, returning to its original position. But during this circuit, its inclination to the plane of the ecliptic remains practically constant, and the effect on the seasons is unchanged, except that they come at different positions of the earth in its orbit round the sun, so that summer and winter alternately come when we are farthest from the sun or nearest to it. At present we are nearer the sun in winter than in summer, and the winter half of the year is shorter than the summer half in the Northern hemisphere. In 11,000 years this position will be reversed, and winter will be shorter than summer in the Southern hemisphere; but there is nothing in these slight changes to affect the general course of the seasons, and as we happen to be now nearer the sun in winter, the effect of any slight change due to precession would rather be to increase the difference between summer and winter heat in high northern latitudes, and so aggravate the difficulty of reconciling the conclusions of the two conflicting sciences. And yet there must be some way of reconciling them. Truth cannot speak with two voices, and the laws of nature cannot give contradictory results.

Let us consider first what the undoubted facts of geology require us to assume. Two things—firstly, that the general temperature of the earth was higher in former times than now; secondly, that it was more uniform. As regards the first condition, astronomy interposes no obstacle but affords no aid, and it must be admitted that we are still in the region of conjecture rather than of certainty. The first obvious guess is that the earth was formerly hotter, and has been gradually cooling. But this guess is contradicted by mathematical calculations as to the cooling of heated bodies, which show that after the earth had cooled down to the point of forming a solid crust, many miles in thickness, of non-conducting rock, internal heat could have had little or no effect on surface temperature. This is confirmed by what we know of the climates of areas where large reservoirs of internal heat lie comparatively near the surface, as in Iceland and other volcanic districts. In the celebrated Comstock lode the heat of the earth increases so rapidly, that it becomes impossible to work the mines below a very moderate depth. Yet in all these cases the temperature at the surface remains the same as that of other regions on the same isotherm, and is determined by the same circumstances of latitude, elevation, aerial and ocean currents, and other known conditions. Nor if the internal temperature of the earth was a factor in the problem, would it be easy to account for our recovery from the cold of the Glacial period, in the face of a continued and progressive diminution of the planet's heat.

Another conjecture is that the sun may have given out more heat formerly. This, however, is a mere guess, confirmed by no theory or experience. On the contrary, theory rather points to the paradoxical conclusion that, as the sun has cooled, it has got hotter; that is, that a volume of gas, in cooling, developes rather more heat by contracting than it loses by radiating. Moreover, as we have already seen, the difficulty is to understand how even the present supply of solar heat can have been maintained long enough for the time requisite to account for the facts of geology; and the improvement in climate since the Glacial period is as inconsistent with solar as it is with terrestrial refrigeration.

The passage of the solar system through warmer and colder regions of space is another explanation which has been invoked. But this—though by no means improbable—is as yet a mere possibility, and based on nothing approaching to actual knowledge.

Of existing known causes there is one which seems, as far as it goes, to be a *vera causa* which might have given the earth's surface a warmer temperature in early ages. Its reality may be proved by the very simple experiment of sleeping on a cold night without a blanket. Evidently, other circumstances being the same, such as the reading of the thermometer and blood heat of the body, the question of blanket or no blanket makes an immense difference in the resulting temperature. Why is this the case? Because the blanket keeps the heat in, or in other words radiates it back to the body instead of letting it radiate out into space. There are other things which do this even more effectually than a woollen blanket, for they let the heat of the sun's rays in, and having let it in, catch it as in a trap, and do not let it out again. Glass, for instance, in a conservatory, is such a trap, and, as we all know, will keep the temperature inside much warmer than it is outside, even without the aid of artificial heat. Many other substances have the same property, and among them two which are essential elements of the earth's atmosphere, water in the form of vapour, and carbonic-dioxide. Tyndall, in his *Heat considered as a Mode of Motion,* has shown clearly what an immense part these gases have in maintaining the temperature of the earth's surface. If the cold is more intense, especially at night, on high mountains, it is not because less heat is received from the sun's rays during the twenty-four hours, but because half the atmosphere is left below, and so the heat-retaining blanket is thin and threadbare. So in deserts where the air is dry and there is little aqueous vapour, the heat by day may be excessive and yet the cold by night well-nigh intolerable. "The removal," says Tyndall, "for a single summer's night of the aqueous vapour which covers England would be attended by the destruction of every plant which a freezing temperature could kill." And such a removal on a winter's night would send the thermometer down far below zero.

This property of retaining heat is not confined to water in the form of vapour; it is common to other gases, and often in a higher degree. Among these is one which is always present in the atmosphere—carbonic-dioxide, a gas formed by the combination of two atoms of oxygen with one of carbon.

The percentage of this gas in the air is very small, only a fraction of one per cent., and yet it constitutes the sole source of supply of the carbon required, directly for vegetable, and indirectly for animal life. At present the balance between the two sorts of life seems to be kept up, as in an aquarium, by animals restoring to the air, in the form of carbonic-dioxide, the carbon which has been abstracted from it by plants. But when we look at the enormous amount of carbon which has been locked up in coal, limestone, and other carboniferous formations of the earth's crust, it is evident that it must be vastly greater than could be derived from such a small percentage of carbonic-dioxide as now exists in the atmosphere. It has been estimated by experienced geologists as many hundred times greater. Where all this carbon could have come from is a question not yet solved. Some have thought that it may have been supplied from the interior of the earth by volcanoes; but although it is certain that some volcanic vents do emit carbonic-dioxide, as in the case of Lake Avernus, and the Grotto-del-cane, near Naples, the quantity is small, and the better opinion seems to be that it is only given out when subterranean fires come in contact with limestone, or some other form of previously deposited carbon. Did the carbon, then, come from the air? If so, there must have been more than one hundred times as much carbonic-dioxide in it in early geological times as there is at present.

This would go some way towards explaining the difficulty of the higher temperature prevailing in past ages, for more carbonic-dioxide would undoubtedly be equivalent to an additional blanket to protect the earth from cold; and the higher temperature thus caused would enable the air to hold more aqueous vapour in solution, and thus increase the thickness of the water-blanket.

It is conceivable that under such conditions a warm and humid climate may have prevailed over a great part of the earth's surface, though this would hardly meet the difficulty of the uniform existence of such a climate in latitudes where the supply of heat from the sun must have been so very different in winter and summer. Nor would this difficulty be removed even if we were to suppose that the earth's axis might have been nearly vertical to the plane of the ecliptic. This might meet the difficulty as to light and actinic rays, for there would be everywhere twelve hours of day throughout the year; but it would not meet the difficulty as to temperature, for if the air-blanket was sufficient to retain heat enough in the Arctic Circle to prevent frosts, from a sun which never rose much above the horizon, it must have retained far too much heat for existing life and vegetation in latitudes nearer to the equator.

There are, however, many grave objections to considering this to be the sole or even the principal cause of the warmer climates of early ages. It is by no means certain that either animal or vegetable life, in anything like known forms, could exist in an atmosphere so surcharged with carbon. Nor is carbon all; we must account also for oxygen. If the whole of the carbon now fixed in the different strata of the earth's crust was derived from carbonic-dioxide originally present in the atmosphere, so also must have been the oxygen, which in various forms of oxides now forms an even larger constituent of that crust. Oxygen is a very active element, which, under moderate conditions of heat and moisture, combines readily with iron, silicon, calcium, aluminium, and all the metallic bases. Many hundred times more oxygen must have been withdrawn from the air than now exists in it to form the rocks which are the principal part of the earth's crust. But an excess of oxygen is as fatal to life as an excess of carbonic-dioxide. Terrestrial life, as known to us, depends on a very delicate adjustment of the quantities of oxygen and nitrogen in the air. A very little excess or deficit of either would destroy all air-breathing animals. With too much oxygen we should be burned up even more rapidly than the drunkard is by too much alcohol; with too little, the fire of life would be choked by ashes and refuse. If there was formerly a hundred, or even ten times more oxygen in the atmosphere than there is now, there must have been a corresponding excess of nitrogen to neutralize it, and if so, what has become of the nitrogen? Nitrogen is an inert element which enters sparingly into combinations, and does not, like oxygen and carbon, get locked up in great masses of the earth's solid crust. Once in the atmosphere it would seem that it must have remained there; and if so, as oxygen was withdrawn in continually increasing quantities, how could the life-sustaining proportion of the two gases have been maintained and continued down to the present day?

It has been said that life may have been so differently organized in past geological ages as to have existed under very different conditions, and the mammoth is appealed to as an instance of an elephant modified so as to resist Arctic cold, and the result of deep-sea dredgings show that molluscs, crustaceans, and other low forms of life may exist in ice-cold water and without light. But we can hardly suppose such profound modifications of existing genera and species of highly-organized plants and animals as would enable them to breathe air of a very different composition.

For we must remember that the evidence for an elevated and uniform temperature is not confined to remote geological ages, but comes down to the close of the Tertiary period, when existing forms, both of animal and vegetable life, were firmly established, and several species have survived to the present day without perceptible change. Thus when the magnolia was growing in Spitzbergen, the dryopithecus was living in Southern France. Can it be supposed that this anthropoid ape breathed a different air from his congeners, the chimpanzee and gorilla; and yet if his lungs required the same air, how could excess of carbonic-dioxide have supplied the extra warm blanket to protect the Spitzbergen magnolia?

A different configuration of sea and land is the explanation which many geologists, following Lyell, have advanced for different conditions of climate. And no doubt aërial and oceanic currents, such as now cause the trade-winds and Gulf Stream, are responsible for great variations of climate, while low lands in low, and high lands in high latitudes must always have had a considerable influence in raising or depressing temperature. But changes of this description can more readily account for the cold of the Glacial, than for the heat of the Tertiary and preceding periods. We have now got the trade-winds and the Gulf Stream in the North Atlantic, and although the diversion of the latter might bring the ice-cap back to London and New York, and make the climate of Scandinavia and Scotland the same as that of Greenland and Labrador, its presence takes us a very short way towards enabling magnolias to flourish in Spitzbergen.

In like manner, even if Croll's theory were established, which it is far from being, and the effect of the obliquity of the earth's axis combined with precession, though imperceptible while the earth's orbit was nearly circular, became great in the two hemispheres alternately, when the orbit was approaching its maximum eccentricity, this would not explain the high and uniform temperature of past geological ages. If this theory were true, what we should look for would be two or three Glacial periods in the course of each geological epoch; for the least time required for any of the great geological formations must have been long enough to include two or three secular variations of the earth's orbit, from minimum to maximum eccentricity. And each of these Glacial periods must have included several changes, alternating, at intervals of 11,000 years, between severe cold and genial heat, owing to the effect of the precession of the equinoxes combined with great eccentricity.

Instead of uniform warmth, there must have been more than 100 Glacial periods during the immense lapse of time between the dawn of life in the Cambrian, and the last of such periods in the Quaternary. It is a moot point with geologists whether traces of a single one of such periods, prior to the last one, have been found. There are a few conglomerates which look very like consolidated boulder-clays, and every now and then we hear of some formation, supposed to be glaciated, being found in the Permian and in other formations in India, South Africa, and Australia; but there is no evidence hitherto which commands the general assent of geologists, for a single Glacial period prior to the recent one which closed the Tertiary period. And there is abundant evidence that during many formations, such as the Carboniferous and Coal-measures, which must have taken millions of years to accumulate, there were no vicissitudes of climate such as must have inevitably occurred if any astronomical cause, such as precession or eccentricity, had been sufficient to bring about great vicissitudes of heat and cold. And what is still more conclusive, the evolution of vegetable and animal life, as shown by fossils, affords no trace of the repeated modifications which must have taken place within the limits of the same geological formation, if there had been such vicissitudes of heat and cold as the theory requires.

It remains to be considered whether any change in the direction of the earth's axis may have been possible. Clearly no such change can have taken place within the earth itself, for its shape is that of an oblate spheroid, revolving round its present axis. Any displacement of the poles must displace the present equator, and tend to establish a new one on a different plane. But the equatorial diameter of the earth is 26 miles longer than the polar diameter, so that any displacement of the poles must have tended to displace this enormous mass of protuberant matter, and send such portion of it as was fluid in a diluvian wave, miles in height, towards the new position of equilibrium; while the solid portion remained in a plane no longer coincident with that of the earth's rotation. There is no trace of anything of the sort having ever occurred, and if the axis has shifted, the whole earth has shifted with it, which is just what astronomers declare to be impossible by any known laws.

But are the whole of the laws really known? There is nothing more difficult than to account for the varying inclinations of the axes of rotation of the different bodies of the solar system. On the nebular hypothesis, which traces the sun, planets, and satellites back to the condensation of a revolving mass of nebulous matter, one might have expected to find the planes of rotation and revolution of planets and satellites, not only in the same general direction from west to east, but nearly coincident. Jupiter, however, is the only one of the planets which fulfils this condition. Its axis of rotation is inclined at an angle of 87°, or very nearly at right angles, to the plane of its revolution round the sun. But there is no certain rule. That of Saturn, which comes next in order on the outside of Jupiter, has an inclination of 64°, while that of the next planet on the inside, Mars, is 61° 18′. The earth's axis is inclined at 66° 33′, while we find its satellite, the moon, rotating like Jupiter in a plane inclined only 1° 30′, and the axis of Venus, on the other hand, is so oblique, that in its winter the Arctic Circle almost extends to the equator.

The case of the moon is most difficult to understand, for on any theory of its origin, whether as a condensed ring left behind as the nebulous matter of the earth contracted, or whether it was ejected from the earth in some eruption of its fiery stages, it might have been expected to retain nearly the same rotatory motion as its parent orb. But if so, clearly some unknown force must have intervened, either to make the earth's axis more, or that of the moon less oblique, than they were originally. No such force is known, nor has any plausible guess been made as to what might have occasioned it; but the same observation applies to many of the phenomena of the solar system. How has the supply of solar heat been kept up for the time required by geology? How does the energy we call gravitation act across space from atom to atom, and from star to star, and how is its supply maintained? Why is the axis of the earth inclined at an angle of 66° 30′ to the ecliptic, while that of Jupiter is almost perpendicular to it, and that of Venus oblique to the extent of nearly two-thirds of a right angle?

These are all problems which depend on natural laws, and must lie within the limits of human reason; but they are pebbles which have not yet been picked up on the shore of the ocean of truth. It may bring home to us the force of Newton's saying that we are but as children picking up such pebbles, when we see what a multitude of the deepest problems, as to the constitution of the earth and of the universe, are raised by the simple fact that Captain Nares brought back a specimen of coral from latitude 81° 40′ in Greenland, and that luxuriant forests, of a sub-tropical or warm temperate vegetation, flourished in Spitzbergen as lately as the period when an anthropoid ape of the stature of man was living in the south of France, and when man himself or his savage progenitors, were possibly or even probably already chipping flints into rude implements.

CHAPTER IV.

THE GLACIAL PERIOD.

Importance of Date of Glacial Period—Its Bearing on Origin of Man—Short Date Theories—Prestwich says 20,000, Lyell 200,000 Years—Croll's Theory—Prestwich's Arguments—Solar Heat—Human Progress—Shown by Palæolithic Remains—Geological Evidence—Advance of Greenland Glaciers—Denudation—Erosion of Cliffs and Valleys—Deposition—Loëss—Elevation and Depression of Land—All show Immense Antiquity—Post-Glacial Period—Prestwich says 8000 to 10,000 years—Mellard Reade 60,000—His Reasons—Inconsistent with Short Date Theories—Causes of Glacial Period—Cooling of Earth—Cold Regions of Space—Change of Earth's Axis—More Vapour in Atmosphere—Lyell's Theory, Different Configuration of Sea and Land—Conditions of Glaciation—Problems Pressing for Solution.

The date and duration of the Glacial period present a problem which is in many respects of the highest interest. It comes nearest to us as inaugurating the recent period in which we live, and for which we have historical data. It affords the best chance of obtaining an approximate standard by which to measure geological time in years or centuries. And it touches directly on the great question of the Origin of Man.

For man is like the mammoth and cave bear, an essential part of the Quaternary fauna, and whatever doubts may be entertained as to his existence in Tertiary times, there can be none as to the fact that his remains are found in great numbers, and widely scattered over the four quarters of the globe, in conjunction with those of the mammoth and other characteristic Quaternary mammals, in deposits which date, probably from the earlier, and certainly from the intermediate and later stages of the Glacial period. A short date, therefore, for that period shortens that for which we have positive proof of the existence of man, and a very short date reduces it to a length during which it is simply impossible that such a state of things as is found existing in Egypt 7000 years ago could have grown up by natural laws and evolution, and therefore brings us back to the old theories of repeated and recent acts of supernatural interference, which, since the works of Lyell and of Darwin, have been generally considered to be completely exploded.

The question, therefore, is one of the highest theological as well as scientific importance, and as such it has too often been approached with theological prepossessions. An extreme instance of this is afforded by Sir J. Dawson, who in his work on *Fossil Man* assigns 7000 years as the probable date for the first appearance of man upon earth, ignoring the fact that at this date a dense and civilized population already existed in Egypt, with a highly-developed language and system of writing and religion; and that the types of the various races of mankind, such as the Negro, the Copt, the Semitic, and the Arian, are as clearly distinguished in the paintings in Egyptian tombs, 5000 years ago, as they are at the present day.

Sir J. Dawson, however, though an excellent geologist as long as the older formations are concerned, is so dominated by the desire to square facts with the account of creation in Genesis, that he becomes totally unreliable when the human era is approaching.

Quite recently, a very different authority, Professor Prestwich, reasoning on strictly scientific grounds, concludes, "that the Glacial period, or epoch of extreme cold, may not have lasted longer than from 15,000 to 25,000 years, and the Post-Glacial period of the melting away of the ice-sheet to from 8000 to 10,000 years or less; giving to palæolithic man no greater antiquity than perhaps about 20,000 to 30,000 years, while should he be restricted to the so-called Post-Glacial period, his antiquity need not go farther back than from 10,000 to 15,000 years before the time of neolithic man."

Prestwich cannot be accused of theological bias, and in fact this estimate is as inconsistent with theological theories of Adam and Noah, as if the figures were multiplied tenfold. But he was influenced by the wish to make geological time accord with the short-date estimates of Sir W. Thomson, as to the possible duration of solar heat. Be this as it may, the fact that an authority like Prestwich reduces to 20,000 years a period to which Lyell and modern geologists generally have assigned a duration of more like 200,000, shows in what a state of uncertainty we are as to this vitally important problem. For even the longest period for man's antiquity assigned by Prestwich would be clearly insufficient to allow for the development of Egyptian civilization as it existed 7000 years ago, from savage and semi-animal ancestors, and still less for the evolution of the human race from earlier types, as is proved to have been the case with the horse, stag, elephant, ape, and other mammals, with whom man is so intimately connected, both in physical structure and in geological association.

It is highly important, therefore, to consider the grounds on which the various theories are based, of the probable cause and duration of the Glacial period. The first natural guess was to attribute it to the precession of the equinoxes. Owing to this cause the North Pole is alternately turned towards the sun every summer, and away from it every winter, the reverse being the case in the southern hemisphere. But owing to the eccentricity of the earth's orbit, the duration of the seasons is not exactly equal, and summer and winter may occur either when the earth is nearest to or farthest away from the sun. At present winter occurs in the Northern hemisphere when the earth is nearest the sun and moving with the greatest velocity, so that it is shorter by some days, and summer longer, than in the Southern hemisphere. Now it is a fact that what may be called a Glacial period prevails at present in the Southern hemisphere, while corresponding latitudes in the Northern hemisphere enjoy a temperate climate. It might be thought that this fact afforded an explanation of the Glacial period; but this conjecture is negatived when it is considered that this revolution of the earth's axis is periodical, and completed in about 22,000 years, so that if it were the sole or principal cause of Glacial epochs, they must have recurred from the beginning of geological time at this short interval, which is altogether inconsistent with the evidence of facts.

Croll expanded this crude theory into one which had vastly more plausibility, viz., that although the effects of precession might be imperceptible while the earth's orbit was nearly circular as at present, they might become very powerful when they coincided with one of the long periods at which the earth's orbit became flattened out into an ellipse of *maximum* eccentricity. He showed by calculation that one such period began 240,000 years ago, attained its *maximum* in 80,000 years, and passed away about 80,000 years before the present era. These figures fitted in so well with those deduced by Lyell and other eminent geologists from geological data, that Croll's theory received very general acceptance. But it is open to the same objection, though in a less degree, that it requires us to assume a periodical succession of Glacial epochs. The oscillations of the eccentricity of the earth's orbit, about its *maximum* and *minimum* limits, though slow as measured by centuries, are not so slow according to the standards of geological time. Croll's calculations have shown that another position, such as is assumed to have caused the latest Glacial period, must have occurred 500,000 years earlier. The calculations have not been carried further back, but it is tolerably certain that, if Croll's theory be correct, at least two or three Glacial periods must have occurred during each of the great geological epochs. This is opposed to geological evidence. The Permian is the only formation in which what look like traces of glacial action have been unmistakably found, and even these are considered doubtful by many geologists. Still more doubtful are the proofs of older Glacial epochs deduced from isolated cases of boulders, as in the Miocene conglomerate of Monte Superga, near Turin, the Flysch of Switzerland, and in some of the conglomerates of the old Devonian. "Not proven" is the verdict which most geologists would return on the few alleged instances of earlier Glacial periods; while if Croll's theory were true, we might expect to find them frequently. Above all, it is difficult to conceive how two or three great changes of temperature could have occurred during each geological formation without showing unmistakable traces in the fauna, and still more distinctly in the flora, of the epoch. Ferns must have died out and been succeeded by mosses; and these in their turn given place to ferns two or three times over or more, during the growth of the Coal-measures, if any changes of climate had occurred at all resembling those of the recent Glacial period.

The confidence, therefore, with which Croll's theory was at first received has been a good deal shaken, and although many geologists still believe that it may have been one among other causes of the last great refrigeration, it can no longer be considered as affording a reliable standard by which to measure the time in historical years, either of the Quaternary, or still less of any previous geological epoch.

We have to fall back, therefore, on the geological evidence of deposition and denudation, of the rise and fall of continents, of the erosion of rivers, valleys, and so forth, in any attempt to decide between the 200,000 years of Lyell, and the 20,000 years of Prestwich. The former period, based on the minute and careful investigations of Lyell, Geikie, Croll, and other eminent geologists, held the field until the recent attempts of Prestwich and others to reconcile geology with Sir W. Thomson's theory of solar heat, by reducing geological time to about one-tenth of the accepted amounts.

Prestwich, in his recently-published works on geology,

states that he has been influenced mainly by two considerations.

1. The wish to bridge over the wide chasm between geologists and physicists as to the possible duration of the supply of solar heat.

2. The difficulty of conceiving that man could have existed for a period of 80,000 or 100,000 years without change and without progress.

And the principal, or rather the sole fact on which he relies is, that the advance of the glaciers of Greenland is found to be much more rapid than that of the Swiss glaciers upon which previous theories had been based of the time required for the advance of the Scandinavian and Laurentian ice-fields over Northern Europe and America.

The two considerations may be briefly discussed. The first, as I have already shown, is based on a theory as to solar heat which is in the highest degree uncertain, and which requires rather to be tested by the positive facts of geology than accepted as an admitted conclusion, to which those facts must be squared. To allow it to distort those facts, or even to influence us in interpreting them, is a prepossession only one degree less mischievous than the theological prepossession which so long retarded the progress of true science.

The second consideration, as to the rate of human progress, is a mere question of what each individual inquirer may think probable estimates, which will depend very much on his habit of mind and previous bias. There are positively no facts on which to base a conclusion as to the rate of progress of isolated savage tribes living in the hunter stage, without contact with more civilized races. The Australian savages, the South African bushmen, the Negritos of the Andaman Islands, may have lived as they were first found by Europeans, any time you like from 1000 to 100,000 years, for aught we know to the contrary. There is, in fact, no record of any such savage race emerging into comparative civilization by any effort or natural progress of its own. Even much more advanced races trace back their knowledge of the higher arts and civilization to some divine stranger, like the Peruvian Manco-Capac, or Chaldæan Oannes, who lands on their shores; or else, like the Egyptians, assign these inventions to gods, which means that they are lost in the mists of antiquity. The neolithic men of Europe were clearly invaders, who brought a higher civilization with them from Asia, and the knowledge of polished stone and metals were diffused by commerce.

It is incorrect, however, to say that palæolithic man shows no signs of change or progress. On the contrary, the evidence of palæolithic deposits shows everywhere a progress which, although it may have been extremely slow, is uniformly in the same direction, viz., upwards. There is no exception in the hundreds, or rather thousands of instances in which palæolithic implements have been found, to the law that the rudest implements are found in the lowest deposits, and that improvements are traced in an ascending scale with ascending strata. This is most markedly the case in caves, where, as in Keat's Cavern, deposits of different ages have been kept distinct and securely sealed under separate sheets of stalagmite. In the rock-shelters also, and river gravels, in which the relative antiquity is proved by their higher or lower levels, the same law prevails. In the oldest, where the cave bear and mammoth are the characteristic fossils, the stone axes, knives and scrapers are of the rudest description. The celts or hatchets are mere lumps of stone, roughly chipped, and with a blunt butt-end, evidently intended to be held in the hand. In the next stage we find finer chipping, and celts adapted for hafting; while arrow and javelin heads appear, at first rude, but gradually becoming barbed and finely wrought. Still later, with the advent of the reindeer in large herds, affording in their horns a softer material than stone, a remarkable improvement takes place, and eyed needles, barbed harpoons, and in some cases engraved and sculptured portraits of animals of the chase, testify to a decided advance in the arts of civilization. Above all these, come the weapons and implements of the Neolithic age which, as already stated, are separated by a sharp line from the earlier records of palæolithic man. No polished stone has ever been found in deposits belonging clearly to the Palæolithic period, and a decided change has taken place in the fauna, which in the Neolithic age corresponds closely with that of recent times in the same locality.

It is impossible, therefore, to deny that both change and progress have existed from the first appearance of man, and there are absolutely no data to enable us to say what may have been the intervals of time required for the successive stages of this progress. All we can say is, that the more nearly primitive man approximated to a state of semi-animal existence, the slower must have been the steps by which he emerged from it into comparative civilization.

We must fall back, therefore, on geology for anything like reliable data on which to base any estimate of the time required for the Quaternary or any preceding geological epoch. Here, at any rate, we are on comparatively certain ground. So many feet of deposition, so many of erosion, so many of elevation or depression; these are measurable facts which have been ascertained by competent observers. How much time required to account for them? This can only be an approximation, based on our knowledge of the time in which similar results, on a smaller scale, have been produced by existing natural laws within the Historical period. Still, if we argue from natural causes, and ignore imaginary cataclysms and supernatural interferences, we may arrive at some sort of *maximum* and *minimum* limits of time within which the observed results must lie.

This was the process by which Lyell and his school of geologists arrived at their estimates of geological time, and it is only by a careful study of their works that it is possible to see how closely the chain is woven, and what a mass of minute investigations support their conclusions. The one solid fact which Prestwich opposes to them is the rapid advance of the glaciers of Greenland. Recent observations by Rink and other explorers have shown that the fronts of these glaciers advance much more rapidly than the rate which had been assumed from the advance of the Swiss glaciers.

The average rate of advance of the great glaciers which discharge themselves into Baffin's Bay is about 35 feet daily, or $2\frac{3}{4}$ miles yearly. Calculating from these data, Prestwich arrives at the conclusion that the old ice-sheets which radiated from the Scandinavian and Canadian mountains to a distance of about 500 miles, might have been formed in from 4000 to 6000 years. The great changes which have taken place since the retreat of the ice-sheets, he accounts for by supposing that with a greater rainfall these changes went on much more rapidly than they have done during the Historical period. These views, however, did not command the assent of other eminent geologists, who were present when Professor Prestwich's paper was read, and they are open to very obvious objections.

The rate of advance of a glacier thrust outwards by such an immense mass of ice as caps Greenland, through a narrow fiord, on a steep descending gradient, into a deep sea which floats off its front in icebergs, affords little test of the advance of an ice-sheet spread out with a front of 1000 miles over a whole continent, unaided by gravity, and obstructed by ranges of mountains 2000 or 3000 feet high, which it has to surmount. Nor does the rate of advance of such a sheet afford any clue to the time during which it may have remained stationary, or been receding. The two latter conditions evidently depend on the climate at the extremity of the ice-sheet, when the ice pushed forward by it is melted by the summer heat. As long as the climate of Switzerland remains the same, the Swiss glaciers will remain at their present level, with slight local and temporary variations; and this must have been equally true of the great Scandinavian and Canadian glaciers. They may have advanced in 5000 years, remained stationary for 50,000 years, and taken 100,000 years to retreat, for anything we know to the contrary, from the Greenland glaciers. Nor is it a question of one advance and retreat only, for there is distinct evidence of several advances and retreats, and of prolonged Inter-Glacial periods.

In the cliffs of the east of England four boulder-clays are found, separated by sands and gravels deposited as each ice-sheet successively receded and melted; and in France there is evidence of at least one Inter-Glacial period, sufficiently warm and prolonged to allow the Canary laurel and fig-tree to supplant the lichen and Arctic willow. The only real test of time is from the amount of geological work that has been done in the way of denudation, deposition, elevation, and depression, since Northern Europe and Northern America were covered by such an ice-cap as now covers Greenland.

Tried by these tests the conclusions point uniformly to a longer rather than a shorter duration of the Quaternary, including the Glacial period. If we take denudation, we may refer to the fact that since palæolithic man left his implements on the banks of the old Solent river above Bournemouth, the level of its valley and of the adjacent land has been denuded by that small stream to a depth of 150 feet, and the erosion of the sea now going on at the Needles has eaten away a wide range of chalk downs, which were then continuous from the Isle of Wight to Dorsetshire. The same action of waves and tides as is now eroding Shakespeare's Cliff has removed the chalk ridge between that cliff and Cape Grisnez, and made England an island. The valleys of the Thames, the Somme, and other rivers of the south of England and north of France, have been excavated to a depth of more than 100 feet and a width of miles, by streams which have produced no perceptible change since the Roman period. And a still more striking proof of the immense time which has elapsed since the Glacial period is afforded by the fact stated in Prestwich's *Geology*, that the great basaltic plateau of the Cascade Range in British Columbia, which is cut through by the Columbia river to the depth of 2000 to 3000 feet, is underlain by the Northern Boulder-drift. Consider what a lapse of time this requires. Since the Boulder-drift, and therefore since the Glacial period, vast sheets of basalt must have been poured out by volcanoes now extinct, and those sheets of hard rock cut down by river action to the levels at which the relics of the old ice-cap now appear.

As regards the erosion of valleys, it is said that there may have been a much greater rainfall formerly than in historical times, and therefore erosion may have gone on much more rapidly. Doubtless there may have been more extensive inundations while great masses of ice and snow were melting under the summer heat of an improving climate, but there seems no adequate reason to account for a much greater rainfall. The maxim "ex nihilo nihil fit" applies to rain as to the other operations of nature, and more rainfall implies more evaporation, brought by warm winds blowing over warm oceans, and deposited when it comes in contact with land at a lower temperature. We already have these conditions in Western Europe, and the Gulf Stream and prevalent westerly winds make the climate more moist and genial than is due to the latitude. To have had it still more moist these conditions must have been intensified, and there is no reason to suppose that in recent times, and with the present configuration of sea and land, the Gulf Stream could have been much warmer than it now is. If the land had extended farther to the westward, the effect must have been to diminish rather than increase the rainfall in the districts where the Somme and the Thames were excavating their valleys; and with more extensive forests and morasses rain-water would be absorbed as in a sponge, and descend more gradually and less in tumultuous floods.

But even if a greater rainfall were granted, it would not affect the erosion of solid chalk cliffs by the sea, and the argument from the disappearance of the downs between the Isle of Wight and Dorsetshire, and between France and England, would remain the same. Sir John Lubbock estimates the rate of erosion of a perpendicular cliff of solid chalk at only a few inches per century, at which rate it must have taken an enormous time to wear away the chalk ridge between the Needles and Ballard downs; but even if we read yards, instead of inches, it must have taken a far longer time than Prestwich assigns for the whole Glacial period. There is nothing upon which reliable data are more wanted than as to the rate of erosion of solid cliffs by the action of the sea, for here the hypothesis of a larger rainfall and greater floods could not be invoked to accelerate the rate, as in the case of the erosion of valleys.

If from denudation we turn to deposition, we find equally conclusive evidence of the immense duration of the Glacial period. The deposit known as "loess" is universally admitted to be one of fine glacial mud, deposited tranquilly from sheets of inundation water, which have overflowed wide tracts during the melting of the ice and snow, as the climate improved and glaciers retreated. It is, in fact, just such a loam as the Arve deposits every summer on the meadows of Chamouni, when the turbid river issues in a swollen stream from the bottom of the mer-de-glace, and overflows its banks. Now this loess covers, as with a mantle, the valley systems of all the great rivers of the Northern hemisphere, whose upper courses lie within the area which was covered by ice and snow during the Glacial period. The Rhone, the Rhine, the Danube, the Mississippi, the Yang-tse-kiang, all run through cliffs of loess, which also fills their tributary valleys and spreads to a considerable height up the slopes of the hills and over the adjoining plateaux. It lies thickest in the valleys, dying off as it ascends the slopes, though it can often be traced to a height of 2000 or 3000 feet. The thin beds of loess at these heights and on the plateaux, are probably the result of the melting of frozen snow; but the great masses in the valleys are evidently the accumulations of mud from the overflows of the existing rivers, as they gradually cut their valley-systems down from higher to lower levels.

These accumulations invariably correspond to the configuration of the existing valleys, and overlie coarser sands and gravels, showing that they have been made since the rivers lost the transporting power which they possessed, when they ran with a more rapid current during the earlier stages of the retreat of the glaciers. The thickness of this accumulation of fine mud is stated by Lyell to be 800 feet or more above the existing alluvial plain of the Rhine, and in other rivers it is even greater. It is impossible that such a thickness could have been accumulated in anything like the shorter time assumed by some geologists for the duration of the whole Glacial period. And yet it represents only one phase of its concluding period, and it not only contains human remains but is itself clearly posterior to many of the sands and gravels in which remains of man and his associated Quaternary fauna have been undoubtedly found.

It is difficult to suppose that the loess can have

accumulated much more rapidly than the alluvium of the Nile, which has been proved to raise the soil of Egypt at the rate of about three inches in a century. At this rate it would require 320,000 years to accumulate the 800 feet assigned by Lyell to the loess of the Rhine valley. Making every allowance for a quicker rate of deposition, it seems impossible that this deposit, which is only an interlude in one of the later stages of the Glacial period, can have been accumulated in anything like the time assigned by Prestwich for the whole of that period.

If we consider the elevations and depressions of land which have taken place since the commencement of the Glacial period, the evidence all points to the same conclusion of immense antiquity. There is distinct evidence that since the first epoch of intense cold a great part of Britain has been submerged, until only a few of the highest mountains stood out from the Arctic Sea as an archipelago of frozen islands, and has been since elevated, with several minor fluctuations, to its present height. Marine shells of an Arctic character have been found on Moel-Tryfane, a hill in North Wales, in glacial drift 1392 feet above the level of the sea, and similar drift is traced continuously, both in Wales and Scotland, to a height of over 2000 feet. It rests on rocks which had been already rounded and polished by glaciers.

It is evident, therefore, that sufficient time must have elapsed during an intermediate phase of the Glacial period, for a depression of more than 2000 feet, followed by a re-elevation of an equal amount. Consider what this means. All we know of these secular movements of large masses of land shows them to be excessively slow. Even the small local elevations and depressions, like those of the temple of Serapis at Pozzuoli, which have taken place principally in volcanic districts, have not exceeded a few feet in historical times.

The deltas of rivers have increased, and the sea has sometimes eroded and sometimes added to the outline of coasts, but there has been no change for more than 2000 years in the general level of sea and land in any of the districts known to the ancient world. The spit of shingle which connects St. Michael's Mount with Cornwall, is still covered at flood and dry at ebb tide, as when the ancient Britons carted their tin across it to barter with Tyrian merchants. Marseilles is a sea-port, as it was when the Phocæan galleys entered its harbour. In Egypt it is evident that no considerable change of level, either of the land or of the Mediterranean, can have occurred since Menes embanked the Nile 7000 years ago.

The only authentic record we have of the rise or fall of masses of land as ascertained by actual measurement, are those of Scandinavia and South America. The Pacific shore of the latter was upheaved five or six feet for a distance of 500 or 600 miles, by the shock of a single earthquake, and remains of human art, such as plaited rushes and string, have been found in a bed of marine shells near Callao, showing that this part of the continent had been elevated eighty-five feet since it was inhabited by man. This, however, gives no clue to the rate of elevation, since we know nothing of the date of man's appearance in Peru, and the whole area is one of volcanic disturbance, which has been raised by successive earthquake shocks, and not by gradual elevation.

In the case of Scandinavia, however, where raised beaches up to the height of 600 feet above the sea level afford proof of much recent elevation, and where there are no signs of volcanic action, attempts have been made to measure the rate accurately by marks cut on rocks. The results, carefully considered by Sir C. Lyell, show a slow, uniform rate of elevation of two or three feet in a century, where the rate is at its maximum at Gefle ninety miles north of Stockholm, which dies out towards the North Cape, and is converted into a slow depression in the south of Sweden. At this rate of three feet per century the depression which carried the hills of Wales and Scotland 2000 feet down would have required 66,666 years, and its elevation an equal period, so that without any allowance for the time the sea-bottom may have remained stationary, this interlude of the Glacial period would have required 133,333 years. Of course, it is not implied that this was the real time, or that the rate both of elevation and depression may not have been faster; but all the evidence points to its having been gradual and not paroxysmal, as there are no traces of any contemporaneous earthquakes or volcanoes in Wales or Scotland. And whatever the rate may have been it is scarcely possible to suppose that it can have been such as to enable us to compress the whole Glacial and Post-Glacial periods, of which this was only one of the intermediate phases, within anything like the limits of from 25,000 to 35,000 years assigned to them by Professor Prestwich. On the contrary, all the evidence from existing known facts points rather to an extension than to a contraction of the times assigned by Lyell and Croll, and if the theory of the latter is correct, it would almost seem as if his first period of *maximum* refrigeration, 700,000 years ago, was that of the formation of the first great ice-cap. And whatever the time may be, it is clear that in its earlier stages man was already widely distributed over the earth, while there is the strongest probability that his origin must have taken place very much further back in the Pliocene or even in the Miocene period.

It must always be remembered that while the date of human origins in years or centuries is a question of great scientific interest, it makes little difference, as regards the religious and philosophical aspects of the question, whether it extends over 50,000 or 500,000 years. In any case, the fact is beyond question, that it is one of immense antiquity, far transcending any period recorded by history or tradition, and that during this immense period the course of humanity has been upward and not downward. Man has not fallen but risen, and arts, morals, societies, and civilization have been slowly developed from an animal-like condition of the lowest savagery.

Perhaps the issue between the long and short dates of the Glacial period can be most closely joined if we take that portion of it which comes nearest to historical times, and is known as the Post-Glacial. Prestwich assigns to this period a duration of "8000 to 10,000 years or less," that is, a duration of not more than 2000 or 3000 years before the time when we know for certain that a dense population and high civilization already existed in Egypt and Chaldæa. I am not aware that he assigns any reason for this highly improbable date, except the conjecture that the erosion of river valleys may have gone on more rapidly, owing to a greater rainfall.

Now the duration of this Post-Glacial period is a question, not of conjecture or theory, but of a vast number of definite and measurable facts. In the British Islands these facts have been carefully examined and ascertained with great accuracy, mainly by the labours of the Geological Survey. An eminent officer of this Survey, Mr. T. Mellard Reade, who has worked for many years at these beds in Lancashire and Cheshire, and is one of the best authorities on the subject, read, as recently as in February, 1888, a paper before the Geological Society, in which he gave a minute description of the successive changes in Post-Glacial times, by which the Mersey valley and estuary were brought into their present condition, with an estimate of the time they may have required. His estimate is "that in round figures 60,000 years for Post-Glacial time is a reasonable one and, as represented by these changes, well within the mark."

This is not a random estimate, but based on a careful calculation of the different changes which are shown by sections and borings to have actually taken place. At the close of the Glacial period the district was submerged, and the valleys of the old Pre-Glacial rivers were levelled up to a height of at least 200 feet by marine boulder-clay. The land then rose until its surface became an undulating upland plain, through which the present rivers began to cut the existing valleys. A mass of boulder-clay 200 feet in depth, and several miles in width, must thus have been removed by sub-aërial denudation before the next stage, which consisted of a general depression of the area, as is proved by the fact that borings show a series of estuarine deposits with marine shells in places fifty feet thick, overlying the boulder-clay, and levelling up the inequalities of its surface due to sub-aërial erosion. Above these silts and clays is a peat-bed, containing stools of trees with their roots running down into the clays below. This is a remarkable deposit, for a similar submerged forest bed is to be traced all round the shores of the British Islands, from Devonshire to the Orkneys. Evidently at a recent period, geologically speaking there has been an age of forests which flourished, and in their decay formed great beds of peat, in localities where no trees have grown within the Historical period. Before these forests could have grown, the marine silts and clays must have been elevated above the sea to a sufficient height to become dry land and covered with trees, and the climate must have been very different from that at present prevailing. It must have been more of a continental and less of an insular climate, and in all probability the German Ocean was then dry land, and the British Islands were connected with an Europe which extended westward up to the 100 fathom line. In no other way can the existence of submerged forests, and vast masses of peat with remains of trees, be accounted for in such isolated islands as those of Orkney and Shetland, now swept by ocean blasts, and where no vestige of a tree has grown for at least 2000 years, when a Roman author described them as "carentes sylva."

But at whatever height the land may have stood during this Forest period, it is evident that it must have subsided, at any rate to the extent necessary to bring the submerged forests to their present level of some feet below low-water mark. Or, indeed, some twenty-four feet more, for there is evidence that a rise to this extent has taken place, quite recently, along a considerable portion of the British coast, as shown by raised beaches. When I say recently, I mean in geological time, for in historical time there has been no appreciable change of level since the occupation of Britain by the Romans, or for nearly 2000 years.

In other regions, however, we have still more conclusive evidence of the great length of time which has elapsed since any appreciable change has taken place in the physical geography of Europe, and in the present relative levels of sea and land. The localities described by Homer in the *Odyssey* can be identified, and the very cave and beach pointed out in Ithaca, on which Ulysses was landed by the Phœnician mariners. The annals of Egypt carry us back still farther, and show that no appreciable change can have taken place in the levels of sea and land in the Eastern Mediterranean, for at least 7000 years, and probably for much longer.

With these facts, even if we had no other evidence than that of the submerged forests, Professor Prestwich's estimate of 8000 to 10,000 years for the whole Post-Glacial period down to the present time seems totally inadequate, and Mr. Mellard Reade's of 60,000 years much more probable. In fact, it seems impossible that changes, such as those demonstrated to have occurred in the Mersey valley, can have been accomplished within a period shorter than that which is shown by historical records to have elapsed in Egypt without perceptible change.

But whether the duration of the Post-Glacial period be more or less, it is evidently a small fraction of the time which is required to account for the work done during the preceding Glacial period, or rather periods, for there is distinct evidence that there were several advances and retreats of the ice-sheets, and alternations of climates, during some of which the winter temperature of Western Europe must have been higher than it is at present. The succession of ice-sheets is clearly shown by the sections afforded by the coast cliffs of the east of England, where four successive boulder-clays are shown, separated by masses of sand and gravel deposited during the melting and retreat of each ice-sheet. The alternations of mild Inter-Glacial with severe Glacial periods, is shown by the frequent presence in caves of a Southern fauna, some of which, like the hippopotamus—which is found as far north as Yorkshire—could by no possibility have lived in a country where the lakes and rivers were bound in ice for a great part of the year. And still more conclusively by the presence, in the south of France, of a vegetation comprising the fig-tree and delicate Canary laurel, in the region over which, at another period of the Glacial age, herds of reindeer roamed, feeding on lichens and Arctic-willows, and accompanied by the musk-ox, the glutton, the lemming, and other exclusively Arctic animals.

But although the evidence for the great antiquity of the Glacial period seems to be conclusive, it must be confessed that we are as far as ever from being able to assign any reliable explanation of the causes which produced it. It came on suddenly, for the interval between the temperate Pliocene and the extreme rigour of the first great ice-sheet is, geologically speaking, very short. Only a few feet of clay and sand separate the Cromer forest, in which the great southern elephant, the Elephas Meridionalis, and other southern mammalia roamed, from the boulder-clay of the Scandinavian ice-sheet, which carried rocks from Lapland and Norway, across the North Sea and over hills and valleys almost to the centre of Europe. This first period was the coldest, and after several oscillations of heat and cold, each apparently less intense than its predecessor, the climate of the Northern hemisphere finally settled down to its present conditions.

These facts seem to negative most of the theories, or rather guesses, which have been hazarded to account for this great and sudden refrigeration. It could not be due to any cooling of the earth, for this must have been gradual and progressive, and the great cold of the first period, instead of decreasing and disappearing, must have gone on increasing. It has been supposed that the solar system, on its journey through space, may have entered into, and emerged from, regions very much colder than those of former ages or at present, but such a cause is at present little more than a conjecture. Nor is it possible that any alteration in the position of the earth's axis can have occurred within the earth, for this would have disarranged its equatorial protuberance, which is precisely that of a fluid mass, rotating about the present axis, and could not be altered without producing a complete cataclysm. No one can suppose that an equatorial protuberance of more than 20 miles can have been shifted through many degrees of latitude, during the short interval between the close of the Pliocene and the commencement of the Glacial period.

Neither can the theories which have been applied to earlier geological epochs, of a warmer blanket of watery vapour and carbonic-dioxide in the atmosphere, account for such a sudden refrigeration and its gradual disappearance. The conditions under which the Pre-Glacial Cromer forest flourished, and those at present existing in the same locality, cannot have been so different as to imply a new order of cosmic or telluric causes.

There remain only two at all plausible theories, the

astronomical one of Croll, and that of Lyell, who explains everything by a different configuration of sea and land. Croll's theory explains many of the facts admirably, but, as we have seen, it cannot be accepted with confidence, in the absence of proof that a succession of Glacial periods has occurred in previous geological epochs. Nor is it very consistent with the fact that the cold period came on suddenly, and was greatest at first; while if due to the eccentricity of the earth's orbit, it ought to have come on gradually, and only attained its *maximum* simultaneously with that of the eccentricity. Lyell's theory is on the whole most generally accepted, as actual experience shows that high land in high latitudes is a cause of glacial conditions, and also that oceanic currents are a main factor in producing climate.

When we inquire under what conditions great glaciers are now formed, we find them to be mainly heavy snow-fall combined with low temperature. Thus the snow-fall is very heavy on the Pacific slope of the Sierra Nevada and coast range of Northern California and British Columbia; but it does not, as formerly, produce glaciers, because the temperature is not low enough to convert the winter snow into the frozen "néve" which is the source of glaciers, and to produce the conditions under which the accumulation finds its way to lower levels by solid rather than by fluid rivers. Again, extreme cold does not of itself produce glaciers, as is seen in Northern Russia and Siberia. The influence of ocean-currents is also apparent from the effects of the Gulf Stream, which gives open winters to the coasts and islands of Western Europe, in a latitude as high as that of the southern extremity of Greenland.

Here, then, are real causes which may account for

such a Glacial period as has been experienced, without invoking utterly unknown and conjectural theories. But there are considerable difficulties in the way of accepting Lyell's theory as the sole and sufficient explanation. The suddenness with which the great cold came on is one of them. It is difficult to suppose that such a great elevation of land in the North Atlantic as would be required, took place, almost at once, in the short interval in which the Pliocene passed almost continuously into the Quaternary. We are tolerably certain, from the similarity of the fauna and flora, that America was connected with the Old Continent during the Miocene period by a land passage across the North Atlantic, and yet there are no traces of a rigorous climate. On the contrary, a climate almost sub-tropical prevailed then in Greenland and Spitzbergen, far within the Arctic Circle.

Again, the Gulf Stream must always have been an important factor in determining the climate, but recent theories as to the great geological antiquity of the Atlantic Ocean make it difficult to conceive how this Stream can have been greatly diverted from its present course, in recent geological times. And the fact that the ice-cap extended much farther to the south in North America than in Europe, makes it almost certain that the influence of the warm Gulf and cold Polar streams must have been felt during the Glacial period, as they are now. How otherwise can we account for the fact that the difference of temperature between Europe and America seems to have been almost the same during the period of extreme cold in both, as it is now under temperate conditions? And the diversion of the Gulf Stream would certainly tend to produce less evaporation in the North Atlantic, and therefore less fall of rain or snow on Northern lands, whereas the contrary is required to account for the ice-caps. We must conclude, therefore, that while Lyell's theory affords the most probable explanation, we are still in a state of great uncertainty as to the causes which may have co-operated in bringing about the last and greatest vicissitude of climate, the Glacial period, which is so interesting to us from its close connection with the origin of man. The causes and duration of the last Glacial period, and whether there have been several, and if so, how many of such periods in former geological ages, are among the problems of the future which are pressing for solution.

CHAPTER V.

TERTIARY MAN.

Antiquity of Man—Man part of Quaternary Fauna—What this Implies —Historical and Neolithic Periods—Palæolithic—Caves and River Gravels—Glacial and Inter-Glacial Deposits—Wide Distribution of Palæolithic Implements in Early Quaternary Deposits—Origin of Species—Evolution and Migration—Diversity of Human Types—Objections to Tertiary Man—Specialization of Type—Survival through Vicissitudes of Climate—Positive Evidence for—St. Prest—Thenay—Tagus Valley—Monte Aperto—Cuts in Bones of Balæonotus—Elephas Meridionalis and Halitherium—Auvergne Worked Flints in Pliocene Tuffs—Castelnedolo—Human Bones in Pliocene—Olmo—Evidence from America—Californian Auriferous Gravels—Tuolumne and Calaveras Skulls—Age of Gravels—Skertchley's Stone Implements—Brazilian Caves—Pampæan Strata —Summary of Evidence.

Of all the discoveries of modern science, that of the antiquity of man has been the most startling. It is not like the abstract discoveries of astronomy and geology, which only indirectly affect the unscientific mass of mankind. It shatters at a blow what had been for centuries the axioms of the whole Christian world, respecting the origin of man, his place in creation, and the course of his development. A literal acceptance of the dates and narrative of Genesis was assumed to be the sole basis of knowledge on the subject, and to question what was told by a Divine revelation was universally considered to be alike ridiculous and impious.

As far as science had a word to say it was thought to confirm theology, for did not Cuvier himself lay down as an axiom that no human remains had been found in a fossil state, or in conjunction with the remains of any of the extinct animals? And although a few scientific men here and there, basing their ideas mainly on the dates of Egyptian monuments, pleaded for a somewhat longer period than the date assigned by Archbishop Usher, there may fairly be said to have been a universal consensus of opinion among all men, learned or unlearned, that the existence of the human race on our planet had not lasted longer than some 6000 or 7000 years before the present period. This was the universal opinion only thirty years ago, when in 1859 Mr. Prestwich read his memorable paper to the Royal Society, confirming the discoveries of M. Boucher de Perthes, and proving beyond a possibility of doubt that flint implements, fashioned by human hands, were found in Quaternary gravels, and brick-earths of the valley of the Somme, in juxtaposition with remains of the mammoth and other extinct animals, which must have been deposited when the river ran at more than 100 feet above its present level. The careful exploration of the Devonshire caves of Brixham and Kent's Hole, by committees of competent geologists, removed the last doubts on the subject, and since then, evidence has accumulated so rapidly from all quarters of the world, that the existence of Quaternary man has become as certain a fact as that the earth revolves round its axis.

Consider what this implies. The Tertiary epoch, in which mammalian life for the first time appears prominently and an approximation is made to existing conditions, is itself but a small fraction of the succession of geological ages since our planet became the abode of animal and vegetable life. At the outside, its three divisions of Eocene, Miocene, and Pliocene, may together represent one-twentieth part of the thickness of fossiliferous strata from the Cambrian to the Cretaceous. The Quaternary period again is but a fraction of the Tertiary; and the recent or existing epoch, including the Historic and Pre-Historic, is but a fraction of the Quaternary. The recent or Historical epoch, characterized by the existing fauna, and, in the main, by the existing climate and disposition of sea and land, is certainly not less than 7000 years old, when Egyptian records and monuments show us a populous and highly civilized nation already existing in the valley of the Nile, and civilized empires of almost as early a date in Chaldæa and China. The Pre-Historic period, characterized by the existing fauna and by neolithic man, must have lasted much longer, before such empires could have been developed from the rude and primitive civilization shown by the Scandinavian Kjokken-middens, the Swiss Lake-dwellings, and other early records of the Neolithic period. Borings in the Nile valley have everywhere brought up rude pottery, and other neolithic remains, from depths below the foundations of the oldest historical monuments, which, at the present rate of silting up by the annual inundations of the river, imply an antiquity of about 18,000 years. This may not be quite accurate as a chronological standard in years, but undoubtedly this, and other similar calculations from physical changes during the Neolithic period, all point to the conclusion that 15,000 or 20,000 years is the shortest time that can have elapsed since its commencement.

Then comes a great break. The climate, geographical

and physical conditions, and fauna, have undergone great changes when we next meet with traces of man, and the Quaternary period stretches back into the Pliocene, through an immense though unknown duration of time. This much however is known, that it embraces two, if not more, great Glacial periods, during the first and most severe of which the northern halves of Europe and America were buried under an ice-cap, in places 5000 or 6000 feet thick, resembling that of modern Greenland, and driving all terrestrial life before it into more southern regions. These Glacial periods alternated with long Inter-Glacial ages, when the ice retreated, and vegetation and animal life again returned to their old abodes, and again advanced and retreated, finally occupying their present stations when the glaciers had shrunk into the valleys of the loftier mountains.

It is certain also that vast changes in the physical geography, and configuration of sea, land, and rivers, occurred during this period. The British Islands, or a large portion of them, were at one time submerged to a depth of certainly 1500, and probably 2000 or 2500 feet beneath an Arctic sea, presenting nothing above it but an archipelago, of what are now mountain peaks; while at another time they were part of an European continent, then connected with Africa, and across which huge extinct lions, tigers, bears, elephants, and rhinoceroses roamed, and left their remains in the caves of limestone districts, and the sands and gravels of rivers, when they flowed 100 feet or more above their present level. During part of this period a southern fauna, and even the hippopotamus, found their way as far north as Yorkshire, testifying to the existence of great rivers flowing from the south across this Quaternary continent.

Now three facts have come out clearly from the latest research.

1. That man is a characteristic member of this Quaternary fauna just as much as any of these extinct animals; or, in other words, that wherever you find the mammoth, cave bear, or woolly rhinoceros, you may expect to find man; and where you find man in old deposits, you may expect to find the mammoth, cave bear, and rhinoceros.

2. That the man whom you thus find is "Palæolithic man," that is, man in such a rude and savage state that he has not yet attained the art of polishing stones, and uses implements roughly fashioned by chipping from flints or other hard stones of the district.

3. That these rude implements are found in the caves and gravels of the Quaternary period in Europe, Asia, Africa, and America; in fact, throughout the whole world, so far as it has been hitherto explored; and wherever they are found, the rudest and earliest implements, such as stone hatchets or celts, and flint flakes and scrapers, are almost identically of the same type.

These facts have such an important bearing on the origin of the human race, that it is desirable to consider them in some detail.

The discoveries, both of implements and of human skulls and skeletons, have now been so numerous, especially in the caves of France, England, Germany, and Belgium, that it has enabled geologists not only to prove the existence of Quaternary man, but to a considerable extent to analyze and classify the successive stages of his progress.

The earliest is that known as the Cave-bear epoch, which occupies the lowest position in the oldest caves, and in which the rudest human implements are found associated with a preponderance of bones belonging to this formidable animal. Thus in Kent's Cavern, in Devonshire, we have in descending order—

1. A layer of black mould, near the entrance, from three to twelve inches thick, containing successively relics of the Historical and Neolithic periods, and bones of existing species of animals.

2. A bed of granular stalagmite from one to three feet thick, securely sealing all below it.

3. Red cave earth, in places five to six feet thick.

4. A bed of older crystalline stalagmite, in places twelve feet thick.

5. Breccia of angular stones; red-clay and bones to the rock floor of the cave.

In the lower deposits (4 and 5) the bones are numerous, but almost exclusively those of the cave-bear, and a few human implements have been found, including a flint hâche or celt in the breccia, which is the oldest deposit of all. In the upper stalagmite, and cave-earth beneath it, were found numerous human implements of various sorts, including a bone needle and barbed harpoon, associated with remains of lion, cave-bear, mammoth, rhinoceros, hyena, reindeer, Irish elk, and other usual animals of the Quaternary fauna, including one tooth of the Machairodus or sabre-toothed tiger, which is characteristic of the Pliocene fauna.

Similar facts have been recorded in such a multitude of caves in France, Belgium, and Germany, especially in those of the South of France, that it is a perfectly well-established fact that the Palæolithic period may be divided roughly into three groups: an upper one, in which the reindeer was very abundant, and human implements showed a considerable advance in civilization; a middle stage, in which the reindeer was scarcer and the mammoth more abundant, with ruder human implements, though still showing considerable design; and the lowest of all, with fewer remains of the mammoth and more of the cave-bear, and with fewer implements, and those exclusively of stone of a very rude type.

This is exactly what might be expected if the theory of evolution applies to the human race. The first dawn of intelligence when primitive man emerged from the animal state, would show itself by his picking up natural stones to use as tools or weapons of offence. He would naturally select stones of the type of the hâche, with a sharp point for crushing in the skull, and a blunt butt-end to give weight to the blow and a firm grasp for the hand. This would hardly require more intelligence than that of the gorilla, who, living in forests, uses branches of trees as clubs; or of apes, who throw stones at enemies. The next stage would be to improve natural stones, or supply them if deficient, by chipping, so as to give a sharper and more solid point or edge, and a similar process would apply to flint chips used as knives or scrapers.

After a while, some genius would discover that by hafting the hâche, and attaching it as a lance to a long handle, he could kill without coming to such dangerous close quarters as was necessary when striking with the hand. This would lead to finer chipping, both to ensure penetration at the point, and to fit the butt-end for attachment. And finally, the invention of the bow would lead to diminished size and still finer chipping for the arrow-head. From this point the progress can be readily traced to the invention of barbs for arrows and harpoons, and the occasional substitution of bone for stone, as being more easily scraped into the desired form; and from these the evolution is uninterrupted up to the beautifully finished weapons of the Neolithic and Bronze periods. But the starting-point is the rude stone hâche, such as is universally found in the oldest deposits of caves and river gravels.

There has been a good deal of discussion as to the purposes for which these implements were employed, but there can be little doubt that their primary use was for killing large game and human enemies. The bushmen of South Africa, who represent most nearly this primitive savage state, use for this purpose implements so closely resembling those of the river drifts, that some of those exhibited at the Colonial Exhibition, and labelled "*pour le gros gibier*," might have been specimens from Amiens or St. Acheul.

A good deal of discussion has also taken place among British geologists as to the exact place, with reference to the great Glacial periods, occupied by the earliest drift and cave implements which have been found in this country. Most of them are Post-Glacial, that is, later than the retreat of the last of the two or more great ice-caps which extended over all except a few of the southern counties of England, during the Quaternary period. Some, however, are clearly proved to be either Inter-Glacial or Pre-Glacial, being overlaid by boulder-clay, as at Brandon, and in the caves of Cae Gwyn in North Wales; while as to the lowest deposits of many caves, as, for instance, the lower stalagmite and bone breccia of Kent's Cavern, there is no distinct evidence except of extreme antiquity, though the presumption is strong that they are either Pre-Glacial or Inter-Glacial. Mr. Pengelley, who has devoted years of research to

Kent's Cavern, expresses an unhesitating opinion that the lowest deposits are Pre-Glacial.

As fresh evidence accumulates, it all points towards the existence of man on British soil in Pre-Glacial, or very early Glacial times, and therefore to carry it back far beyond the period assigned to it by Post-Glacial geologists.

Thus, quite recently, rude palæolithic implements of unmistakable human design have been found near Wye, in Kent, at an elevation of upwards of 300 feet, in a gravel which does not correspond with the existing valleys, but which overspread the chalk plateau of the North Downs, and was drained by rivers running southwards in a directly opposite course to that of the present streams. Professor Prestwich, whose bias, as we have seen, is towards shortening the period of man's antiquity, after a personal examination of the locality, came to the conclusion that this drift was immensely older than the ordinary high-level gravels of existing rivers, and in all probability was Pre-Glacial.

Since Professor Prestwich's paper was read, similar palæolithic implements have been found by Mr. Worthington Smith, on the Chalk downs near Dunstable, up to a height of 759 feet above Ordnance datum, and some of them embedded in the brown clay which, with gravel, covers the chalk. But the question of the evidence afforded by England is comparatively unimportant, for the wider induction of continental experience settles conclusively the general relations of palæolithic man to the Quaternary period. It is absolutely certain that in the later stages of the palæolithic record, when man had already made considerable progress, and was able to draw and carve figures of the contemporary animals with a good deal of artistic skill, vast herds of reindeer roamed over the plains of Southern France and Germany, accompanied by a group of Arctic animals, such as the musk-ox and the lemming, which are found even on the Italian side of the Alps. When this was the case in Southern Europe, it is evident that all its northern portion and higher mountains must have been covered by ice and frozen snow, and one of the great Glacial periods must have been in full force. All earlier deposits, therefore, in which ruder implements and a more temperate or even African fauna are found, must of necessity have been either Inter-Glacial or Pre-Glacial, and there is no reasonable doubt that the earliest of such deposits date back at least to the earlier stages of the Quaternary period. We must recollect that when we talk of geological periods, there was no real break in the succession of time. We merely use a convenient expression to distinguish those formations, between which the evidence of the regular progression of development has been lost for such a long period, that when we find it again the characteristic fauna and flora have undergone a marked change. But the idea of cataclysms and of repeated destructions and miraculous renovations of the whole vegetable and animal worlds, is completely exploded, and every day affords fresh evidence of the gradual process of transition from one so-called epoch or formation to the succeeding one. Thus types and even species appear sparingly in one formation, become abundant in another, and finally die out and disappear, or persist with slight modifications, as we see in the first appearance of fish in the Silurian, and of reptiles in the Carboniferous eras, in each case in one or two geological periods before they became the predominant type. This applies specially to the relation of the Quaternary to the Pliocene and Miocene periods. It is difficult to say definitely where one begins and the other ends. Thus not only do most of the great Mammalian genera persist from the Miocene, through the Pliocene and Quaternary, down to the recent periods, but some specific forms, such as the tapir, have continued unchanged; while the ox, bear, horse, wild boar, and other species first found in the Pliocene, survive through the Quaternary to the present day.

The gravels and sands of St. Prest, the forest bed of Cromer, and other Pre-Glacial formations, contain such a mixture of characteristic mammals, that some geologists have considered them to be Pliocene, while others have pronounced them to be Quaternary.

What we really can affirm with certainty is, that as soon as we find a Quaternary fauna firmly established, we find man forming an essential and characteristic part of it. Can he be traced further back into the Tertiary? The question involves points of the highest interest, for, as in the issue between short-time and long-time geologists as to the duration of the Glacial period, the issue really is between evolution and miracle.

Even if the Glacial or Quaternary periods were extended to the 200,000 years assigned to them by Lyell, Croll, Geikie, and other leading geologists, the difficulty as to man being a product of evolution would be only postponed, and not removed. By no possibility could such conditions of the human race as are found at the commencement of the Quaternary period have been produced by the natural laws applicable to the rest of the animal creation, unless man can be carried back into the Tertiaries.

For under what circumstances do we find undoubted

traces of the existence of man upon earth, early in the Quaternary period? Not in small numbers, or in some limited locality, in which we may suppose the human species to have originated, and from which we can trace the different races slowly developing and radiating out to more distant regions. No, when we find them lowest in the Quaternary, we find them in large numbers, and practically all over the world, from China to Peru, and from Northern Europe to South Africa.

This is so important that I proceed to state the facts in some detail, and specify the localities in which stone hatchets and knives, of the rude type of the oldest river drifts and lowest cave deposits, have been found in Europe, Asia, Africa, and America.

The list is doubtless incomplete, and every day is adding to it, but it is already amply sufficient to prove the general proposition.

In England they have been found in the river drifts and deposits of the Thames, the old Solent river, and all the existing and Quaternary valley systems south of a line drawn across it, a little to the north of the Bedford Ouse; and in the caves of all the limestone districts of Yorkshire, Derbyshire, North and South Wales, Somersetshire, and Devonshire, and they are absent only in those northern districts which were covered with ice during the successive phases of the Glacial period. In France and Belgium they are met with in the oldest drifts of the valleys of the Seine, Somme, Meuse, Loire, Rhone, Garonne, and other rivers, and in almost innumerable caves and rock-shelters in all the limestone districts, from Liege and Maestricht to the Pyrenees, and on the Mediterranean coast at Mentone. In Spain and Portugal they appear in the drifts of the Tagus and Ebro, and in Italy in those of the Tiber and Arno. In Central and Southern Germany and Switzerland they are found in numerous caves and river drifts, often deeply buried under thick beds of the loess, or fine glacial mud, which was deposited during the melting of the great ice-fields.

In Asia these palæolithic implements, associated with extinct animals, have been found almost everywhere, where search has been made for them. They have been found in Asia Minor and Syria, in the Caucasus, in Mongolia, China, and Japan. India, which has been examined by competent geologists, affords the most authentic and complete record. Here they have been found in large numbers, both in the river drifts of the Nerbudda, Godavery, and other rivers, and in the laterite of Madras and other places, which is a loamy land-deposit similar to that of the loess of Europe and China. Implements almost exactly of the type of those of St. Acheul, though made of quartzite, as flints were wanting, have been found in Bengal, Orissa, the Deccan, Scinde, Assam, and other provinces; and some of them in deposits which, from the extinct animals associated with them, experienced geologists are doubtful whether to consider as upper Pliocene or as the lowest Quaternary.

In Africa, well characterized palæolithic implements have been found in Algeria and in the valley of the Nile; and at the other extremity of the continent, at Natal and other places in Cape Colony.

America furnishes some of the most conclusive proofs, both of the extreme antiquity, and of the wide diffusion of man. Human implements, human skulls and bones have been found associated with the mastodon and other extinct animals, over nearly the whole area of the United States; in Mexico, Brazil, and in the pampas of Buenos

Ayres and Patagonia; associated in South America with the Glyptodon and other extinct mammals of its peculiar fauna. In one instance, in Buenos Ayres, a human skull was found under a huge carapace of this extinct armadillo, which it was conjectured might have been used as a roof for a hut. In these South American cases, however, as well as in those which will presently be referred to from California, the geological age is uncertain, and they are considered by some to be evidences of Pliocene, by others of early Quaternary man; while in other instances they are probably Post-Glacial, or at latest, Inter-Glacial. But in one typical case, that of the discoveries of Mr. Abbott in the drift of the Delaware valley at Trenton, in New Jersey, there can be no hesitation in referring them to the same early Quaternary period as the corresponding finds in the oldest river drifts of Europe and Asia. The Trenton implements are of a granular argillite, exactly resembling in size and form the flint implements of the valley of the Somme; and they are found sometimes twenty feet deep, in an old bed of gravel, with large boulders, which is exposed in the cliffs of the river's banks. A portion of a human lower jaw was found at a depth of sixteen feet in the gravel, and also a human skull of a peculiar type, being small, long, and very thick.

We are able, therefore, to affirm as an undoubted fact, that at the earliest stage of the Quaternary period the human species not only existed, but was already widely diffused over four continents, and occupied nearly the whole surface of the habitable globe. How did man get there? Evidently by the same process by which other fauna become distributed over wide distances and extensive zoological provinces, that is, by migration from one or more centres, where the different species were first developed in the course of evolution. In the case of land mammals, this implies where there has been an uninterrupted land connection within recent geological periods.

There is no fact better established by geological and zoological research, than that the existing fauna are not uniformly alike throughout the world, but are located in separate provinces, bounded by some barrier of sea, mountain, or desert, insurmountable by the ordinary animal species. The most signal instance of this is that of the absolute separation of the two totally dissimilar faunas of Southern Asia and Australia, by the narrow strait of Lombok, not above twenty miles wide, which is a deep sea fissure or channel, dating back to very remote geological times. On the other hand, in the north temperate zone of Europe and Asia one may travel from the Atlantic coast of Western Europe to the Eastern coast of China, without observing any marked change in the familiar fauna and flora, the extension of which to the British Islands and Japan, leaves no doubt that they recently formed part of the same continent; while the existence of so many of the same forms in North America, makes it certain that there was a land connection, at no distant geological date, between the Old and New Worlds, by what is now the North Atlantic, and probably also by Behring's Straits. The familiar instance of the absence of snakes in Ireland, shows clearly how this extension of a fauna was accomplished by gradual migration. Ireland was connected with England and with continental Europe long enough to enable most forms of the European fauna to occupy it. Herds of Irish elk, deer, oxen, wolves, and other animals roamed over it; but some of the slower moving reptiles had not had time to reach it before it became finally separated from England by St. George's Channel.

The only alternative to migration is the special miraculous creation of every separate species which has ever existed throughout the vast range of geological time, and this idea is as thoroughly exploded as that of the absence of snakes in Ireland being due to the prayers of St. Patrick in the seventh or eighth century. It breaks down under the weight of the innumerable instances of special miracles, which must be invoked on the most trivial occasions. Thus it has been shown that more than 160 miraculous creations must have taken place to account for the separate species of land-shells alone, which are peculiar to the little group of the Madeira Islands.

Admitting, then, evolution to be the cause of the origin of species, and migration for their diffusion, it must be observed that the human species is specially organized for extensive migration. The structure of man, and his intelligence, even in the most rudimentary form, enable him to overcome obstacles and resist changes of climate and environment, which would be fatal to most of the brute creation. And, as a matter of fact, in historical times we know that New Zealand and the Pacific Islands have been peopled by migration; and that races like the Bushmen, Esquimaux, and Australians, which come nearest to the state of primitive men, are essentially migratory. If the population of America were annihilated, with the exception of the Esquimaux and Fuegians, there is little doubt that they would creep onwards along the sea-coast, accumulating their Kjokkenmiddens as they went, until they had occupied the whole continent. But the process must necessarily have been a very slow one, and there must have been already a considerable population and pressure on the means of subsistence, before these Quaternary men could have spread over nearly the whole habitable globe, and left their remains where we now find them. The fact that they are so found makes it certain that they must have had a long series of ancestors, and that the first origins of the human race must be sought in a vastly more remote antiquity. The immense time required for such migrations will be apparent when we consider that it is not only a question of traversing such great distances, but much more of becoming gradually acclimatized during the passage from Arctic, or temperate, through tropical regions. Evidently the existing Esquimaux or Laplanders could not reach Patagonia or South Africa, without passing through a wide extent of hot and pestilential country, in which the northern immigrants could only live by the gradual survival of new types adapted to the altered conditions.

Another well-established fact points to the great antiquity of the human race when those early palæolithic implements were so widely distributed. A sufficient number of skulls and skeletons have been found associated with these implements to enable ethnologists to classify them as belonging to essentially different races. Thus the skulls found in America all present distinctive characters of the high and narrow type now existing among the various native races of that continent. In Europe, those of the Canstadt type, which is considered to be the oldest, and of which the celebrated Neander-thil skull is an extreme instance, are very dolicocephalic, or long-headed, with markedly projecting brows, differing essentially from those of the Cro-Magnon type, which represent an exceptionally tall race with a good cranial development, equal to that of many modern European races; while the Furfooz type again, is that of a dwarfish race, with small round heads, resembling the modern Laplanders. This diversity of race argues for a long departure from the original type, involving development through a long series of ages. We know from the Egyptian monuments that a period of 5000 years has been insufficient to produce any perceptible change in the type of the Negro, the Copt, the Semite, and other races of Africa and Western Asia.

It is remarkable, however, that while this diversity of race type is thus early found, there is almost perfect identity among the early palæolithic implements found in regions the most distant from one another. Rude stone hatchets, knives, and scrapers, are of the same form and fabricated in the same way, whether they come from the gravels of the Delaware, the Thames, the Tagus, the Godavery, or the Yang-tse-Kiang; from the caves of Devonshire, the deserts of Mongolia, or the plains of Patagonia and South Africa. The only apparent exception is afforded by the stone implements found in the auriferous gravels of California, which consist mainly of rude stone mortars and pestles, resembling those used for pounding acorns by modern tribes of Digger Indians, inhabiting the same districts. This uniformity of industrial type over such wide spaces shows that the peopling of the earth by migration must have been effected while the human race was still in that uniform state of rudimentary intelligence, which had not got beyond the first stage of supplementing natural stones by rude chipping.

Thus far we have been going on ascertained facts, admitted by all competent geologists, but in taking the next step and carrying man back into the Tertiaries, we enter on new ground, where positive evidence is scanty and disputed, and where probabilities and theoretical preconceptions are to a great extent invoked to supply its want. Among English geologists especially, there still remains a strong desire to abridge as much as possible the time of man's existence upon earth. The evidence furnished by England, which has been almost entirely covered during recent geological times by two or more successive ice-sheets, is comparatively weak to carry back the evidence for palæolithic man, even into Pre-Glacial times, and some good authorities still contend for all such remains in this country being Post-Glacial. Others, again, of less weight, and the general public who have a smattering of science, have a vague fear that every extension of man's antiquity carries them further away from the old theological standpoint, and brings them nearer to the proof that man is the product of evolution from an animal ancestry. The evidence of facts has, however, become too strong to maintain this ground, and the Quaternary line of defence being broken through, the defenders of old ideas have fallen back on their next entrenchment, and insist that man, if not Post-Diluvian, or Post-Glacial, is at any rate Post-Tertiary.

We pass here from the region of facts universally admitted, into that of probabilities, and statements of facts which although probable in themselves, and apparently well authenticated, are still disputed by competent authorities. Let us first deal with the probabilities for and against the existence of Tertiary man. It is objected that an animal so highly organized and specialized as man, can hardly have come into existence in geological periods characterized by a fauna, so much nearer the primitive and generalized type of Mammals, as those of the Pliocene, and still more of the Miocene and Eocene eras. The answer to this is that such a highly specialized specimen of the anthropoid type as the Dryopithecus undoubtedly did exist in the Middle-Miocene. This, which was an anthropoid ape, as highly organized as the chimpanzee or gorilla, and of a stature equal to that of man, has been found in that formation in the South of France and in Germany. Now, looking at man simply as an animal, the anthropoid ape is just as much a specialized development of the primitive quadrumanous type as man. Monkeys and apes are specialized for life in forests and climbing trees, as man is for life on the earth and walking, but in their anatomical structure they correspond bone for bone and muscle for muscle. If there is any truth in evolution they must have descended, not necessarily one from the other, but both from a common ancestor.

Again, it is said that man could not have survived for such a succession of geological periods during which so many other species have died out and disappeared. But here again the answer is, that many of the animals which are associated with man as part of the Quaternary fauna, have in fact survived unchanged from the Pliocene, and with slight modifications from the Miocene periods, and that man's larger brain, and consequently greater intelligence, must have given him a better chance of survival than in the case of elephants, rhinoceroses, oxen, and horses. If man could survive, as we know he did, the severe and extreme fluctuations of the different Glacial, Inter-Glacial, and Post-Glacial periods, what was there in the milder and more equable conditions of the Pliocene and Miocene to have prevented his existence?

The theoretical objections, therefore, to Tertiary man seem to be of the weakest and vaguest character, while on the other hand, the probabilities in its favour are so cogent as almost to amount to demonstration. How could man, early in the Quaternary period, have already found his way to the remotest regions of the globe, and developed a variety of types and races, if his first appearance on earth lay within the limits of that period? One might as well suppose that elephants, horses, and all the other mammals associated with man in the Quaternary period, sprung suddenly into life along with him by some act of miraculous creation, in the teeth of all the accumulated and irresistible evidence which shows them existing in the upper Tertiary, and traces their ancestry and lines of progressive development through the Miocene into the earliest Eocene period.

Having thus cleared the ground of probabilities, I proceed to state the positive evidence for discoveries of human remains in Tertiary formations, premising that it is nearly all the result of the last few years, and is rapidly accumulating; and that there is no reason to expect that it will ever be abundant, as the more nearly we approach to the time and place of man's origin, the narrower must be the area, and the fewer the stations, at which we can hope to find his traces, and the greater the effect of denudation in obliterating those traces.

The first well-authenticated instance is that of St. Prést, near Chartres, on the Eure, one of the tributaries of the Seine. Here the lowest gravels of the present river rest on gravels of what Lyell, after personal examination, considered to be an earlier Pliocene river, and which are characterized by the older forms of elephant and rhinoceros; the Elephas Meridionalis, and Rhinoceros Leptorhinus, instead of by the Quaternary Mammoth and Rhinoceros Tichorinus. In these older gravels have been found stone implements, and bones of the Elephas Meridionalis with incisions evidently made by a flint knife worked by a human hand. This was disputed as long as possible, but Quatrefages, a very cautious and competent authority, states in his latest work, published in 1887, that it is now established beyond the possibility of doubt. It is contended, however, by some geologists, that this formation, though always considered to be Pliocene until human remains were found in it, is in reality a very low stage of the Quaternary, or a transition bed between it and the Pliocene. The instance, therefore, cannot be accepted as absolutely conclusive for anything more than the existence of man at the earliest commencement of the Quaternary period, though the evidence all points to the gravels being really Pliocene. The same uncertainty applies to the celebrated discovery by the Abbé Bourgeois, of flint knives and scrapers in the Miocene strata of Thenay, near Blois. When these were first produced, the opinion of the best authorities was very equally divided as to their being the work of human hands, but subsequent discoveries have produced specimens as to which it is impossible to entertain any doubt, especially the flint knife and two small scrapers figured by M. Quatrefages at p. 92 of his recent work on *Races humaines*. They present all the characteristic features by which human design is inferred in other cases, viz.: the bulb of percussion and repeated chipping by small blows all in the same direction, round the edge which was intended for use.

The human origin of these implements has been

greatly confirmed by the discovery that the Mincopies of the Andaman Islands manufacture whet-stones or scrapers almost identical with those of Thenay, and by the same process of using fire to split the stones into the requisite size and shape. These Mincopics are not acquainted with the art of chipping stone into celts or arrow-heads, but use fragments of large shells, of which they have a great abundance, or of bone or hard wood, and the scrapers are employed in bringing these to a sharper point or finer edge. The main objection, therefore, at first raised to the authenticity of these relics of Miocene man, that they did not afford conclusive proof of design, may be considered as removed, and the objectors have to fall back on the assumption, either that the implements were fabricated by some exceptionally intelligent Dryopithecus, or that the Abbé Bourgeois may have been deceived by workmen, and mistaken in supposing that flints, which really came from overlying Quaternary strata, were found in the Miocene deposit. This hardly seems probable in the case of such an experienced observer, and had it been so, the implements might have been expected to show the usual Quaternary types of celts, knives, and arrow-heads, fashioned by percussion, whereas the specimens found all bear a distinct type, being scrapers and borers of small size, and partly fashioned by fire. The other supposition is based on no evidence, and contrary to all we know of the limited intelligence of any anthropoid ape. If it were true we might at once say that the missing link had been discovered, as a Dryopithecus, able to do what the Mincopics are now doing, might well have been the ancestor of man. On the whole, the evidence for these Miocene implements seems to be very conclusive, and the objections to have hardly any other ground than the reluctance to admit the great antiquity of man, which so long opposed itself to the recognition of the discoveries of M. Boucher de Perthes.

The same class of objection apply to the palæolithic hatchets found by M. Ribiero, in beds of the valley of the Tagus, at Olta, in Portugal, which have always been considered as being of the upper Miocene. It is thought possible that they may have fallen at some distant period from overlying Quaternary gravels, and become mixed up with the upper bed of the Miocene. The congress of geologists, therefore, who met at Lisbon three years ago, thought it wise to suspend their opinion as to the Tertiary age of M. Ribiero's implements.

Other discoveries, however, of the same nature, seem to be absolutely conclusive for man's existence, at least as far back as into the Pliocene era. An Italian geologist, M. Capellini, has found in the Pliocene strata of Monte Aperto, near Sienna, bones of the Balæonotus, a well-known species of a sort of Pliocene whale, which are scored by incisions obviously made by a sharp cutting instrument, such as a flint knife guided by design, and by a human hand. At first it was contended that these incisions might have been made by the teeth of fishes, but as specimens multiplied, and were carefully examined, it became evident that no such explanation was possible. The cuts are in regular curves, and sometimes almost semi-circular, such as a sweep of the hand could alone have caused, and they invariably show a clean cut surface on the outer or convex side, to which the pressure of a sharp edge was applied, with a rough or abraded surface on the inner side of the cut. Microscopic examination of the cuts confirms this conclusion, and leaves no doubt that they must have been made by such an instrument as a flint knife, held obliquely and pressed against the bone while in a fresh state, with considerable force, just as a savage would do in hacking the flesh off a stranded whale. Cuts exactly similar can now be made on fresh bone by such flint knives, and in no other known or conceivable way. It seems, therefore, more like obstinate prepossession, than scientific scepticism, to deny the existence of Tertiary man, if it rested only on this single instance.

As regards the evidence from cut bones it is very conclusive, for experienced observers, with the aid of the microscope, have no difficulty in distinguishing between cuts which may have been made accidentally or by the teeth of fishes, and those which can only have been made in fresh bone by a sharp cutting instrument, such as a flint knife. In fact, the best authorities on the subject, such as M. Mortillet, the Curator of the Museum at St. Germain, M. Hamy, and M. Quatrefages, while admitting the authenticity of the cuts submitted to them in a few cases, have rejected it in numerous others, as in the well-known instance of the grooves on the bones of a rhinoceros, which Delaunay had found in a Miocene deposit at Billy.

The only incisions on bones from very early strata, which these experts have admitted as undoubtedly made by sharp cutting instruments held by a human hand, are those above-mentioned, viz.: on the Elephas Meridionalis of St. Prest, and the Pliocene Balæonotus of Monte Aperto; and in the humerus of a Halitherium from the Upper Miocene of Pouancé (Maine et Loire). This shows with what caution and scrupulous good faith the experts have worked, who bear testimony to facts, which, if admitted, are a conclusive demonstration of the existence of Tertiary man.

But in addition to these instances from cut bones, there are others equally certain and well-authenticated. In the region of the extinct volcanoes of Auvergne, in which the celebrated fossil man of Denise was discovered under a stream of lava, embedded in a volcanic tuff, which, however, was considered to be probably Quaternary, there are older lava streams overlaying tuffs and gravels, which, from the fossils contained in them, are undoubtedly Tertiary. From one of these Tertiary gravels at Puy-Courny, M. Rames, a competent geologist, assisted by MM. Badoche, Chibret and Grandvaux, obtained at three different points a considerable number of flint implements, which, if found in any Quaternary deposit, would have been accepted without hesitation as of human origin. They comprise small and rude specimens of the types found in the lowest Quaternary gravels, such as celts, knives, and scrapers, and present all the characters by which artificial are distinguished from natural flints in those formations, viz.: bulbs of percussion, and chippings in a determinate direction on the sides and points intended for use; while no such chippings appear on other parts of the flint, as must have been the case if they had been the result of casual blows on natural flints.

M. Quatrefages, by whom the subject is fully discussed, and the objects figured in his recent work, lays great stress on the fact that while the beds contain five different sorts of flints, those which present traces of design are confined exclusively to one description of flint, which is most easily manufactured, and best adapted for human use. He observes with much force that a torrent capable of tearing flints from their bed and rolling them on, with collisions violent enough to imitate artificial chipping, could not have exercised a selection, and confined its operations to one only, out of five different descriptions of flints. He shows also that the worked edges exhibit, when closely examined, both intentional chipping and fine parallel striæ, as from repeated use in cutting or scraping, while nothing of the sort is to be seen on the sides left in the natural state, though they are often as sharp, or even sharper.

It only remains to add that these specimens were submitted by M. Rames to two Congresses of French geologists, the first at Blois, when doubts were expressed in some quarters; the second one, last year, at Grenoble, when the Congress decided that the existence of Tertiary man was in this case fully established.

Italy supplies the next instance, and it is a very remarkable one, for here competent geologists have found, not merely implements or cut bones showing human design, but man himself, including skeletons of several individuals. The discovery was made on the flank of the hill of Castelnedolo, near Brescia, in a bed which is identified by its fossils as belonging to the Lower Pliocene. The excavations were made with the utmost care, in undisturbed strata, by M. Ragazzoni, a scientific man of good reputation, assisted by M. Germani, and the results confirmed by M. Sergi, a well-known geologist, who visited the spot and inquired minutely into all the circumstances. According to their united statement some human bones were found in this deposit by M. Ragazzoni as far back as 1860. This led to further excavations, made at different times, and with all the precautions pointed out by experience. The deposit was removed in successive horizontal layers, and nowhere was the least trace found of the beds having been mixed or disturbed. At a considerable depth in it, were found the bones of four individuals, a man, a woman, and two children, which presented the same appearance of fossilization as the bones of extinct animals found in the same deposit. The female skeleton was almost entire, and the fragments of the skull were sufficiently perfect to admit of their being pieced together so as to show almost its whole form.

This preservation of the entire skeleton might lead to the conjecture that it had come there as the result of a subsequent burial, but this supposition is negatived by the undisturbed nature of the beds, and by the fact that the other bones were found scattered in the same stratum, at considerable distances from the perfect skeleton. M. Quatrefages sums up the evidence by saying, "that there exists no serious reason for doubting the discovery of M. Ragazzoni, and that if made in a Quaternary deposit, no one would have thought of contesting its accuracy. Nothing, therefore, can be opposed to it but theoretical *à priori* objections, similar to those which so long repelled the existence of Quaternary man; objections which have long since been refuted, and shown to be absolutely inconsistent with a multitude of established facts."

If we accept this conclusion this remarkable consequence follows: that man, so far back as the Early Pliocene period, was perfectly human, for the skull and bones present no marked peculiarity, or approximation to an animal type. The skull is of fair capacity, and very much what might be expected of a female of the Canstadt race. But if this be so, it necessarily puts back the origin of the human species to a vastly more remote antiquity, which can hardly be less than that of the Early or Middle Miocene, in which the remains of the great anthropoid Dryopithecus have been found.

A skull very similar to the above has also been found in Italy, in a lacustrine deposit at Olmo, near Arezzo, on the flank of the Apennines; but although it was found at a depth of nearly fifty feet from the surface, and some feet lower than a layer of clay containing a tooth of the Elephas Meridionalis, a species which in Northern Europe scarcely survived the Pliocene period, the whole formation is considered, from other remains found in it, as probably belonging to an early Quaternary age, and therefore not affording satisfactory evidence of Tertiary man. It can only be quoted as affording some corroboration of the discoveries of Capellini and Ragazzoni, by showing that man has existed in Italy for an immense period, and is found in deposits between which and the Pliocene there is no abrupt line of demarcation.

This completes the evidence from the Old World. Turning to the New World, we find, both in North and South America, numerous proofs of the existence of man from a very remote antiquity, but there is some difficulty in arriving at definite conclusions as to their Tertiary date, from the fact that the succession of geological periods does not exactly correspond on the two sides of the Atlantic. America has been said to be, in some respects, a whole period behind Europe and Asia in this succession. Thus the mastodon, which in the Old World is a characteristic Miocene and Pliocene species, and did not survive into the Quaternary, is found in America in the latest drifts, and even in peat masses associated with neolithic flint arrows, and not impossibly survived into the Historical period. The bear family, on the other hand, which is so conspicuous in the old formations of

Europe, is not found in America until the Quaternary. The extinct fauna also of South America is, like the present, that of a distinct zoological province from either North America or Europe, so that we cannot assume that the Zenglodon and other huge ancestral types of armadillos and ant-eaters, were necessarily of an age corresponding to our Tertiary.

With this reservation I proceed to state some of the leading instances which have been referred to by American geologists as establishing the existence of Tertiary man on that continent.

The most important case is that of the skulls and stone implements which have been found in the auriferous gravels of California, the evidence for which, and for other ancient remains in North America, has been very carefully summed up by the distinguished naturalist, Mr. Alfred Wallace, in an article in the *Nineteenth Century* of November, 1887. These gravels are the result of an enormous denudation of the Sierra Nevada, which has filled up all the great valleys on its Pacific slope with thick deposits of *débris*, forming in some cases detached hills, and even mountains, of considerable height. While this was going on, there were repeated volcanic eruptions in the higher range, giving rise to beds of lava, tuff, and ashes, which are frequently interstratified with the gravels; and finally, the close of the volcanic period was marked by a great flow of basaltic lava, which spread in a nearly level capping over the whole surface of the country. This, and the subjacent beds of gravels and tuffs, has since been cut down by the action of the present rivers, to a depth of sometimes 1500 or 2000 feet, leaving a series of isolated, tabular hills composed, on the upper part, of a horizontal layer of basalt, varying from 50 to 200 feet in thickness, and in the lower part, of 800 to 1500 feet of gravels, lava-beds, and tuffs. Thus what was once a single lava stream, or succession of lava streams, is now a series of detached hills, the tops of which form parts of one gently inclined plane, sloping from the mountains towards the plains, and now, in some cases, 1000 feet or more above the adjacent valleys.

The present rivers have in some places cut down the lavas and gravels to the beds of ancient rivers, which flowed in different courses from the existing ones, and it is in the beds of these ancient rivers that the principal accumulations of gold are found. Hence an enormous amount of the oldest gravels has been excavated in working for gold, and in some of these workings human remains have been found, associated with animal remains, which are all of extinct species, entirely distinct from those that now inhabit any part of the North American continent. Some of the genera, such as Hipparion, Auchenia, and Elotherium, would, if found elsewhere, undoubtedly be taken to denote a Pliocene, if not a Miocene formation. The vegetable remains also indicate a totally different flora from that now prevailing in California, and which Professors Lesqueraux and Whitney—the latter the geologist of the State, and well-known from his *Report on the Auriferous Gravels of the Sierra Nevada*—consider to be of Pliocene age, with some affinities to Miocene. Numerous stone implements have been found associated with this extinct fauna and flora in nine different counties, and human bones in five widely-separated localities. The two most remarkable instances of the latter are—

1. The Tuolumne skull. A fragment brought up

from a shaft in Table Mountain, at a depth of 180 feet below the surface, beneath a bed of three feet of consolidated volcanic tuff, with fossil leaves and branches, over which is a deposit of 70 feet of clay and gravel.

2. The Calaveras skull. This was found in 1866, under four beds of lava, and in the fourth bed of gravel from the surface, embedded in a rounded mass of earthy and stony matter containing bones. The cemented gravel was removed with great difficulty, and disclosed a human skull, nearly entire, with several bones of the human foot and other parts wedged into the cavity of the skull, the whole being in a fossilized condition, like that of the animal bones in similar formations. Human bones have been found in two other instances—one by an educated observer, under a bed eight feet thick of lava; and more recently a discovery has been announced of rude stone implements in Tertiary gravels of Stone Creek, Colorado, associated with shells which are considered by conchologists to be not later than of the older Pliocene.

The Calaveras case is, however, the typical one, owing to its having been extracted from the matrix by Professor Wyman, and all the circumstances of the find thoroughly investigated by Professor Whitney. When the discovery was first announced, it was objected that the skull was possibly taken by the miners from some Indian grave. But this objection disappears before the fact that it was fossilized, and embedded in a matrix which no forger could have counterfeited, and even more conclusively from the great number of instances in which human bones and implements have been discovered at different localities in similar formations. Even the polemical imagination of the Duke of Argyll could hardly invent a conspiracy of so many groups of Californian miners, at different times, and in different localities, to hoax scientists, or to supply proofs for or against the Darwinian theory of the Descent of Man. Nor would men intent on such a fraud have buried fragments instead of whole skulls, and stone implements of a type different from that which, if they had known enough on these subjects to conceive the fraud, they must have been aware would have been expected. For the nature of these implements is an exception to the general rule, that the oldest type found throughout the world, from South Africa to China, is everywhere the same, consisting of rudely-chipped celts, knives, and scrapers, the Californian implements consisting of stone plates or mortars, and pestles or pounding stones, very like those used by some living tribes of Indians for crushing acorns.

Quatrefages, assuming that these implements were used for pounding corn, justly considers it highly improbable that agriculture could have been known at such an early period, and that Pliocene man in California could have been so far in advance of his Quaternary brother on the Atlantic side of the continent, as shown by the rude celts, and knives of the Trenton gravels. But if they were used for crushing acorns, the argument is not so clear, for a tribe of primitive savages, living among oak forests, might use flat stones and pounders for the purpose, while hunting tribes might use rude celts, as the bushmen do at the present day. Either form seems equally within the range of the early dawn of human intelligence, and not much in advance of that of the gorilla or chimpanzee.

Equally futile is Sir J. Dawson's surmise that the

skull may have been dropped into some old mining shaft. There is no evidence for any prehistoric mining for gold in California, such as is found in the copper region of Lake Superior, and it is certain that, if any such had existed, it must have been confined to the superficial deposits. Nothing but an intrepid determination to ignore facts could have led to such a supposition. The Calaveras skull is not a solitary instance, but one of several human bones, and hundreds of human implements, which have been found, at wide distances apart, in these auriferous gravels, and often underneath beds of dense basalt, which could by no possibility have been pierced without the aid of metal tools and blasting powder. Objections like these prove nothing except that the objector is in the theologico-scientific frame of mind, which sees everything relating to the origin of man through the medium of the first chapter of Genesis.

The only serious objection to assuming these Californian discoveries to be a conclusive proof of the existence of Tertiary man, arises from the fact that several good American geologists dispute Professor Whitney's conclusion that these auriferous gravels are of Tertiary origin. They consider that such an enormous accumulation could only have been formed during a Glacial period, when frost and ice were grinding down the mountains, and swollen rivers, from melting snow and glaciers, sweeping the *débris* down the valleys into the plains. This leaves doubt as to their origin in the comparatively mild and equable climate of the Pliocene period, but as regards the question of the great antiquity of man, it does not much signify to which period we assign them. Any time subtracted from the Pliocene has to be added to the Quaternary, for the fact remains unquestioned that, since man existed in California, valleys have been filled up by drifts from the waste of mountains to a depth in some cases of 1500 feet; these covered by a succession of tuffs, ashes, and lava streams, from volcanoes long since extinct, and finally cut down by the present rivers through beds of solid basalt, and through this accumulation of lavas and gravels. Such an operation corresponds in time with that by which the great river systems of the Old World were sculptured out from a table-land, standing, in some cases, many hundred feet higher than at present, as shown by the deposit of the loess, which is universally recognized to be an accumulation of fine glacial mud.

The latest contribution towards the antiquity of human remains in California is contained in a paper read to the Anthropological Society by Mr. Skertchley, the well-known geologist, to whom we are indebted for the discovery of palæolithic implements beneath the chalky boulder-clay at Thetford, in Norfolk.

During a visit to the Spring Valley gold-mine, in one of the tributary valleys of the Sacramento River, he ascertained the following facts: This mine is worked by hydraulic jets directed on the sands and gravels of an old river which once flowed in an impetuous course down a steep gradient from the Sierra Nevada. It has long since ceased to flow, and the bed of the old river is now buried under 500 feet of its own deposits, capped in places by 100 feet of basalt, which has flowed in wide sheets from long-since extinct volcanoes. The section given by Mr. Skertchley is—

1. Basalt cap	25 to 100	feet.
2. White sands and gravels	450	„
3. Blue gravel, with boulders	2 to 15	„
4. Blue gravel, with large boulders ...	50	„
5. Bed rock—metamorphoid cretaceous slates.		

Stone mortars, rudely chipped, occur abundantly in the white sand (No. 2), about 300 having been found, and one is said to have occurred in No. 3. There can be no question of their occurring *in situ*, as they are washed out of the gravel by powerful hydraulic jets, from the working face of the mine, which forms an artificial cliff of 400 to 600 feet in height.

Nor can there be any doubt as to their human origin, for the specimen produced by Mr. Skertchley to the Anthropological Society was universally admitted to have been artificially wrought. Their use was probably for pounding acorns, which then afforded a great part of the food of the savages who inhabited the district, as they did recently of the Digger Indians.

The question, therefore, is entirely one of the age of the gravels, as to which American geologists differ, some assigning the upper, or white gravels, to the Pliocene, others to the early Quaternary period. As Mr. Skertchley says, "If the human remains had not been found in them, geologists would never have doubted their Tertiary age. At any rate they must be of immense antiquity. Since they were deposited, the present river system of the Sacramento, Joaquim, and other large rivers has been established; cañons 2000 feet deep have been excavated by these later rivers through lava, gravels, and into the bed rock; and the gravels, once the bed of a large river, now cap hills 6000 feet high."

This definite information, conveyed by an experienced

geologist like Mr. Skertchley, gives confirmation and precision to what has been stated from a variety of other sources, as to the frequent discovery of human implements, and even, in a few instances, of human skulls, from similar auriferous gravels over a wide range of country in California. Whether Tertiary or not, it is evident that they must carry back the date of man's existence, in the north-west of America, to a period vastly older than that of 25,000 or 30,000 years assigned to him by the latest guess of Professor Prestwich.

The other instances from America are open to the same doubt as to their geological age. The cavern of Semidouro, in the plateau of Lagoa-Santo, in Brazil, has yielded sixteen human skulls, associated with bones of extinct species, such as Glyptodon, Machærodus, Hydrochærus, Scalidotherium, and others, which, if found in Europe, would undoubtedly be taken to imply a Tertiary fauna. But there remains the doubt as to the real succession of geological periods in America, and if the Mastodon lived on there until recent times, for which there is a good deal of evidence, there is no conclusive reason why the Machærodus and other Tertiary forms might not have survived from the Pliocene or Miocene into the Quaternary. The human implements also found in these Brazilian caves seem, in many cases, of too advanced a type to be readily accepted as of such extreme antiquity.

The same doubt also applies to the numerous human remains found by two competent observers, M. Ameghino and M. Burmeister, at different points in the pampas of Buenos Ayres. They both recognize two distinct beds in this pampean formation—an upper one, in which these remains have been found, and a lower one, in which nothing of human origin has yet been discovered. Ameghino, relying on the fossil remains of extinct animals, considers the upper bed to be Tertiary; while Burmeister considers the lower one only to be Pre-Glacial, and the upper one to be Quaternary. While these doubts continue, we must hold our judgment in suspense as to the evidence from America, though undoubtedly it tends as far as it goes to confirm the rapidly accumulating evidence from the Old World of the existence of Tertiary man; and the discovery of his traces at so many widely-separated places, at such a remote antiquity, adds to the irresistible force of the conclusion that his first origin, and subsequent diffusion by migration, must be sought in one of the geological formations preceding the Quaternary.

To sum up the evidence, there are at least ten instances of the alleged discovery of human remains in Tertiary strata, of each of which it may be safely said that if the remains had been those of any other Mammalian species, no doubt would have been entertained of their Tertiary origin by any geologist. Four of these are in France, those of St. Prest and of Puy-Courny in the Pliocene, and of Thenay and Pouancé in the Miocene; three in Italy, in the Pliocene of Monte Aperto, St. Olmo, and Castelnedolo; one in Portugal, in the Miocene of the Tagus; in North America, the skull of Calaveras and other numerous human remains in the presumably Pliocene auriferous gravels of California; and in South America, in the pampean remains of Buenos Ayres. Of these, the discoveries at Puy-Courny, Monte Aperto, St. Olmo, and Castelnedolo, seem to be undoubted, both as regards the human nature of the remains, and the Tertiary character of the deposits. Those of St. Prest and of the Californian gravels are doubtful only as regards the question whether the deposits may not be of the earliest Glacial or Quaternary period, rather than Tertiary, the evidence from the associated fossil remains being strongly in favour of their Tertiary origin. There remain three cases of alleged discoveries in the Miocene, viz.: at Thenay, Ponancé, and in Portugal, the evidence for which, especially for the two former, is extremely strong and almost conclusive, while the objections to them are obviously based on a reluctance to admit such an extension of human origins, rather than on scientific evidence.

In none of these cases, as further evidence has accumulated, has it tended to shake the conclusions of the first discoverers as to the human character of the implements and the Miocene age of the formations. On the contrary, the most cautious authorities, such as M. Quatrefages, who held their judgment in suspense when the first implements were produced, have been converted by subsequent discoveries, and expressed their conviction that doubt is no longer possible. And the latest Congress of French geologists has expressed the decided opinion that the existence of Tertiary man is fully proved.

On the whole, we may say with confidence of the problem of Tertiary man that, if not completely solved, it is very near solution, and that there is little doubt what the solution will be.

The next generation will probably accept it as an obvious fact, and wonder at the doubts now entertained, very much as we wonder at the incredulity with which the discovery of palæolithic implements in the Quaternary gravels of the Somme by M. Boucher de Perthes was received by the scientific world, when it was first announced.

CHAPTER VI.

THE MISSING LINK.

Human Origins—Evolution or Miracle—First Theories Miraculous—Conception of Natural Law—Law proved to be Universal in Inorganic World—Application to Life and Man—Darwin and Evolution—Struggle for Life and Survival of the Fittest—Confirmed by Discovery of Missing Links—Professor Cope's Summary—M. Gaudry—Instances of Missing Links—Bears and Dogs—Horse—Pedigree of the Horse from Palæotherium and Eohippus—Appearance and Disappearance of Species—Specialization from Primitive Types—Condylarthra—Reptiles and Birds—Links between other Genera and Orders—Marsupials and Mammals—Monotremata—Ascidians and Fish—Evolution of Individuals and Species from Primitive Cell—Question of Missing Links applied to Man—Man and Ape—Resemblances and Differences—Specialization of Human Type—For erect posture—How Man differs from Animals—Mental and Moral Faculties—Language—Tools—Progress—Mental Development—Lines of Research for Missing Links—Inferior Races—Fossil Remains—Point in direction of Tertiary Origin.

Of all the problems which have been raised, but not solved, the most important is that of the origin of man. It is important, not only as a question of the highest scientific interest, but from its bearings on the deepest mysteries of philosophy and religion. Is man, like the rest of the animal creation, a product of evolution acting by natural laws, or is he an exception to the general rule, and the product of some act of secondary supernatural interference? Or to put it in theological language, is man a consequence of that "original impress," which Bishop Temple pronounces to be more in accordance with the idea of an omniscient and omnipotent Creator; to whom "a day is as a thousand years, and a thousand years as a day," than the traditional theory of a Creator constantly interposing to supplement and amend His original creation by miracles? Or is he an exceptional supplement and amendment to such original creation, miraculously introduced at one of its later stages? It is a question which has to be solved by facts, and not by theories or prepossessions.

As regards the physical universe, and the whole of the world of life, with the possible exception of man, it may be taken as already solved in the sense of evolution and original impress. But in the case of man, the question is still *sub judice;* the missing links have not yet been discovered which connect him with primitive forms, and scientific authorities are not yet agreed whether the time which has elapsed since his first appearance on earth is sufficient to afford a possibility of his being a creature of evolution. The problem is of such importance that it may be well to state its conditions in some detail.

When I say that evolution has become the accepted law of the whole animate and inanimate universe, with the possible exception of man, why do I say this? The old theory of special miraculous interpositions to account for all unexplained phenomena was the most natural and the most obvious. It was, in fact, the inevitable result of the first attempts of the human mind to connect effects with causes, or, in other words, to reason. Take the case of thunder. What could the first savage who reasoned on the subject infer except that, the noise being like the roar of an angry wild beast or enemy, and the flash like that of the darting of an arrow or javelin, there was probably a sort of magnified man like himself in the clouds, full of wrath and very capable of doing him an injury? The savage who reasoned thus, and the early priests and astronomers who, whenever they saw motion in the sun and planets, inferred life, were natural philosophers, who reasoned correctly from their premises, only their premises were wrong. In course of time, it came to be demonstrated that phenomena formerly supposed to be isolated miraculous acts of an Anthropomorphic power, were linked together by that invariable sequence which we call law, and that their real first cause or origin must be pushed vastly further back in space and time, and relegated more and more from the known to the unknown.

The establishment of Newton's law of gravity as the pervading principle of all celestial movements, gave the first great blow to the old miraculous theory, and introduced the conception of Natural Law. Geology did for time what astronomy had done for space, and since the publication of Lyell's *Principles* no serious thinker has doubted that the successive stages by which the earth was brought to its present state were due to evolution, acting by natural laws over immense periods of time. The discoveries of modern chemistry have confirmed the impression of the uniformity and invariability of Law, by showing it extending from the infinitely great to the infinitely small, from stars to atoms; while the spectroscope shows the identity of matter and energy throughout this extreme range. Above all, the establishment of the laws of the indestructibility of matter and energy, and their mutual transformation into new forms and new modes of action, have placed special causes altogether out of court, and reduced all the phenomena of the inorganic universe to one law of universal simplicity and generality. Instead of speculating with ancient sages who may be the God who flashes lightnings from the skies, or drives the chariot of the sun; or even as late as Kepler, assigning a spirit to each planet to direct its harmonious movement, the question for modern science is reduced to the ultimate stage of—What mean these atoms and energies into which everything can be resolved? Whence came they, and how did they become endowed with those laws which have enabled them to build up the universe by an irresistible evolution?

But the miraculous theory died hard. Based as it was on popular apprehension and on theological prepossession, when driven from the outwork of the inorganic universe, it held out stoutly in the inner citadel of life. Were not species distinct, and if so how could they have come into existence unless by a series of special acts of miraculous creation? Above all, was not man a miracle, with his high faculties, "only a little lower than the angels;" and did not all records and traditions describe him as a recent creation, who had fallen from a high state of perfection by an act of original sin? Nay more, did not science itself confirm this view, and had not Cuvier laid down the axiom that no human remains had been found in connection with any extinct animals, or in any but the most superficial deposits? The discovery of innumerable human implements and remains in all quarters of the globe, in caves and river drifts of immense antiquity, and associated with extinct animals, has shattered this theory into fragments, and it is now as impossible to believe in man's recent origin and fall, as it is in the sun's daily journey round the earth, or the notion that it might be as big as the Peloponnesus.

Still, the difficulty as to the creation of distinct species remained, and until the publication of Darwin's celebrated work on the *Origin of Species*, the miraculous theory, though driven back, could hardly be said to be routed. But evolution was in the air and Darwin's book produced the effect of a fragment of crystal dropped into a saturated solution. In an incredibly short time, all the floating elements crystallized about it, and the speculations of science took a definite form, the evidence for which has gone on strengthening and increasing from that day to this, until, as I have said, with the solitary exception of human origins, evolution or original impress has become the axiom of science, and is admitted by every one who has the slightest pretensions to be considered a competent authority.

This predisposition to accept Darwin's views arose from various causes. The establishment of evolution as a fact in the material universe had familiarized men's minds with the idea of Natural Law, and the discoveries of astronomy and geology had proved to demonstration that the accounts of creation, formerly taken to be inspired truths which it was impious to question, could only be considered as vague poetical versions of the ideas which were current among Eastern nations in the infancy of science. The last remnant of respect for these narratives as literal records of actual events vanished when the discoveries of M. Boucher de Perthes were confirmed, and it became apparent that man was not a recent creation who had fallen from a high estate, but the descendant of palæolithic savages, who had struggled slowly up to civilization through immense periods of time. As a knowledge of natural history increased, it became apparent that the earth had not been peopled recently from a single centre, but that it was divided into numerous vegetable and zoological provinces, each with its own separate flora and fauna; and a better acquaintance with the zoological record showed that this had been the case for millions of years, and through the vast succession of strata of which the earth's crust is composed. Finally, the multiplication of species, both now existing and in past geological ages, reached a point which, on any theory of separate supernatural creations, required an amount of miracle which was plainly absurd and impossible. When it came to this, that 160 separate miracles were required to account for the 160 species of land shells found to exist in the one small island of Madeira, and that 1400 distinct species of a single shell, the Cerithium, had been described by conchologists, the miraculous theory had evidently broken down under its own weight and ceased to be credible.

In this state of things, Darwin not only supplied a vast number of instances, drawn from his own observation, of graduation of species into one another, and the wide range of varieties produced and rendered permanent by artificial selection, but what was more important, he showed the existence of a *vera causa* operating in nature, which could not fail to produce similar effects. If a pigeon-fancier could, by pairing birds which showed a tendency to variation in a particular direction, produce in a few generations races as distinct from the original blue-rock as the fantail or the pouter, it is evident that nature could do the same in a longer period. Nay, not only that nature *could*, but that nature *must*, do this, for in the struggle for existence, variations, however slight, which gave an advantage to individuals, must tend to survive, and become extended and fixed by the operation of heredity. This was the famous theory of "Natural Selection" and "Survival of the Fittest," which at once converted the chaos of life into a cosmos, and extended the domain of harmonious law to the organic as well as the inorganic universe. Attractive, however, as the theory was from the first to thinking men, its universal acceptance at the present day is due mainly to the immense amount of confirmation which it has since received. This confirmation has come from two independent sources—the discovery of Missing Links and Embryology.

When Darwin's theory was first propounded, the objection was raised that if species were not created distinct, but gradually evolved from one another by slight variations, geology ought to show us the intermediate forms which must have existed before the permanent types were established. The objection was reasonable, and Darwin was the first to admit it, but he pleaded the imperfection of the geological record, and predicted that with fuller knowledge of it, the gaps would be filled up and the missing links discovered. The truth or falsehood of his theory was thus staked on the discovery of missing links. The case was almost similar to that of the truth of Halley's calculations as to the orbit of his comet, being staked on its return at the predicted period. The comet did return, and the missing links have been discovered, or so many of them that no doubt remains in the minds of scientific men that evolution has been the real law of the animal and vegetable kingdoms.

In fact, the discovery of missing links has gone so far, that Professor Cope, one of the latest and highest authorities on the subject, and who has done so much for it by his discoveries of the wonderfully rich fossil fauna of the Tertiary formations of the Rocky Mountains and California, says—"We have attained the long-since extinct ancestor of the lowest vertebrates. We have the ancestor of all the reptiles, of the birds, and of the mammals. If we consider the mammals separately, we have traced up a great many lines to their points of departure from very primitive types. Thus we have obtained the genealogical-trees of the deer, the camel, the musk, the horse, the tapir, and the rhinoceros; of the cats and dogs, of the lemurs and monkeys, and have important evidence as to the origin of man."

M. Gaudry, the celebrated discoverer of the fossil treasures of the Upper Miocene of Pikermi, repeats the same thing. He says—"If we take a skeleton of a fossil mammalian species, and compare it with one of an analogous living species"—as for instance a Mammoth or Mastodon with a modern elephant—"placing the heads, vertebræ, humerus, radius, femurs, feet, &c., of the one, side by side with those of the other, the sum of the likenesses will appear so much greater than that of the differences, that the idea of family relationship will impose itself on the mind. In vain would sceptics try to throw doubts on this relationship by pointing out some slight shades of difference. We see too many points of resemblance to admit that they can be all fallacious." And again he says, "Where our predecessors saw ten or one hundred distinct beings, we see only one; and instead of creations thrown, as it were, into the world at haphazard, without law and without connection, we follow the trace of a few types whose essential characters are so similar as to enable us to comprehend them in still simpler types, and thus hope to arrive some day at understanding the plan which God has followed in producing and developing life in the world."

This is almost identical with Bishop Temple's profession of faith "that it seems something more majestic, more befitting of Him to whom a thousand years are as one day, thus to impress His will once for all on His creation, and provide for all its countless varieties by this one original impress, than by special acts of creation to be perpetually modifying what He had previously made."

A clear, popular conception of this question of "missing links" is so important for all who desire to understand the latest conclusions of modern science, that it may be well to illustrate it by a homely example. Fifty years ago, the popular belief respecting the animal creation was summed up in the simple words of Dr. Watts' hymn:

> "Let dogs delight to bark and bite,
> For 'tis their nature to;
> And bears and lions growl and fight,
> For God has made them so."

Science could only shrug its shoulders and say, "So it seems; I have no better explanation to give."

How different are the terms in which science would now reply. "Made, if you like, but how made? As *individuals*, each from a cell not distinguishable from any other microscopic cell of the lowest animal and vegetable organisms, but endowed with such an impress of evolution that it develops through the stages of fish, reptile, and mammal, into the special mammalian form of its parents. As *species*, traceable through a similar progression backwards from the living form, through intermediate ancestral forms graduating by slight distinctions into one another, up to the generalized Eocene type of the Placental mammal, and thence backwards by less definite but still traceable variations, to the types of the marsupial, the reptilian, the fish, the vertebrate, and so up to the primitive cell in which the individual living animal originated."

Thus the dog and bear, now so distinct, can be traced up to Amphicyon and Hyænarctus, which combined the qualities of both; the former being rather more dog than bear, the latter rather more bear than dog; and these again, either through the Creodonta to the Bunodonta of the early Eocene, or through the Ictitherium to the Cynodictis, or weasel-like dog of the same formation, which is clearly a descendant of the insectivorous Marsupials of the Secondary age.

The horse affords the best example of this progressive evolution, the specialization from the generalized Eocene type of a five-toed and tubercular toothed mammal being clearly traced, step by step, down to the present one-toed horse. The evolution took the course of adapting the original form to the requirements of an animal which had to live on wide prairies or desert plains, where a bulky body had to be transported at high speed, by leaps and bounds, over great distances, both to find food and to escape from enemies by flight.

For this purpose, evidently, one solid toe, protected by a single enlarged nail or hoof, was preferable to five or three weak toes terminating each in a separate nail or claw; and in like manner, teeth adapted for cutting and masticating grass were better than the more mill-stone-like tubercular teeth adapted for grinding down shrubs and branches of trees. Accordingly, we find the evolution of the horse constantly following this line. In Europe, the Hipparion, who is the immediate ancestor of the horse whom it closely resembles, has already the two lateral toes so rudimental as to have become wholly useless; in the Anchitherium the tips of the outer toes just touch the ground, while the Palæotherium is a distinctly three-toed animal, though the middle toe is larger than the two side toes. We have thus a complete progression from a slow, heavy animal, adapted for living on marshy ground, like the tapir, to the courser of the plains, whose latest development, under artificial selection, is seen in an Ormonde or a Donovan.

In America, the links in the pedigree of the fossil horse are still more numerous, and the transitions closer. The line begins in the Early Eocene with the Eohippus, an animal of the size of a fox, which in addition to four well-developed toes of the forefoot, had the remnants of the hoofed fifth toe. In the Upper Eocene, the Eohippus was replaced by the Orohippus, in which the rudimentary first digit had disappeared, and the fifth was reduced to a splint. In the Lower Miocene the Mesohippus, which was about as large as a sheep, had only three toes with a rudimentary splint on the foreleg, and in its teeth and other particulars approached more closely to the horse. In the Upper Miocene, Mesohippus is replaced by Miohippus, which approaches closely to the Anchitherium of Europe; while in the Lower Pliocene this gives way to the Protohippus, which approached the horse very closely, and was about the size of an ass. Like the Hipparion of Europe, which in many respects it resembles, it had three toes, of which only the middle one reached the ground. In the Middle Pliocene we have the Pliohippus, which has lost the small hooflets on the rudimentary toes, and is in all respects very like a horse; and finally in the Upper Pliocene we have the true horse. This progression gives rise to two important remarks. First, that size cannot be accepted as of much importance in tracing lines of descent, as might indeed have been anticipated from the wide variations in the size of dogs and other domestic animals introduced by artificial selection. Secondly, that the extinction of widespread and apparently unexhausted races of animals is a fact which has to be reckoned with. The total disappearance of the horse in America, where it and its ancestors had existed in such numbers from the Early Eocene down to quite recent times, is a most perplexing problem. There is no appearance of any great change of environment since the horse roamed in countless numbers over the continent of America, and we know from the experience of Europe that it was a hardy animal, capable of resisting both the torrid heat of Arabia, and the intense cold of the Glacial period. And so many other species survived in America, from the Pliocene to the Quaternary and recent periods, as to show that the extinction of the horse was an isolated phenomenon. And as of extinction, so of creation. We do not fully understand the exact process by which types and species have either appeared or disappeared, and this affords the only ground left to those who, from theological or other prepossessions, are hostile to Darwinism. They say his theory of natural selection from spontaneous variations does not account for everything, and does not explain fully all the laws of these variations. This may be partly true, but it in no way affects the truth of evolution, which is a *fact* and not a *theory*, and is quite independent of the subsidiary question, whether natural selection can account for all, or only for a principal part of the facts which, in some way or other, have to be accounted for. Thus, whether the long neck of the giraffe was developed by natural selection taking advantage of accidental variations in this direction, or partly by this, and partly by heredity fixing variations induced by use and disuse of organs in stretching to reach the branches of palms, in no way affects the question whether the animal is a product of evolution or a miraculous creation.

To return to the pedigree of the horse, which may be taken as the typical instance of descent traced by progressive specialization. What is a horse? It is essentially an animal specialized for a particular object, that of the rapid progression of a bulky body over open plains or deserts. When mammalian life first appears abundantly in the lower Tertiaries, it is in the primitive generalized type, in which nature seems always to make its first essays, as if it were trying its 'prentice hand on a simple sketch, to be gradually developed into a series of finished pictures. The primitive sketch in this instance took the form of what Professor Cope calls a "pentadactyle, plantigrade, bunodont," by which formidable collocation of words we are to understand an animal which had five toes at the extremities of each of its limbs; which walked on the flat of its feet, and whose molar teeth presented a flat surface, with four, or in the very earliest form, three little cones or tubercles, to assist in grinding its food. It may give some idea of the precision and certainty to which such researches have attained, to say that this primitive form was predicted by Professor Cope in 1874, from the progress towards it traced in following backwards various lines of later descent; and that seven years later, in 1881, the prophecy was fulfilled by the discovery that such a type of mammals, now known as the Condylarthra, actually existed in large numbers in North America, in the early Eocene period.

Consider now what the specialization from this original type to the horse implied. The first step was to walking on the toes instead of on the flat of the foot, a change which, whether owing or not to the lady Condylarthra having adopted the modern fashion of wearing high-heeled boots, became general in most lines of their descendants. For galloping on hard ground it is evident that one strong and long toe, protected by a solid hoof, was more serviceable than four short and weak toes, protected by separate nails. Accordingly, coalescence of the toes is the fundamental fact in the progress of structural changes through successive species, by which the primitive Bunodont was converted into the modern horse. Corresponding with this are other progressive changes in the articulation of the joints, especially those of the bones corresponding to the ankle and wrist joints, which are modified from a contact of plane surfaces into a system of tongues and grooves, which give freedom of action in direct progression, but secure them against the dislocations from shocks and strains, to which they would be exposed in galloping or jumping. So in other types the specialization takes different forms, but always towards the sharper distinction of species formerly more united and generalized. Thus the half-bear, half-dog, and half-cat original type of the Eocene, becomes differentiated into the three distinct types of the wholly bear, dog, and cat of later formations.

Nor is this tracing back of existing mammalian species to ancestral forms in the Early Tertiary all that recent science has accomplished. The course of geological discovery for the last twenty, and specially for the last ten years, may almost be summed up as that of the discovery of "missing links," until gap after gap, which seemed to separate not only species, but genera and orders, by insurmountable barriers, has been bridged over by intermediate forms. Thus to take one of the most striking instances—What can, at first sight, appear more unlike than reptile and bird, and who would have ventured to predict that any relationship could be traced between a tortoise and a swallow? And yet nothing is more certain than that the Reptilia pass over into the Aves, by successive gradations, which make it difficult to pronounce where one ends and the other begins. The pterodactyle, or flying dragon of the lias, approaches in structure and habits towards the bird type; the ostrich retains some resemblance to the pterodactyle, but the complete transitional type, or "missing link," has been found in those feathered reptiles, or birds with reptilian heads and teeth, whose remains have fortunately been preserved in a fossil state. The Archæopteryx, from the Œningen slate of the Upper Oolite, in the museum of South Kensington, is a beautiful specimen of such a missing link, and would certainly be taken for a bird by any casual observer, though comparative anatomists find many of its essential features to be reptilian.

The Archæopteryx and other transitional types which have been discovered in Europe and America between birds and reptiles, afford perhaps the most obvious and universally intelligible instances of what recent geology has done in the way of the discovery of "missing links," between genera and orders now widely separated; but similar discoveries have gone a long way towards establishing the continuity of life from the earliest periods in which it appears, down to the present day, and showing the kind and progress of the changes in structure, which, in the course of evolution, have linked the various orders and species of living forms together. Thus the higher form of Placental mammals which became predominant in the Early Tertiary, differs from the Marsupials, which extend into the trias of the Secondary period, by the greater extension of the allantoid or membrane which surrounds the fœtus. In the Placentals this completely surrounds it, so that the fœtus remains part of the mother until birth; while in the Marsupial the young are born incomplete, and take refuge for a time in a pouch which is attached to the mother's stomach. But there are fossil animals in the Eocene which combine the two characters, showing a Marsupial brain and dentition, with a Placental development. They are, in effect, Marsupials in which the allantoid, instead of being arrested at an early stage, has continued to grow.

Again, the Marsupials are linked on to still lower forms of animal life through the Monotremata, of which a few specimens survive in Australia, typified by the Ornithorynchus, or water-mole, which has the bill of a duck, and lays eggs. This order has only one opening, called the cloaca, for the purposes which, in higher orders, are performed by separate organs, and it is remarkable that this stage is passed through by man and the higher mammals in the course of their embryonic development.

Going still further back, the lines of demarcation between orders are, as in the case of birds and reptiles, more and more broken down every day by the discovery of intermediate forms, and we can almost trace the evolution from the Ascidian or lowest vertebrate type into the fish, the amphibia, the reptile, and so upwards. And it is remarkable that this course of evolution invariably corresponds with the general progressive evolution of types through geological ages, and with the embryonic evolution of individual life from the primitive cell. It is not too much, therefore, to assume evolution to be the demonstrated law of the world of life as well as of that of matter, and to confine ourselves to the question whether man is or is not a solitary exception to this law.

We are now in a position to examine more closely the bearing of this question of "missing links" on that of human origins. Geologically speaking, man is one of the order of Primates, which includes also the catarrhine apes and monkeys of the Old World, the platyrhine apes and monkeys of America, and the lemurs or half-monkeys which are found principally in Madagascar and a few districts of continental and insular Asia and Africa.

Of these, the anthropoid apes—the chimpanzee, gorilla, and orang approach most closely to man in their structure.

In fact, considered as mere machines, the resemblance between them and man is something wonderful. It is much closer than is suggested by a mere comparison of outward forms. One must have read the results arrived at by the most distinguished comparative anatomists, to understand how close is the identity. Not merely does every bone, every muscle, and every nerve in the one, find its analogue more or less developed in the other, but even in such minute particulars as the direction of the hairs on the forearm converging towards the elbow, there is an absolute correspondence.

It is in the brain, however, which is the most important organ, as being that on which the specially human faculty of intelligence depends, that the close physical resemblance between man and the other quadrumana is most striking. The brain of all quadrumanous animals is distinguished from that of quadrupeds by certain well-defined characters. Those of lemurs, monkeys, baboons, and apes, show a progression of these characters from the lemurs, whose brain differs little from that of rodents, up to the anthropoid apes, the chimpanzee, the gorilla, and the orang, who have a brain which in its most essential particulars closely resembles that of man. In fact, the brain of these apes bridges over much more than half the interval between the simplest quadrumanous form of the lemur and the most advanced—that of man; while in like manner the brains of some of the inferior races of mankind, and of idiots, where the development of the brain has been arrested, bridge over the interval between man and ape, and in some extreme cases, approach more nearly to the latter than to the former type both in size and structure.

Attempt after attempt has been made to find some fundamental characters in the human brain on which to base a generic distinction between man and the brute creation, but such attempts have invariably broken down under a close investigation. Thus, in the celebrated controversy between Owen and Huxley, the former distinguished anatomist thought that he had found such a distinction in the hinder part of the human brain, but it turned out that he had been misled by relying on the plates in the work of the Dutch anatomists Camper and Vrolik; and Huxley, confirmed by them, proved by actual dissection that all the characters on which Owen relied were to be found equally in the brain of the chimpanzee and other higher quadrumana.

The distinction also on which the very term quadrumana is founded is proved to be fallacious, for Huxley has shown that the termination of the hinder limbs of the anthropoids is really a foot with a prehensile great toe, and not a hand, and there are many instances, both of human individuals and races, in which this toe has considerable flexibility, and is used in climbing trees or picking up small objects. And so in innumerable other cases in which anatomical observations, supposed to be specifically human, have either been found wanting in some individual men, and present in some individual quadrumana, or have been traced in both in some undeveloped or fætal condition.

And yet with this close identity of anatomical conditions there is, as Huxley emphatically asserts, a wide gap between man and the highest ape, which has never been bridged over, and which precludes the idea of direct lineal descent from one to the other, though it implies close relationship. The differences are partly physical and partly intellectual. Of the former, it may be said that they may be all summed up in the fact that man is specialized for erect posture.

Speaking broadly, it may be said that man is a member of the order of Primates, specialized for erect posture; while monkeys are specialized for climbing trees, and anthropoid apes are a sort of intermediate link, specialized mainly for forest life, but with a certain amount of capability for walking erect and on the ground.

Thus, to begin at the foundation of the human structure, the foot, with its solid heel bone, arch of the instep, and short toes, is obviously better adapted for walking and worse for climbing than that of monkeys. The upright basis of the foot corresponds with longer, stronger, and straighter bones of the leg, and a greater development of muscles to move them. The erect posture determines the shape of the pelvis and haunch bones, which have to support the weight of the vertebral column and intestines in a vertical direction. The vertebral column, again, is arranged with a slight double curvature, so as to enable the body to maintain an upright posture, and to afford a vertical support for the head. And finally, the larger brain is rendered possible by its weight being nicely balanced on a vertical column, instead of hanging down and being supported by powerful muscles requiring strong processes for lateral attachment in the vertebræ of the neck.

Again, the fore-limbs being entirely relieved from the necessity of being used as supports, acquire the marvellous flexibility and adaptability of the human arm and hand; a specialization which has doubtless a good deal to do with man's superior intelligence, for as we see in the case of the elephant, the intelligence of an animal depends not merely on the mass of the brain, but very much on the nature of the organs by which it is placed in relation with the surrounding environment. In this respect there is no animal organ comparable to the human hand, and we may probably trace its influence in other divergencies of the human from the bestial type. Thus, the greater development of the jaws and bones of the face in animals, giving rise to a projecting muzzle, is no longer requisite when the arm and hand afford so much better an instrument than the mouth for seizing objects, and for attack or defence; while from the same cause the canine teeth tend to diminish. In fact, the specialization of improved types from the early generalized type, takes very often the form of a reduction of the number of teeth to that required for the relations of the new types to their environment. Thus, in the pure carnivora, like the cats, the molars disappear and the canines and sectorial premolars assume a great development. In the Herbivora, on the other hand, the molars are developed at the expense of the flesh-cutting teeth; and in civilized man there is a progressive diminution in the size of the jaws, which hardly leaves room for the normal number of teeth, some of which are probably destined to disappear, as the so-called wisdom-teeth have already almost done.

Thus, from the single point of view of specialization for erect posture, we arrive at all the physical characteristics which distinguish man from the monkeys and anthropoid apes. At the same time, it is a difference only of adaptation and not of essence. The machine man differs from the machine ape, much as the modern railway locomotive differs from the old-fashioned pumping steam-engine. The essential parts, boiler, pistons, cylinders, valves, are the same, but differently modified; those of the locomotive being vastly better adapted for condensed energy and rapid motion in a smaller compass. Still, no one can doubt their affinity and common origin, or suppose that while the Newcomen engine owed its existence to human invention, the Wild Irishman or Flying Scotchman could only be accounted for by invoking supernatural agency.

This is precisely the case as regards man in his physical aspect. It is difficult to imagine that the combination of bones, muscles, and nerves, which make a man, originated in any different manner than did the combination of the same identical bones, muscles, and nerves, which make a chimpanzee or gorilla. If one originated by evolution, the other must have done so also; and conversely, if one came into being by special miraculous creation, so also must the other, and not only the other, but all the innumerable varieties of distinct species, now, and in past geological times, existing upon earth.

It is only when we come to the higher intellectual and moral faculties, that the wide gulf appears between man and the animal creation, which it is so difficult to bridge over. It is true that all or nearly all of these faculties appear in a rudimentary state in animals, and that not only apes and monkeys, but dogs, elephants, and others of the higher species, show a certain amount of memory, reasoning power, affection, and other human qualities; while, on the other hand, some of the inferior races of mankind show very little of them. The chimpanzee Sally, in the Zoological Gardens, and Sir John Lubbock's dog Van, can count up to five; while it is said that three is the limit of the counting power of some of the Australian tribes. The gorilla, in his native forests, according to the accounts of travellers, lives respectably with a single wife and family, and is a better husband and parent than many of our upper ten who figure in Divorce Courts. Still, there is this wide distinction, that even in the highest animals these faculties remain rudimentary, and seem incapable of progress, while even in the lowest races of man they have reached a much higher level, and seem capable of almost unlimited development. No human race has yet been discovered which, however savage, is entirely destitute of speech, and of the faculty of tool-making in the widest sense of adapting natural objects and forces to human purposes. As regards speech, no animal has advanced beyond the first rudimentary stage of uttering a few simple sounds, which by their modulations and accent give expression to their emotions. They are in the first stage of what Max Muller calls the "bow-wow and pooh-pooh theory," and even in this they have advanced but a little way. They have a very few root-sounds, and those are all emotional. A dog or ape can express love, hatred, alarm, pain, or pleasure, but has not risen even to the height of coining roots imitating sounds of nature, such as "crack" and "splash," and still less to that which all human races have attained to, of multiplying these primitive roots indefinitely, by extending them by some sort of mental analogy, to more abstract ideas; and connecting them by some sort of grammar, by which they are made to express a variety of shades of meaning and modifications of human thought. Animals understand their own simple language perfectly well, and to a certain extent some of the higher orders, such as dogs and monkeys, can be taught to understand human language; but no animal has ever learned to speak in the sense of using a series of articulate sounds to convey meaning, though, as in the case of the parrot, the vocal organ may be there, capable of uttering imitation words and sentences.

As regards tool-making, no human race is known which has not shown some faculty in this direction. The rudest existing tribes, such as Bushmen or Mincopies, chip stones, and are acquainted with fire and with the bow and arrow, spear, or some corresponding weapon for offence and defence. The highest apes have not got beyond the stage of using objects actually provided for them by nature, for definite objects. Thus monkeys enjoy the warmth of a fire and sit over it, but have never got the length of putting on coals or sticks to keep it up, much less of kindling it when extinguished. Sally and Mafuca perfectly understood the use of the keeper's key, and would steal and hide it, and use it to let themselves out of their cage; but no chimpanzee or gorilla has ever been known to fashion any implement, or do more than use the sticks and stones provided by nature, for throwing at enemies or cracking nuts. Their nearest approach to invention is shown in constructing rude huts or nests from branches and leaves, for shelter and protection; an art in which both apes and savages are very inferior to most species of birds, to say nothing of insects. The difference is a very fundamental one, for in the case of man, we can trace a constant progression from the rudest form of palæolithic chipped stones, up to the steam-engine and electric telegraph; but in the ape, we can discern no signs of progress, or of a capacity for progress. It is conceivable that by taking a certain number of Bushmen or Australians when young, placing them in a favourable environment, and breeding selectively for intelligence, as we breed race-horses for speed, or short-horns for fat, we might, in a few generations, produce a race far advanced in culture; but it is not readily conceivable that we could do the same with orangs or chimpanzees. It would be a most interesting experiment to try how far we could go with them in this direction, but unfortunately it cannot be tried, as we have no sufficient number of specimens to begin with, and the race cannot be kept alive, and much less perpetuated in our climate. Even if it could, there is no reason to expect that it would succeed up to the point of making a race of apes or monkeys who could speak a primitive language or make primitive tools. For the fundamental difference between them and man may be summed up in the words "arrested development."

At an early age the difference between a young chimpanzee and a young negro is not very great. The form and capacity of the skull, the convolutions of the brain, and the intellectual and moral characters are within a measurable distance of one another; but as age advances, the brain of the negro child continues to grow, and its intelligence to increase up to manhood; while in the case of the ape the sutures of the skull close the growth of the brain is arrested, and development takes the direction of bony structure, giving rise to a projecting muzzle, protuberant crests and ridges, and generally a more bestial appearance; while the character undergoes a corresponding change and becomes less human-like.

It is evident, therefore, that these two branches of the Primates, man and ape, follow diverging lines of development, and can never be transformed into one another, and that the "missing links" to connect the human species with the common law of evolution of the animal kingdom, are to be sought in other directions than that of direct descent from any existing form of ape or monkey.

There are three lines of research which may be followed in looking for traces of such missing links.

1. We may compare the higher with the lower varieties of the existing human species, and see if we can discover any tendency towards a lower form of ancestral development.

2. We may observe the results in the cases of arrested development which occur in those unfortunate beings who are born idiots or microcephali, that is, with deficient brains.

3. We may explore the records of the past, of which we have now numerous remains preserved in the fossil state.

The first and second of these lines give us a certain amount of clear and positive result. Comparing civilized man with the Negro, Australian, Bushman, and other inferior races, we invariably find differences, which all tend in the direction of the primitive "pentadactyle, plantigrade, bunodont." The brain is of less volume, its convolutions less clearly marked, the bony development of the skull, face, and muzzle more pronounced, the legs shorter and frailer, the arms longer, the stature less. The most primitive savage races known to us are apparently those Pygmies who, like the Akkas and Bushmen of Africa, the Negrillos of Asiatic islands, some of the hill tribes of India, and the Digger Indians of North America, have been driven everywhere into the most inaccessible forests and mountains by the invasion of superior races. The average stature of many of these does not exceed four feet, and in some instances falls as low as three feet six inches; and in structure, as well as in appearance and intelligence, there is no doubt that they approximate towards the type of monkeys.

In the case of idiots, the resemblance to an animal type is carried much further, so far, indeed, that they may be almost described as furnishing one of the missing links. As Vogt says, "we need only place the skulls of the negro, chimpanzee, and idiot side by side, to show that the idiot holds, in every respect, an intermediate place between them."

Thus the average weight of the brain of Europeans is about 49oz., while that of Negroes is 44¼oz., and in some of the inferior races it is still lower, descending to about 35oz. in the case of some skulls of Bushwomen. This approaches very closely to the limit of 32oz. which Gratiolet and Broca assign as the lowest weight of brain at which human intelligence begins to be possible, but in many cases of small-headed idiots the weight descends much lower, and has even been observed as low as 10oz. The average weight of the brain of the large anthropoid apes is estimated at about 20oz., and in some cases is even higher, so that the brains of some of the inferior human races stand about half-way between those of the superior races and of the anthropoids, which latter again differ more from those of the lemurs and inferior monkeys than they do from those of man.

The approximation towards primitive conditions shown by a comparison of superior with inferior races, and of normally developed men with idiots and apes, might have been expected to derive further confirmation from tracing back the third line of inquiry, that of fossil remains.

And yet it is just here, where we might expect to find conclusive evidence, that it has hitherto failed us. Not only have we found no fossil remains which stand to modern man in something of the same relation as the Hipparion does to the horse, but nothing has yet been discovered which seems to carry us so far in that direction as is done by a comparison with some of the existing savage races. The number of skulls and skeletons dating back to early Quaternary times, distant from us certainly not less than 50,000 years, and probably much more, is now so great as to enable us to speak confidently as to their character, and even to classify their different types. The oldest is that known as the Canstadt type, the next oldest that of Cro-Magnon. Now the Cro-Magnon type is not only not a degraded one, but, physically speaking, that of a fine race, tall in stature, with large and symmetrical brain-structure, and on the whole on a par with some of the best modern races.

The Canstadt type is somewhat more rude, and in extreme cases, like that of the celebrated Neanderthal skull, so simious in the low forehead and massive bony ridges, that at first sight it was thought that one of the missing links had really been discovered. But further inquiry showed that this was only an extreme instance of a type which is presented by numerous other skulls of a character entirely human, certainly not inferior to that of existing savages, and which may be traced as surviving among many of the best European races. Even in the extreme case of the Neanderthol skull, the brain was of fair capacity, and a modern skull, that of Lykke, a Dane of distinguished intellectual capacity, is preserved in the museum at Copenhagen, which closely resembles it in all its principal peculiarities.

If the Tertiary skulls of Olmo, Castelnedolo, and Calaveras are accepted as genuine, they carry us back much further in the same direction. Everything about these remains is entirely human, and in the female skull of Castelnedolo, M. Quatrefages thinks he can discover a specimen of one of the milder and less savage forms of the Canstadt type.

Reports occasionally reach us of discoveries of alleged missing links, but they lack confirmation. The nearest approach to a scientific fact is afforded by a human jaw found in the Cave of La Naulette, in Belgium, in which Mortillet and other good authorities assert that the genial tubercle is wanting. This is a small bony excrescence on the chin, to which the muscle of the tongue is attached, and is said to be necessary for the movements of the tongue which render speech possible. It is absent in the monkey and all non-speaking animals, and Mortillet asserts that in the Naulette skull the bone is absent, and its place shows a hollow. He argues that the primitive men of the Neanderthol or Canstadt type were incapable of speech, and his conclusion is thought probable by several good authorities. But the induction seems too wide to be drawn from a single instance, and as far as I am aware, it has not been confirmed by any other undoubted specimen of early palæolithic man.

We are still therefore without any conclusive evidence

of human evolution from fossils, and the negative evidence remains, that while so many Pliocene and Miocene formations have been explored, and so many missing links of other animal forms have been discovered, no such links have yet been found in the case of the human species.

What can be said to these facts? Only this, that if the missing links exist, they must be sought much further back. From the wide diffusion of mankind over nearly the whole of the habitable globe in early Quaternary times, it is clear that if the race originated like other animal races from evolution, the origin must be sought in a much more remote antiquity. The existence of the Dryopithecus and other anthropoid apes in the Middle Miocene, shows that the development of another branch, so closely allied to man in physical structure, had been completed in the first half of the Tertiary period. Unless we assume direct descent, and not parallel development, for the two species, why should the starting-point of man be later than that of the Dryopithecus? The horse, whose ancestral pedigree is the best established of any of the existing mammals, was already in existence in the Pliocene period, and the Hipparion, which is the first of the links connecting him with the primitive mammal, is first found in the Miocene and not later than the Pliocene. Why should the development of man have begun later, and followed a more rapid course than that of the horse? Man, as M. Quatrefages observes, must, from his superior intelligence and knowledge of fire and clothing, have been more able to resist changes of climate and environment than many of the animals which undoubtedly outlived the change from the Tertiary to the Quaternary period, and even survived the excessive rigour of the Glacial epoch.

If, as seems almost certain, the first origins of man are to be sought as far back as the Miocene, we can hardly expect to find many specimens of the missing link. If we find such an abundance of palæolithic remains early in the Quaternary period, it must be because the human race had long existed, and been driven by the pressure of increasing population to diffuse themselves over nearly the whole of the habitable globe. But this radiation from the original birth-place must have been extremely slow, and immense periods must have elapsed before it reached the countries which have been the fields of scientific research. Again, great geological changes have taken place since the Miocene period, and it is quite probable that the earliest scene of man's development may be now submerged beneath the Atlantic or Pacific Oceans.

In Miocene times, when Greenland and Spitzbergen supported a luxuriant vegetation, such a continent would be found to the north, possibly in that submerged northern continent which afforded a bridge for the passage of so many forms of animal life between the Old and New Worlds. In fact, many geologists incline to the conclusion that the more recent forms of animal and vegetable life have migrated southwards from this circumpolar Miocene land, and not northwards from tropical regions.

In any case the conclusion seems certain, from the failure to discover any missing links in the later formations, that either a vast period of time must have elapsed since man first began to be specialized from the primitive mammalian type, or that he is an exception to the general law of evolution, and owes his origin to some miraculous act of secondary supernatural interference. The solution of this question must be sought in two directions: firstly, the probable duration of the Quaternary period, during which the existence of man as a component part of the Quaternary fauna is no longer doubtful; secondly, the evidence for his existence much farther back into the Tertiary period in common with many of the animal types with which he is associated. This evidence is accumulating, and any day may bring us conclusive proof by the discovery of some "missing link" in the Miocene or Eocene formations, bearing the same relation to man as the Hipparion and its ancestors do to the horse. In the meantime, the attitude of the scientific world must be described as one of eager expectation rather than of assured knowledge, and this most important and interesting of all problems must be relegated among the problems of the future.

CHAPTER VII.

ANIMAL MAGNETISM AND SPIRITUALISM.

Binet and Féré's Volume—School of Salpetriere—Dr. Braid—Hypnotism—How Produced—Effects of—Lethargy—Catalepsy—Somnambulism—Hallucination—Dreams—Hypnotic Suggestion—Instances of—Visible rendered Invisible—Emotions Excited—Acts Dictated—Magnet—Trance—Alternating Identity—Thought Reading—Clairvoyance—Spiritualism—Slate Writing—Scybert Commission—All Gross Imposture—Dancing Chairs and Tables—Large Field opened up by French Investigations—Point to Materialistic Results.

THE volume by Messrs. Binet and Féré, published in the *International Scientific Series,* gives a lucid view of the recent researches by which the mysterious subjects comprised under the cognate heads of animal magnetism, hypnotism, somnambulism, catalepsy, hallucination, and spiritualism have been, to a considerable extent, brought within the domain of experimental science. The existence of extraordinary phenomena in this misty region had been known since the time of Mesmer, and at times professors of what seemed to be something very like the black art, had excited a temporary sensation, which died out as their tricks were exposed, or as folly changed its fashion. But there was such an atmosphere of imposture, delusion, and superstitious credulity about the whole subject, that rational men, and especially men of science really competent to make experimental inquiries, turned from it in disgust.

The first step towards a really scientific inquiry was made by Dr. Braid, a well-known surgeon in Manchester, about forty-five years ago. He proved conclusively that the state known as mesmerism, or artificial somnambulism, could be produced by straining the eyes for a short time to look at a given object.

A black wafer stuck on a white wall could do just as much as a Mesmer with his flowing robes and magic wand. This led to the further conclusion that anything that strained the attention, or in other words excited certain sensory centres of the brain abnormally, threw it, so to speak, out of gear, and caused both sensory and motor nervous centres to behave in a very extraordinary and unusual manner.

Thus it produced a state of anæsthesia, and if chloroform had not proved a more generally efficacious and manageable agent, hypnotism would probably have been employed to this day in surgical operations. Healing effects also were produced, which bordered very closely on what used to be considered as miraculous cures, and in several cases Braid literally made the blind to see and the lame to walk, by directing a stream of vital energy to a paralyzed nerve.

Still more extraordinary were the effects produced in exalting the faculties and paralyzing the will. Muscular force could in certain cases be so increased that a limb became as rigid as a bar of iron, and memory so stimulated that words and scenes scarcely noticed at the time, and long since forgotten, started into life with wonderful vividness and accuracy.

Thus in one of Dr. Braid's experiments, an ordinary Scotch servant-girl startled him by repeating a passage from the Bible in Hebrew. It turned out that she had been maid to a Scotch minister who was learning Hebrew, and who used to walk about his study reciting passages from the Hebrew text.

Another instance shows the remarkable obliteration of the will in hypnotized subjects. A puritanical old lady, to whom dancing was an abomination, was sent capering about the room by playing a reel tune on a piano, and telling her to join in the dance.

Dr. Braid's experiments, however, did not carry the subject much farther than to make people believe that there was really something in it, and the subsequent rise of spiritualism, with its vulgar machinery of table-turning and spirit-rapping, and frequent exposures in police-courts, once more repelled rational men and consigned the subject to oblivion.

But within the last few years a school has arisen of French medical men, connected with the hospital of Salpetriere, at Paris, who have taken up the subject in a thoroughly scientific spirit, and have arrived at truly wonderful results. This hospital, affording as it does a constant supply of hysterical and epileptic patients, presents peculiar facilities for conducting a series of experiments. In cases of individual experiments there is always danger of error from simulation on the part of the patient, or delusion on that of the operator. But here the experiments were conducted by a body of scientific and sceptical men, selected from the flower of French surgeons and physicians, and the patients were so varied and numerous, that by proper precautions it was possible to eliminate the element of conscious imposture. This supply of a large number of patients, suffering from hysteria and other nervous disorders, was an essential element for success, for it is with this class of patients, and especially of female patients, that the phenomena can be produced with most completeness and certainty. It is a moot point whether all human organisms are subject more or less to the influence of hypnotism; but it is certain that with healthy adults not more than one out of every five or six subjects can be hypnotized at the first attempt, and a great majority of those who can, are only so in a slight degree.

The liability, however, to hypnotic influence increases rapidly by practice, so that nervous patients on whom the process is repeated, may be soon brought into a state in which the slightest hint or suggestion is sufficient to produce the abnormal condition. Thus a highly sensitive patient may be hypnotized, if led to believe that an operator is making passes in an adjoining room, although he is not really there; while, on the other hand, the weight of evidence is against any effect being produced by real passes, if the patient is totally unaware of anything of the sort going on, or being expected.

But with the class of patients at the Salpetriere, the various effects can, in many cases, be produced with as much precision and certainty, as when a bar of iron is magnetized or de-magnetized by turning on or off an electric current through a coil of copper wire surrounding it.

These effects may be classed under two heads—physical, and mental or psychical. Not but that the latter depend ultimately on mechanical movements of nerve-centres of the brain, but they are connected with will, consciousness, and other phenomena which we are accustomed to consider as mental. The purely physical efforts, again, may be classified under three heads, viz.: those of lethargy, catalepsy, and somnambulism. The divisions shade off into one another, but the typical states are sufficiently distinct to justify this classification, which is due to M. Charcot, the Director of the Salpetriere.

In lethargy the patient appears to be in the deepest sleep. In fact, all the functions of mind and body, except the bare life, seem to be suspended. The eyes are closed, the body is perfectly helpless; the limbs hang slackly down, and if they are raised they drop heavily into the same position. The characteristic feature of this state is that any excitement of the muscles either direct or through a stimulus applied to the connecting motor nerve produces what is called a contracture. Thus if the ulnar nerve is pressed, the third and fourth fingers of the corresponding hand are forcibly contracted, and so for every other nerve and corresponding muscle of the body. This evidently affords a perfect security against simulation, for no one who was not a skilled anatomist would know what muscles were connected with a particular nerve.

One of the most remarkable phenomena connected with these contractures is that they may be produced by a magnet not in physical contact with the nerve or muscle excited, and still more wonderful, that it may be transferred by a magnet from one side of the body to the other. Thus if the fingers of the right hand have been contracted by pressure on the ulnar nerve of the right arm, and a magnet is brought close to that nerve, both hands become agitated with slight jerking movements, and soon the contracture of the right fingers ceases, and is transferred to the same fingers of the left hand. We shall see later that in more advanced stage of hypnotism still more marvellous effects are produced by the magnet, even to the extent of transferring moral emotions into their opposites, as love into hatred, or hatred into love.

In the meantime, it may be sufficient to observe that these experiments with the magnet seem to point out the most likely way of bringing these mysterious phenomena within the domain of accurate science, and here the researches of the Salpetriere school seem to be deficient. We are merely told that the magnet produces certain effects, but we want to know at what distance does it produce these effects. Do the effects and distance vary with the power of the magnet? are they produced differently by the presentation of the positive or negative pole? are they produced by an electro-magnet or by electric currents? is there any and what reaction by the nerve or muscle on the magnet? and other similar questions. When these are certainly known and can be expressed in terms of weight and movement, we shall have made the first solid and secure step in advance towards a solution of the more complicated problems.

The next stage is that of catalepsy, into which lethargy may be made to pass by simply opening the eyelids. But although so closely allied to lethargy, the states are very different. In catalepsy all power of movement, or of resistance to movement, is absolutely suspended, and the body is like a lump of plastic clay, which may be moulded into, and will retain, any form given to it by the operator. In fact the subject becomes a lay figure, with this difference only, that he remains so only for some ten or fifteen minutes, after which the constrained positions give way to natural ones. But that he is a *bonâ fide* lay figure for the time is proved by registering the movements of the extended arm and the regularity of the respiration, by means of tracing instruments, and comparing them with those of a healthy man voluntarily assuming the same position. The contrast of the tracings is most remarkable. That of the arm extended by catalepsy is a straight line showing absolutely no tremors; while that of the arm voluntarily extended, shows such a series of abrupt and increasing oscillations, as to make it quite conceivable how thought-reading may be possible by contact between persons of exceptionally delicate nervous organization.

Another remarkable feature in catalepsy is that the position in which the body is placed seems to react on the mind, and call up the emotions, and their reflex muscular motions, which are habitually associated with the attitude. Thus if the head is depressed the face assumes the expression of humility; if elevated that of pride.

The most extraordinary phenomena known are those of somnambulism and of the artificial somnambulism which is produced by animal magnetism or hypnotism. These are of various stages, graduating from that of ordinary waking dreams to that of profound hypnotism in which will, consciousness, memory, and perception, are affected in a way which at first sight appears to be truly magical or supernatural. The symptoms may be classed for convenience as physical or psychical, although the latter are really physical, depending ultimately on movements of nerve-centres.

The direct physical effect seems to be the exact opposite of that of lethargy, viz.: that the senses, instead of being asleep, have their sensibility exalted in an extraordinary degree. Thus subjects feel the heat or cold produced by breathing from the mouth at a distance of several yards. The hearing is so acute that a conversation may be overheard which is carried on in the floor below.

The amount of this exaltation of the senses can almost be measured. There is a familiar experiment in which the impression of two points, as of separate pencils near one another, is felt as one, and an instrument has been constructed, known as Weber's compasses, which measures the amount of deviation necessary to produce a twofold sensation. This deviation appears to be six times greater in the waking than in the somnambulistic state, whence it may be inferred that the sensibility of the sense of touch has been exalted sixfold.

A similar exaltation is produced in the faculty of memory, as shown in the instance already quoted, in which an ignorant servant-girl recited a long passage in Hebrew. As in dreams, perceptions long since photographed on the brain and completely forgotten seem to be revived with all the vividness of actually present perceptions, when recalled by some association with the dominant idea which has taken possession of the mind. This arises doubtless, in a great measure, from the mind being closed against the innumerable other impressions which, in the waking state, wholly or partially neutralize any one suggested idea, and weaken its impression. Thus a somnambulist walks securely along a narrow plank, because no other outward impressions of surrounding objects confuse his mind with suggestions of danger.

It is, however, when we come to the partly psychical phenomena of hallucination and suggestion, that the results are most startling and most opposed to ordinary experience. What is an hallucination? It may be described in one word as seeing the invisible and not seeing the visible. And the same of the other senses. They not only deceive us, but give evidence directly contradictory of that of the waking senses. We hear the inaudible, and are deaf to the audible; we touch the intangible, and lose touch of the tangible; bitter tastes sweet, and sweet bitter. The fundamental fact seems to be, that if certain conditions or molecular movements of certain sensory nerve-centres of the brain are caused, no matter how, the corresponding perceptions, with their train of associated ideas and reflex movements, inevitably follow. In the normal waking state these conditions are created by real objects conveyed to the brain through the senses. We see a man, and we conclude him to be a real man because our other senses confirm the testimony of sight. If he speaks we hear him, if we touch him we feel him, and the evidence of all other people who see and hear him confirms our experience. But in dreams we have the commencement of a different experience, for we see and hear distinctly for the time, though in a fleeting and imperfect manner, scenes and persons which have no real objective existence. In hallucinations we have the same thing, only in a waking or partially waking state, and the impressions made are vastly more vivid and permanent.

Take the following as instances of positive hypnotic hallucinations, or seeing the invisible, recorded by Messrs. Binet and Féré from their experience at the Salpetriere. A patient told to look at a butterfly which had just alighted on the table before her, immediately said, "Oh, what a beautiful butterfly," and proceeded cautiously to catch it and impale the imaginary butterfly with a pin on a piece of cardboard. Another patient being shown a photographic plate with an impression of a scene in the Pyrenees, and told that it was a portrait of herself in a very unbecoming dress, or rather want of dress, immediately saw it so, and was so enraged, that she threw the plate on the ground and stamped on it. And what is remarkable, as showing the intensity and persistency of these hallucinations, for nearly two months afterwards, when shown in her waking state photographs of this landscape which had been taken from the plate, she saw her own portrait, and fell into fits of passion. In another case, a patient being told that one of the hospital doctors would be present at a ball to be given next night among the inmates of Salpetriere, saw, conversed, and walked about with this imaginary doctor, who was not really present, and when she saw the real man the day after, could not recognize him until she had been again hypnotized, and the hallucination dispelled.

The negative experiences of making the visible invisible are even more extraordinary. Take the following case. "We suggested to a hypnotized patient that when she awoke she would be unable to see F——. She could not see him, and asked what had become of him. We replied, 'He has gone out; you may return to your room.' She rose, said good morning, and going to the door knocked up against F——, who had placed himself before it. We next took a hat, which she saw quite well, and touched it so as to be sure that it was really there. We placed it on F——'s head, and words cannot express her surprise when she saw the hat apparently suspended in the air. F—— took off the hat and saluted her with it several times, when she saw it, without any support, describing curves in the air. She declared the hat must be suspended by a string, and even got on a chair to feel for it."

Numerous other instances equally remarkable are recorded, and there is a whole class of cases in which suggestions impressed on the subject's mind in a state of hypnotism may long afterwards, and when totally forgotten, be revived at predicted periods, with irresistible force, in the waking mind and produce the effects corresponding to the idea as by an inevitable piece of machinery. This brings the subject within the domain of criminal jurisprudence, for there is abundant evidence that a normally moral person may obey a hypnotic suggestion which had been totally forgotten, even to the extent of committing the greatest crimes, as attempting to stab or administer poison. Thus M. Féré relates that having ordered a subject in a state of somnambulism on awakening to stab M. B—— with the pasteboard knife he put into her hand, as soon as she awoke she rushed on him and struck him in the region of the heart. M. B—— feigned to fall down. The subject, being asked why she had killed him, replied with an expression of ferocity, "He is an old villain and wished to insult me."

It is evident that if these phenomena are real, hypnotism ought to be regulated by law as much as the far less dangerous practice of vivisection. The practice of it should be confined to licensed medical practitioners, and under conditions requiring the presence of at least two or more witnesses, one of whom, especially in the case of females, should be some respectable friend or relative. I prefer, however, not to dwell on this branch of the question, but to return to its purely scientific and philosophical aspects.

The purely mechanical origin of these hallucinations is shown by a number of interesting experiments. An hallucinatory image can be reflected, refracted, or made to appear double, in precisely the same manner as a real one. Thus in what is known as Brewster's experiment, where an image is duplicated by a slight lateral pressure on one eye throwing it out of focus with the other, the same effect is produced. A case is recorded where an hysterical patient who had a vision of the Virgin Mary appearing in great glory, saw two Virgins directly this lateral pressure was applied. Complementary colours also appear to an hallucinatory image of a red or green spot on a sheet of white cardboard, just as they would in the waking state if the spot were real. The magnet also, by a purely mechanical action, transfers unilateral hallucinations which affect one eye only, from the right to the left eye, and *vice versâ*, and it may be made to destroy an hallucination, as when X—— was made invisible to an hypnotic subject, on applying a magnet to the back of the head, X—— again became visible.

And what is still more wonderful, the magnet is capable of transferring emotions. Thus the idea was impressed on a hypnotized subject, that on awaking she would feel a desire to strike F——. A magnet was placed near her right foot. On awaking, she jumped up and tried to give F—— a slap, saying, "I do not know why, but I feel a desire to strike him." In another moment, her face assumed a gentle and endearing expression, and she said, "I want to embrace him," and tried hard to do so. Consecutive oscillations between love and hatred were then observed.

Another most remarkable phenomenon is recorded.

It was suggested to a subject X—— that she had become M. F——. On awaking, she was unable to see M. F——, who was present, but she exactly imitated his gestures, put her hands in her pockets, and stroked an imaginary moustache. When asked if she was acquainted with herself, X——, she replied with a contemptuous shrug, "Oh, yes, an hysterical patient. What do you think of her? She is not too wise."

There are two experiments recorded which throw a good deal of light on the phenomena of what is known as spiritualism. In slight hypnotism, the subjects assert, on awaking, that they have never for a moment lost consciousness, and that they have been present as witnesses at the phenomena of suggestion developed by the magnetizers. In another case, the furniture of the room seemed to the subject to be noisily moved about by invisible hands, being really displaced by F——, who had been rendered invisible by suggestion. It is evident that if there is any real residue of facts in the phenomena of spiritualistic *séances*, after deducting what is due to legerdemain and imposture, the above experiments would go a long way to account for them. The preliminaries of a *séance*, such as darkened rooms, contact of hands, and excited imagination, are almost identical with those employed by Mesmer, and it would be contrary to experience if they did not frequently produce, on susceptible subjects, hypnotic effects which made them susceptible to hallucinating suggestions. If so, there is no doubt that they might see tables move and Mr. Home float in the air, with a full conviction that they were awake all the time and in possession of their ordinary senses.

This much I would observe, that all these attempts

to escape from the inexorable laws of nature invariably fail. Spiritualism is grasped at by many because it seems to hold out a hope of escaping from those laws and proving the existence of disembodied spirits. But when analyzed by science, spiritualism leads straight to materialism. What are we to think of free will, if, as in the case of Dr. Braid's old lady, it can be annihilated and the will of another brain substituted for it, by the simple mechanical expedient of looking at a black wafer stuck on a white wall? Or what becomes of personal consciousness and identity if, as in the case above quoted, a young woman can be brought to refer to herself with contemptuous pity as a strange girl who "was not over wise"? These cases of an alternating identity are most perplexing. Smith falls into a trance and believes himself to be Jones. He really is Jones, and Smith has become a stranger to him while the trance lasts; but when he awakes he is himself, Smith, again, and forgets all about Jones. He falls into another trance, and straightway he forgets Smith and takes up his Jones existence where he dropped it in the previous trance, and so he may go on alternating between Smith and Jones. I often ask myself the question—If he died during one of his trances which would he be, Smith or Jones? and I confess that it takes some one wiser than I am to answer it.

Again, what can be said of love and hate, if under given circumstances they can be transformed into one another by the action of a magnet? It is evident that these phenomena all point to the conclusion that all we call soul, spirit, consciousness, and personal identity, are indissolubly connected with mechanical movements of the material elements of nerve-cells, and that if we want any further solution, we must go down deeper and ask what this matter, and what these movements, or rather the energy which causes them, may really mean. Can the antithesis between soul and body, spirit and matter, be solved by being both resolved into one eternal and universal substratum of existence? When Shakespeare said—

> "We are such stuff as dreams are made of,"

he enunciated what has become a scientific fact. The "stuff" is in all cases the same—vibratory motions of nerve-particles.

The researches of the French school of physiologists throw a good deal of light on the mysterious regions of phenomena, or alleged phenomena, which are classed under the general heads of thought-reading, clairvoyance, and spiritualism. Those of thought-reading and clairvoyance may be summed up in the question whether or no it is possible for one brain to communicate with another, otherwise than through the ordinary medium of the senses. It is certain that in the immense majority of cases it is not possible. Consider how the ideas or perceptions of A are communicated to B. Certain movements of the brain-cells of A which are, if not the cause, the invariable concomitants of those ideas and perceptions, send currents along the nerves, which at their extremities contract muscles and cause movements. These are transmitted, in the case of hearing, by sound-waves of air; in that of sight by light-waves of ether, to the nerve-endings of B, and along those nerves to his brain, where they originate cell-movements corresponding to the original movements in the brain of A, and which are accompanied by the same train of ideas and perceptions. In the sense of touch, there is no intermediate medium between the nerve-endings of A and B, and the movements of the former are communicated directly to those of B by contact. The senses of taste and smell are hardly used by the human species as means of communicating ideas, though in many animal species, as in the dog, the latter sense is one which is greatly used in placing them in relation with their environment.

This also may be affirmed respecting the different senses, that they are capable of being brought to an exceptional degree of susceptibility by necessity and practice, as is well illustrated by the facility with which the blind substitute the sense of touch for that of sight, and read fluently books printed with raised letters. The sense of sight also may be brought to a degree of unusual acuteness, enabling the observer to read indications in the face and expression so slight as to be invisible to the ordinary sense, and of which the person observed is himself unconscious. A remarkable instance of this is given by Sir John Lubbock, of a dog who could pick out from a series of numbers on cards laid on the floor the correct answer of sums in arithmetic, and even extract cube-roots, doubtless by observing unconscious indications in his master's face when he touched the correct card.

This, no doubt, goes a long way towards explaining the phenomena of what is called thought-reading. It is quite conceivable that, with contact, an exceptionally delicate sense of touch, exceptionally cultivated, may enable a man to read the insensible tremors which are unconsciously transmitted to nerve-ends and superficial muscles, the existence of which is a necessary consequence of all brain-motion or thought, and which is proved to exist as a matter of fact by the irregularities in the line traced by a pencil under suitable conditions. And it is to be remarked that keeping the mind fixed on the idea, in other words, making the corresponding brain-motions and nerve-currents stronger and more persistent, is the condition usually required for a successful experiment in thought-reading.

Thus far, and Mr. Cumberland the most successful thought-reader of the day carries it no farther, there is nothing impossible, or even *à priori* improbable, in the assertion that thought may be thus read. It is a question of evidence, and here the weight of the negative evidence is so great that it requires extremely strong proof to establish exceptions. It is a matter of notoriety that persons, even of delicate temperaments, may lie in the closest contact, clasped in each other's arms, without either having the remotest idea of what is passing in the mind of the other, unless it is conveyed by the ordinary channels of sight or hearing. On the other hand, the evidence for a few rare exceptions is strong, especially in the case of some of Mr. Cumberland's experiments, which are all the stronger because he does not pretend to any supernatural power, and shows none of the ordinary signs of an impostor. All we can say, therefore, is that where there is contact, or where unconscious indications may be read by the eye, there is nothing in thought-reading inconsistent with the known laws of nature; but that the evidence, though strong, is hardly strong enough to enable us to accept it as an established fact.

But when we come to thought-reading at a distance, and to the analogous alleged phenomena of clairvoyance, fulfilled dreams and visions, and communications across the globe, mostly from the dead and dying, such as are so plentifully recorded in the annals of the Psychical Research Society, the case is different. If they are true, we must assume either a reversal of the known laws of nature, or an otherwise unknown and unproved addition to them. Vibrations cannot be transmitted without a medium, and in the supposed cases the medium is certainly not the air which transmits sound-waves, or the ether which transmits waves of light, heat, and chemical energy, or any modification of it which transmits magnetism or electricity. It must either be some sort of personal aura, or a universal aura which pervades space, and is specially adapted for transmitting brain and nerve vibrations, and those only. But the evidence is overwhelming against the existence of such a medium. In the case of the real mediums, air and ether, they respond invariably and uniformly to the same stimuli; but we may point our fingers to the end of time to a magnet without making it vibrate, and think for ever of absent friends without conveying to them the slightest intimation. It is only in the rarest exceptional cases that the contrary is even alleged, and that only under conditions which may either be accounted for by coincidence or imposture, or which not only lie outside of, but directly in conflict with, known laws of nature. This is most apparent in the cases which fall under the heads of clairvoyance or supernatural communications. Consider the enormous number of dreams, 300,000,000 at least, of civilized human beings dreaming for most nights of the year, and these dreams all made up of fragments of actual scenes and persons, which have been photographed on the brain. The wonder is not that there should be occasional coincidences between dreams and contemporaneous or subsequent occurrences, but that there should be so few of them. How many anxious brains must have dreamt of absent friends or relations dying or in danger, and in how many millions of cases must the dream not have been verified. And how many vivid dreams, or dreams in the dozing state, between sleeping and waking, must have passed into the stage of hallucination, and been taken for actual visions. And how weak is memory, and how strong the myth-making propensity of the human mind to convert these dreams and visions into waking realities. Of the many cases of distant communications collected by the Psychical Research Society, I do not know of one which may not be thus accounted for; and in some the proof is conclusive, as where visions have been seen or impressions felt of events before they occurred, owing to the difference of time due to longitude.

In the case of spiritualism it is remarkable that it is only the more vulgar and grotesque forms which there is any difficulty in explaining. We understand how spirits are materialized, for the apparatus has been frequently exposed in the police-courts; there is nothing very mysterious in the way in which slight hints and clues are followed up by professional mediums. And there is this conclusive consideration that the spirits never say or know anything which has not passed through the mind of the medium. If he is illiterate, the spirits would be plucked for their spelling; if he is weak in his h's, so are they; if he makes a mistake or is entrapped into a contradiction, they follow suit. In no single instance has any communication of the slightest use or novelty been made by these visitors from another world.

In short, the whole affair is obviously. legerdemain in rapping or writing on slates, answers to questions known to the medium, supplemented by any hints or clues he may possess, and in the absence of these by such commonplaces as "we are happy," "we are with you." I saw a conclusive proof of this in the only experience I ever had with a professional medium, one of great repute. The question put was, "What was my mother's Christian name?" This was written on a slate out of sight of the medium, and turned down, and apparently held by one of his hands under a table, while the other hand was held by the questioner. Nothing occurred for a while, but then began a series of groans and twistings by the medium, which I took to be part of the usual conjuror's patter to divert attention; but looking closely, I distinctly saw a corner of the slate reversed under the table, with the writing on it uppermost, followed by the scratching of a pencil, after which the answer was produced, alleged to have been written by the spirits. But mark what the answer was! The "m" of "mother" had been written not very legibly, with the first stroke too long, so that at a hasty glance in a constrained position it might be easily read as "brother." And sure enough the answer came, "Your brother's spirit not being here we do not know his Christian name." This was my first and last experience of omniscient spirits, and it was perfectly apparent that it was only a piece of very simple and very clumsy legerdemain. No doubt things more marvellous are done by superior legerdemain, but nothing that I have ever heard of that is beyond the resources of legerdemain, or which is so wonderful as the mango and other tricks of Indian jugglers. No one who has not studied the art of legerdemain can be aware how great its resources are, and how completely the senses may be deceived by a skilful operator.

Nor is it at all difficult to understand how slight clues may be used by an experienced operator, to give what are apparently astounding answers. Thus if a medium happens to know that a death has at any time occurred in the family of the questioner, the answer rapped or written out is sure to profess to come from the spirit of the deceased relative.

If any doubt had remained as to the nature of these spiritualistic experiences, it would have been removed by the report made in 1887 by the Scybert Commission. In this case Mr. Scybert, an enthusiastic spiritualist in the United States, bequeathed a considerable sum of money to the University of Philadelphia, on the condition that it should appoint a Commission to investigate modern spiritualism. Ten commissioners were appointed, including several professors and well-known men of science; some of whom, including their chairman, Dr. Furness, confessed "to a leaning in favour of the substantial truth of spiritualism." They took great pains with the investigation, which was conducted with scrupulous fairness, and examined many of the most famous mediums, among whom was the well-known Dr. Slade. Their unanimous report was that the whole thing was based on "gross, intentional fraud." They saw distinctly how the tricks were effected, and a professional conjuror, Mr. Kellar, who had been at first baffled by the phenomena of slate-writing, having turned his attention more closely to this branch of conjuring, was able not only to repeat the processes of the best mediums, but to do so with far greater skill, and produce effects which they could not imitate; while he has given a challenge to the spiritualistic world that he will reproduce by sleight-of-hand any alleged spiritualistic phenomena which he has witnessed three times.

This report is so conclusive to any reasonable mind, that it is scarcely necessary to refer to the mass of corroborative evidence to the same effect; such for instance as the confession of the Fox family, that the rappings, in which the spiritualistic faith originated, were produced by a knack they had of half-dislocating toe and knee joints, and replacing them with a sudden snap, a knack which, singularly enough, is also possessed by Professor Huxley; the confessions of Home and other exposed mediums; and the experiences of Mr. Davy, Mrs. Sedgwick and others, related in the last volume of the Psychical Research Society.

Those who are not convinced by such proofs as these are impervious to reason, and it would be a waste of words to argue the matter any farther. It may be assumed as a demonstrated fact, that all the phenomena which profess to be based on a communication with a spiritual world are, in the words of the Scybert Report, simply instances of vulgar legerdemain, and of human credulity.

It is only when we come to what may be called the tomfoolery of spiritualism, such as unmeaning tricks of dancing chairs and tables, that we are left in doubt how some of the appearances are produced. There is a good deal of evidence from persons whose good faith cannot be doubted, that they have seen pieces of furniture move at the end of a room, without any contact or apparent cause, and that this took place in private houses, where there was no possibility of prepared machinery.

The mediums say it is done by spirit-hands. This is obviously absurd, for it is not a case which lies outside of known laws of Nature, but one which radically conflicts with them. As long as the law of motion holds "that action and reaction are equal and opposite," there can be no action without a solid point of resistance. Archimedes said that he could move the world if you gave him a *που στω*, or fulcrum, on which to rest his machinery, and the ghost of Archimedes, if summoned from the Elysian fields at the bidding of a seedy professional medium, could say no more. Spirit-hands must be attached to a solid spirit body, standing on solid feet on a solid floor, to lift a weight. And the same thing applies to any supposed magnetic or psychic force enacted by the medium. If the medium pulls the chair, the chair must pull the medium, and it becomes a case of "pull devil, pull baker." If a magnet lifts an iron bar, it is because the magnet is fixed to some point of attachment.

The question therefore resolves itself into one either of hallucination or legerdemain. Do the chairs and tables really move, or only seem to move? There seem no trustworthy evidence as to this fundamental point, and yet it is one easily determined. Does the housemaid when she comes into the room next morning, or any one who has not been under the influence of the *séance*, find the furniture where it was originally, or where it seemed to be? If it was really moved, who moved it? Here, also, hallucination might come into play in another form, for if, as described in the experiment of Binet and Féré, already mentioned, the medium could release his hands without being perceived, and render himself invisible by suggestion, or perform the trick in a dark room, he could easily move the chairs himself without being seen. This seems the more probable, as in all the accounts I have read, the articles moved do not exceed the weight which the medium might move, either in his natural condition, or with his muscular strength excited by hypnotism. Assuming a state of hypnotism to be induced in the spectators, the explanation would be easy, and, in fact, identical with many of the scientifically recorded experiments of Binet and Féré. And it is remarkable that the preliminary conditions of the *séance*, such as darkened rooms, clasped hands, and strained attention, are identical with those employed, from Mesmer downwards, in producing real hypnotism.

At the same time, it would seem that the hypnotism (if it be so) introduced at *séances* differs from ordinary hypnotism. The subjects retain the fullest convictions that they have been wide-awake all the time, and in full possession of their ordinary senses. Can there be a state of semi-hypnotism in which the brain, while retaining its full consciousness, is rendered susceptible to suggested hallucinations? If so, the whole matter is explained. If not, it is very singular that the same preliminary operations which produce hypnotism, where hypnotism is expected, should make chairs and tables dance, and bodies float in the air, where that is what the spectators expect to see. But the problem could easily be solved so far as the medium is concerned, by connecting him with an electric current, which would be broken and ring a bell if he moved hand or foot, and seeing whether, under such circumstances, the furniture could be moved.

It is singular that the men of really scientific attainments who profess a belief in spiritualism, such as Professor Crookes and Mr. Wallace, do not seem to have proceeded in this way of accurate experiment pursued by the French school of Salpetriere, even as regards the first rudimentary alleged facts of moving heavy bodies at a distance without apparent contact. Nor do they seem to have thoroughly studied and mastered the resources of legerdemain, which are obviously one of the principal, and in many cases the sole cause of the so-called spiritualistic manifestations, and without a knowledge of which no one is really competent to form an opinion. Indeed, it is questionable whether, when all the more refined tricks of spiritualistic mediums have been so thoroughly exposed, it is worth while to seek for any other hypothesis than that of ordinary conjuring, to account for those mere childish and unmeaning manifestations, the *modus operandi* of which has not yet been fully explained.

It is evident, however, from the well-attested experiments of the French school, that there really is opening up a most interesting field of inquiry as to the relations of mind to matter under certain exceptional conditions, and the extent to which illusions may appear as realities under the influence of excited imagination. Hypnotism, somnambulism, dreams, and hallucinations are becoming exact sciences; and researches pursued in the same manner into the alleged phenomena of spiritualism and thought-reading, would end either in exposing imposture, or in reducing such residuum of truth as they may contain, to known laws analogous to those which prevail in other branches of physiological and psychological investigation.

In the meantime, I conclude by saying that, so far

as we have yet gone, the whole of what is called "spiritualism" seems to be quite dreadfully "materialistic." The one fact which comes out with demonstrated certainty is, that definite ideas are indissolubly connected with definite vibrations of brain-cells; and that however these vibrations are induced, the corresponding ideas and perceptions inevitably follow. In the ordinary course of things these vibrations are induced by what are called realities acting through the senses, and by the normal action of the brain-cells on the perceptions thus received and stored up.

But this applies only to about two-thirds of our existence, viz., the waking state. In sleep and dreams, the vibrations set up are from former perceptions, photographed on the brain, and grouped together in unreal and often fantastic pictures. In somnambulism this is carried to a further point, and we act our dreams. In hypnotism it is carried still farther, and the vibrations are excited by a foreign will, and by foreign suggestions. In the ultimate state, madness, the hallucinations have become permanent. But what strange questions does it raise when we find that, in certain abnormal conditions, all that is most intimately connected with what we call soul, individuality and consciousness, can be annihilated, or exchanged for those of another person, by the mechanical process of exciting their corresponding brain-motions in another way. What are love and hate, if a magnet applied to a hypnotized patient can transform one into the other? What is personal identity, if the suggestion of a third person can make an hysterical girl forget it so completely, as to make her talk of herself as a distant acquaintance "who is not over wise"? What is the value of the evidence of the senses, if a similar suggestion can make us see the hat, but not the man who wears it, or dance half the night with an imaginary partner? Am I "I myself, I," or am I a barrel-organ, playing "God save the Queen," if the stops are set in the normal fashion, but the "Marseillaise" if some cunning hand has altered them without my knowledge? These are questions which I cannot answer. All I can say is, that practically the wisest thing I can do is to keep myself, as far as possible, in the sphere of normal conditions, and assume its conclusions to be real; avoiding, except as a matter for strict scientific investigation, the various abnormal paths which, in one way or other, all converge towards the ultimate end of insanity.

CHAPTER VIII.

THE RELIGION OF THE FUTURE. AGNOSTICISM AND CHRISTIANITY.

PART I.

Are they reconcilable—Definitions of Agnosticism and Christianity—Christian Dogma—Rests on Intuition, not Reason—Descartes, Kant, Coleridge—Christian Agnostics—Tendency of the Age—Carlyle, George Eliot, Renan—Anglican Divines, Spurgeon—*Robert Elsmere.*

Is Agnosticism reconcilable with Christianity, or are they hopelessly antagonistic? That depends on the definition we give to the two terms. That of Agnosticism is very simple. It is contained in the sentence of Professor Huxley's, "That we know nothing of what may be beyond phenomena," and "that a man shall not say he knows or believes that which he has no scientific grounds for professing to know or believe." This is not a positive or aggressive creed, and is reconcilable with any form of moral, intellectual, or religious belief which is not dogmatic—*i.e.*, which does not attempt to impose on us some hard-and-fast theory of the universe, based on attempts to define the indefinable and explain the unknowable. The definition of Christianity is by no means so simple. Practical Christianity resolves itself very much, and more and more every day, into a sincere love and admiration of the life and teaching of Jesus, the son of the carpenter of Nazareth, as depicted in the narratives which have come down to us respecting them, mainly in the Synoptic Gospels. This love and admiration translates itself into a desire to imitate as far as possible this life and to act upon these precepts; to be good, pure, loving, charitable, and unselfish even to the death.

With this form of Christianity the Agnostic has no quarrel; on the contrary, if he is not dwarfed and stunted in his faculties, if he has a heart to feel and an imagination to conceive, he recognizes as fully as the most devout Christian all that is good and beautiful in the true spirit of Christianity and its author. Nay, more, he will not quarrel with the mass of humble and simple-minded Christians who show their love and admiration by piling up adjectives until they reach the supreme one of "divine," and who, in obedience to the ineradicable instinct of the human mind to personify abstract ideas and emotions, make Jesus of Nazareth their Ormuzd, or incarnation of the good principle, and author of all that is pure, righteous, and lovely in the universe.

But there is another definition of Christianity of a totally different character—the dogmatic or theological definition, which, commencing with St. Paul and St. John, and culminating in the Athanasian Creed, has been accepted from the early ages of Christianity, almost until the present day, as the miraculous revelation of the true theory of the universe. It teaches how a personal God created the universe, how He deals with it and sustains it, how He formed man in His own image, and what relations He has with him. It professes to explain mysteries such as the origin of evil, man's fall and redemption, his life beyond the grave, the conditions of his salvation, and a variety of other matters which, to ordinary human perception and human reason, are absolutely and certainly hidden "behind the veil."

With this definition of Christianity Agnosticism has nothing in common. It cannot be both true that we know certain things and that we do not and cannot know anything about them. Theology asserts that we are quite capable of knowing the truth respecting these mysteries, and that, in point of fact, we do know it, either by intuition or by historical evidence. Philosophy traverses the assertion that we know it by intuition; Science shatters into fragments the scheme assumed to be taught historically by a miraculous revelation.

To begin with intuition. It rests on Cardinal Newman's celebrated theory of the "Illative sense," or a complete assent of all the faculties, which gives a more absolute proof than any that can be attached to proofs of science, which are only deductions from certain limited faculties, such as experience and reason. This is very clearly put by Father Dalgairns in a discussion on "The Uniformity of Laws of Nature" at the Metaphysical Society. He says: "I believe in God in the same sense in which I believe in pain and pleasure, in space and time, in right and wrong, in myself. If I do not know God, then I know nothing whatever." That is, the idea of such a being as the God of theology, a personal creator of the universe, with faculties like, though transcendently like, those of man, appears to him a necessary postulate, or rather a fundamental instinct or mould of thought, as universal and imperative as those of space and time. Now is this so? It is at once refuted by the fact that it is not universal and not imperative. The immense majority of mankind, both now and in all past ages, have had no such intuition. It is the refined product of an advanced civilization, confined to a few exceptional minds of high culture, acute intellect, and tender conscience. Even in Christian countries it is an affair of education and authority, rather than of necessary intuition; and even those who assert most loudly that it is a fundamental category of thought, complain that ninety-nine men out of every hundred in modern England live practically as if there were no God. Not so with the real categories of thought and perception. No man, past or present, in Monotheistic, Pantheistic, or Polytheistic countries, has ever lived practically as if there were no such things as space and time, or as if such primary perceptions as those of pain and pleasure had no real existence. These have never deceived us; but the instances are innumerable in which the "illative sense," the complete, earnest, and conscientious assent of all the faculties, has deceived us, and has led to conclusions which a wider knowledge has shown to be not only erroneous, but, in many cases, absurd and noxious.

When closely analyzed, the theological idea of God may be clearly seen to be an attempt to define the indefinable. The primary idea is that of a creator. But what is creation? Making a thing, in the sense in which alone man makes anything—that is, transforming existing matter and energy into new forms—we can understand. As we make a watch or a steam-engine, we can conceive how a Being, with faculties like our own, but indefinitely magnified, might make a universe out of atoms and energies, and make it so perfectly that it would go for ever. But how He could make something out of nothing, which is what creation really implies, altogether passes our understanding. We have absolutely no faculties which enable us to form even the remotest conception of what those atoms and energies really are, how they came there, or what will become of them.

The more closely we examine, the clearer it will appear that these theological intuitions are, in effect, nothing but aspirations; or reflections, like Brocken spectres, of our earnest longings, fears, and hopes on the back-ground mists of the Unknowable; and that all the attempted definitions are mere juggles with words which convey no real meaning. We talk of creation, but when it comes to the point we find that we really mean transformation, and that of creation, properly speaking, we have no more idea than the babe unborn. We talk of immortality, but what we were before we were born, or what we shall be after we die, what soul, consciousness, personal identity, really are, how they came to be indissolubly connected with matter, and what they will be when that union is dissolved, are mysteries as to which we can only make guesses, like the Brahmins and Buddhists, whose guess is transmigration, or the Red Indians, whose guess is a happy hunting-ground beyond the setting sun.

The greatest philosophers have come to this as the ultimate fact of their metaphysical reasonings. Descartes says, "that by natural reason we can make many conjectures about the soul, and have flattering hopes, but no assurance." Kant confesses that reason can never prove the existence of a God. Even great theologians, in the midst of their dogmatic definitions, let drop admissions which show that, at the bottom of their hearts, they feel their ignorance of the high mysteries of which they talk so confidently. The Athanasian Creed, the very essence and incarnation of dogmatism, says "the Father incomprehensible" in the midst of a long series of articles, every one of which is absolutely devoid of meaning unless on the assumption that He is comprehensible, and that St. Athanasius rightly comprehended him. St. Augustine writes, "God is unspeakable," and then proceeds, in a long treatise on "Christian Doctrine," to speak of Him as if he knew all about His personality, attributes, and ways of dealing with the world and man. Even St. Paul says, "O the depths of God! how unsearchable are His judgments, and how inscrutable are His ways!"

What more have Huxley and Herbert Spencer ever said? Only they have said it deliberately, consistently, and knowing the reason why; while theologians, admitting the premises, have preferred to act and argue as if a totally different set of premises were true. The cause is obvious: Reason failing, they have fallen back on Revelation. They had an assured belief that an inspired volume, attested by miracles, taught things respecting these mysteries which otherwise must have remained unknown. Thus Coleridge, who, of those who have attempted to base Christian theology on abstract reason, occupies a foremost place, arrives at this conclusion, that "a Christian philosophy or theology has its own assumptions, resting on three ultimate facts—namely, the reality of the law of conscience, the existence of a responsible will as the subject of that law, and, lastly, the existence of God. The first is a fact of consciousness; the second, of reason necessarily concluded from the first; the third, a fact of history interpreted by both." He clearly sees that any certain knowledge respecting the existence of God, and the various conclusions deduced from it by Christian theology (such as the creation of man, his fall and redemption, the origin of sin and evil, atonement, grace, and predestination), if a fact at all, is a *fact of history*—that is, depends on a conviction that these mysteries were actually revealed as recorded by the Bible, and that the Bible is an inspired book attested by historical facts; that it contains prophecies which really were fulfilled, and describes miracles which actually occurred.

This assumption has turned out to be a broken reed. In face of the discoveries of recent science, no reasonable man doubts that, beautiful and admirable as the Bible, and especially the New Testament, may be in many parts, it is not a true, and therefore not a Divine, revelation of the scheme of the universe. It is not true that the world was created as described by Genesis; that man is a recent creation made in God's image, who fell from his high estate by an act of disobedience; or that the course of things is regulated by a special personal providence, frequently interfering by miracles with the course of evolution and the uniformity of the laws of nature. The cause of miracles may be considered as out of court when even enlightened advocates who hold a brief for them, like Dr. Temple, a Bishop of the Anglican Church, throw it up and declare "that all the countless varieties of the universe were provided for by an original impress, and not by special acts of creation modifying what had previously been made."

Dogmatic theology, therefore, having no solid foundation either in abstract reason or in historic facts, and being in hopeless conflict with science, is bound to disappear; and even now, in addressing enlightened and impartial men, it may be taken as "*une quantité negligeable.*" This being the case, the barrier which separates Agnosticism from Christianity is to a great extent removed. The term "Christian Agnostic" is coming more and more to the front in the thoughts and utterances of enlightened Christian men. I notice these with pleasure, for it is always more profitable to find points of agreement rather than of difference with sincere and reasonable men. A Professor of Divinity, preaching in the University of Oxford a short time ago, said: "The field of speculative theology may be regarded as almost exhausted: we must be content henceforward to be Christian Agnostics." Canon Freemantle, in an article in the *Fortnightly Review*, quotes this with approval. In the course of a very able argument on the changed conditions of theology, he says that "theologians, in defiance of Aristotle's axiom, that you must not expect demonstration from a rhetorician, have begun with axioms and definitions and proceeded to demonstrations. They have said or 'proved' that God is just or good, God is personal, God is omniscient and omnipotent; and they have used these phrases, not in a literary, but in a quasi-scientific manner, and have proceeded to draw strict inferences from them. But, in doing this, they have not only acted in the way of unwarrantable assumptions; they have often produced what St. Paul termed the vain janglings of a science falsely so called; have enslaved the Divine to their own puny conceptions, and have provoked violent revolt."

This is precisely what Agnostics contend for. They

do not deny that, in the course of evolution, certain feelings and aspirations have grown up and come to be part of the mental furniture of civilized nations, which find a poetical expression in the ideas of God and of immortality. They simply deny that we have, or ever can have, any certain, definite, and scientific knowledge respecting these mysteries. To take an instance, that of the pre-existence of the soul before birth. We recognize a certain poetical truth in Wordsworth's noble ode when he asserts this pre-existence, and tells us that in infancy

> "Trailing clouds of glory do we come
> From God, who is our home."

But we do not accept it as a known or knowable fact. We have absolutely no experience of any consciousness or personal identity before birth, or as existing otherwise than in association with the matter and energy of our corporeal body. No more have we of any continuance of that identity after death. It is "behind the veil," in that great region of the "Unknowable," where nothing is known, and therefore all things are possible. Here Agnosticism comes in as a powerful auxiliary to those emotions and aspirations which constitute what is called "religion." It is the best of all arguments against Atheism and Materialism, for, if we cannot prove an affirmative, still less can we prove a negative. No man who understands what knowledge really means can affirm that any conception of what may exist in the great Unknowable which compasses us about on every side, is impossible. He can only call it impossible when it conflicts with known facts and laws; but as long as it remains in the region of poetical imagination or moral emotion he cannot disprove it, and may even, if he finds consolation or guidance from it, give it a sort of provisional assent. Thus no Agnostic can deny that, if he had faculties to see Him, there might be in the Unknowable a Divine spirit or substratum, bearing some resemblance to what enlightened men understand by the term "God"; that there may be a Divine eye watching his every thought and recording his every action; and he will not be acting unwisely if he endeavours to mould his life as if this were a true supposition. Only he does not pretend to know this as a dogma or certain truth, and therefore he does not quarrel with any brother-man who thinks differently, or who fancies that he has more certain assurance. Christian morality he recognizes fully, not as taught by the later inventions of churches and casuists, but as displayed in the life and teachings of Jesus, the son of the carpenter of Nazareth, as they stand out, when stripped of their mythical and supernatural attributes, in the narrative of the Gospels. He looks on these moral precepts as the results of a long process of evolution in the best minds of the best races, and not as arbitrary rules, invented for the first time, and imposed from without by miraculous teaching; and he sees in Jesus simply the brightest example and best model of a large class of the virtues which are most needed to make practical life pure, lovely, and of good repute. In this sense may we not all shake hands in the near future and be "Christian Agnostics"?

The tide is already running breast-high in this direction. During the last half-century how many of the foremost men of light and leading have drifted towards orthodox Christianity, and how many away from it? Darwin, Herbert Spencer, Huxley, Carlyle,

Mill, all the great thinkers who have influenced the currents of modern thought, are men who had renounced all belief in the traditional theories of miracles and inspiration, and who, a few centuries earlier, would have been burned as heretics. The conversions have been all one way. Darwin, greatest of all, was an orthodox believer in his early life, and had even contemplated taking orders before he embarked on his mission of naturalist to the expedition of the *Beagle.* In his case no violent impulse or sudden crisis changed his views; but the theological mists simply melted away as the sun of Science rose higher above his horizon. Patiently he worked out his great work, guided solely by his unswerving allegiance to truth, until his conception of the universe as the product, not of innumerable supernatural interferences, but of evolution by natural law, became the creed of all men of all countries who are able to appreciate scientific facts and evidence.

But Darwin and men of scientific training are not the only ones who have exchanged the old for the new standpoint. Conversions have been even more remarkable among eminent leaders in literature and philosophy, who were brought up in the strictest traditions of the old religious beliefs. In another work[1] I have called attention to the fact that, if ever there were three minds trained under the strongest influences binding them to typical though different forms of faith in Christian theology, they are Carlyle, George Eliot, and Renan. Carlyle was a Puritan of the Puritans, bred in a farmhouse, whose inmates might have been Covenanters who

[1] *Modern Science and Modern Thought.*

fought against Claverhouse at Drumclog; George Eliot was, in her surroundings and early life, a typical representative of middle-class English Evangelicalism; Renan of the simple Catholic piety of Breton peasants, developed in an ecclesiastical seminary. How came they, all three, to break away, with a painful wrench, from old ideas and associations, and become leaders of advanced thought? How, indeed, except that they were sincere searchers after truth, and that truth compelled them? If the case for miracles and the inspiration of the Bible had been convincing or even plausible, is it conceivable that Carlyle, George Eliot, and Renan should have all three rejected it? Where are the conversions that can be shown in the opposite direction? Where the leading minds which, bred in the doctrine of Darwinism, have abandoned it for the doctrine of St. Athanasius or of Calvin? The few eminent men who still adhere to the old theology, such as Cardinal Newman and Mr. Gladstone, are all of the old generation which is passing away. Where are their successors? Where are the rising naturalists who are to refute Darwin? where the young geologists who are to dethrone Lyell? where the Biblical critics who are to answer Strauss?

Perhaps the best proof of the irresistible force of the movement is afforded by the attitude of those who still remain within the pale of the Church and are among its most distinguished members. Three eminent Bishops of the Anglican Church preached sermons in Manchester Cathedral, during the meeting of the British Association there in 1887, which were published in a pamphlet, under the title of *The Advance of Science.* They adopt the doctrine of Evolution and the conclusions of modern science so frankly that Huxley, reviewing them in the *Nineteenth Century*, says that "theology, acting under the generous impulse of a sudden conversion, has given up everything to science, and, indeed, on one point, has surrendered more than can reasonably be asked." Other bishops, it is true, denounce this as "an effort to get up a non-miraculous invertebrate Christianity," and assert that "Christianity is essentially miraculous, and falls to the ground if miracles never happened." Perfectly true of the old theological Christianity; but, if this is the only Christianity, it is its sentence of death, for it is becoming more and more plain every day that it is as impossible for sincere and educated men to believe in Scripture miracles as it is to believe that the sun stood still in the Valley of Ajalon, or that the world was peopled from pairs of animals shut up, a few centuries ago, in Noah's Ark.

These truths are rapidly passing from the schools into the streets, and becoming the commonplace possessions of the rank-and-file of thinkers. Thus, in a lower plane of thought and among the strictest sect of believers, we find Spurgeon complaining that, whereas "twenty years ago there was no question of fundamental truth (brethren used to controvert this or that point; but they were at least agreed that whatever the Scripture said should be decisive), now, however, it did not matter what Scripture said; it was rather a question of their own inner consciousness." And, again, that "the position of sitting on the fence is the popular one. There are two or three very learned men who are trying to get down on both sides of the fence at once."

There is something touching in the spectacle of a

man like Spurgeon thus finding the solid earth giving way and heaving under his feet, and even the preachers of his own persuasion lapsing into views inconsistent with his own rigid orthodoxy. But does it never occur to him to ask himself why the landmarks are thus drifting steadily past him all in one direction? Is it a question of inner consciousness and human perversity, or is it not rather that a flood-tide of advancing knowledge and allegiance to truth is really setting in and running with increasing velocity?

Another significant symptom of the times is that the popular novel of the day, *Robert Elsmere*, is a life-history of the conversion of a clergyman of noble nature and cultivated mind from orthodoxy to a faith which I have endeavoured to explain in these pages and elsewhere as "Agnostic Christianity," or "Christianity without miracles." The gifted authoress describes the process by which his belief in miracles is gradually undermined, and, while his love and admiration for the human Jesus comes out stronger than ever, he feels it impossible to remain in a Church which demands assent to such dogmas as those of the Logos, the Resurrection, and the Atonement. Accordingly, he resigns his living, and devotes himself to a life of active charity in the East-end of London, where he labours to found a new religion which shall satisfy reason by rejecting revelation, while it satisfies emotion by dwelling on the lovely character of the carpenter's son of Nazareth. The hero dies, and the new religion remains a pious aspiration; but it is a sign of the altered atmosphere of the times that, instead of being received with a howl of execration, the book is favourably accepted by so many readers as a true picture of the course of modern thought, and as presenting an ideal of what may possibly become the religion of the future. It is a significant symptom of that drift which is setting in from so many lines of thought, irresistible as that of the stars of heaven, away from orthodoxy and towards Agnostic Christianity.

CHAPTER VIII. (*continued*).

PART II.

Effect on Morals—Evolution of Morality—Moral Instincts—Practical Religion—Herbert Spencer and Frederic Harrison—Positivism and the Unknowable—Creeds and Doctrines—Priests and Churches—Duty of Agnostics—Prospects of the Future.

ASSUMING as I do that some form of liberal and reverent Agnosticism is certain to supersede old theological and metaphysical creeds in our conceptions of the universe, it remains to consider how this will practically affect the machinery and outward form of religion, and, what is of more importance, the interests of morality.

In stating the results of my reflections on this subject I am far from wishing to dogmatize, or, like Comte, to build up any positive religion of the future, which, like his, might be comprehensively summed up as "Catholicism without Christianity." I know too well that religions, like other social institutions, are evolved and not manufactured, and that religious rites and institutions only flourish when they are a spontaneous growth. Nevertheless, I think the time has come when the intellectual victory of Agnosticism is so far assured that it behoves thinking men to begin to consider what practical results are likely to follow from it.

The first question is as to the effect on morals. Those who cling to old creeds make great use of the argument that religion is the best of policemen, and that, if faith in a future state of rewards and punishments, as taught by an inspired Bible, were once shaken, all security for life and property would be at an end. This, if it were true, would be no argument, any more than the fact that a nurse may occasionally quiet a naughty child by the threat of a bogey, would prove the existence of a black man with horns and a tail in the cupboard. But it is distinctly untrue. The foundations of morals are fortunately built on solid rock, and not on shifting sand; they are based on ideas and feelings which, in the course of the evolution of the human race, have gradually become instinctive in civilized communities, and passed beyond the sphere of abstract reasonings or speculative criticisms. So far from morality being a thing altogether apart from human nature, and which owes its obligation solely to its being a revelation of God's will, it may be truly said in a great many cases that, as individuals and nations become more sceptical, they become more moral. Thus, for instance, an implicit belief in the inspiration of the Old Testament perverted the moral sense to such an extent that the most monstrous cruelties were inflicted in the name of religion. Murders, adulteries, witchcraft, religious wars and persecutions, all found their origin and excuse in texts either expressly enjoining them, or showing that they formed part of the character and conduct of men "after Jehovah's own heart." We no longer burn heretics, torture old women, or hew captives in pieces before the Lord. Why? Because we have become sceptical, and no longer believe in the Bible as an infallible record of

God's word. When we find anything in it contrary either to the facts of science or to the moral instincts of the age in which we live, we quietly ignore it; and, instead of trying Science and Morality, as our forefathers did, at the bar of Inspiration, we reverse the process and bring Religion before the bar of Reason.

Is the world better or worse for this latest phase of its evolution? Is it more or less tolerant, humane, liberal-minded, charitable, than it was in the ages of superstitious faith? The answer is not doubtful, and it confirms my position that, as a matter of fact, as we have become more sceptical we have become more moral.

If there is one fact more certain than another in the history of evolution, it is that morals have been evolved by the same laws as regulate the development of species. They were no more created, or taught supernaturally, than were the various successive forms of animal and vegetable life. Take, for instance, the simplest case—the abhorrence of murder. It is not an implanted and universal instinct, for even at the present day we find sections of the human race among whom murder is honourable. The Dyak maiden scorns a lover who has not taken a head; the Indian squaw tests a suitor's manhood by the number of scalps in his wigwam, and the more they were taken by stratagem and treachery the more honourable are they esteemed. The priest and prophet of ancient Israel considered it an act of duty towards Jehovah to hew Agag to pieces before the Lord; and Jael was famous among Hebrew women because she drove a nail into the head of the sleeping refugee who had sought shelter within her tent. David, the man after God's own heart, committed the most treacherous and cold-blooded murder in order to screen a foul act of adultery. Where in those cases was either the implanted instinct or the recognition of a divine precept commanding "Thou shalt do no murder"? Millions of Brahmins and Buddhists, who never heard of Moses or of the commandment inscribed on the table of stone at Sinai, have carried the abhorrence of murder to such an extreme as to shrink from destroying even the humblest form of animal life, while millions of savages have killed and eaten strangers and captives without scruple or remorse.

Evidently moral ideas are, like other products of evolution, the result of the interaction of the two factors, heredity and environment, determined in the course of ages by natural selection. They may be seen in the simplest form in the instinct of all social animals, from ants and bees up to man, which makes them abstain from injuring those of the same nest or herd, and prompts them to act together for the common good. Those who had this instinct strongest would be most likely to survive in the struggle for existence, and each successive generation would tend to fix the instinct more strongly by heredity. What is instinct? In the last analysis it is motion, or tendency to motion, of certain nerve-cells, which have become so fixed, by frequent practice or by heredity, that they become unconscious, and follow necessarily on impulses from without, as in the act of breathing or swallowing. The simpler instincts, as in the case of animals, are the most spontaneous and inevitable. The duckling swims, to the alarm of the mother hen, because it is the descendant of generations of ducks which have taken to the water as their natural element. The sight of water sets up certain motions in the duckling's brain which, by reflex action, impel it to swim.

But, in higher organizations and more complicated instincts, what is inherited is not so much absolute motion as tendency to motion. The almost infinitely complex molecules of the higher brain do not move mechanically, so as to produce a definite result from a definite impulse, but they move more readily in certain directions than in others, those directions being determined partly by the ancestral channels in which they have run for generations, and partly by the action of the surrounding environment. Thus it may be accepted as certain that a child born and educated in England in the nineteenth century will, as a rule, grow up with an instinctive abhorrence of murder; but it is not so certain as that it will breathe and eat. A very violent outward impulse, such as greed or revenge, may overcome the instinct; and if the child had been kidnapped in infancy and brought up among Dyaks or Indians, its notions would probably have been the same as theirs as to the taking of heads or scalps. But, speaking generally of modern civilized societies, there is such an enormous preponderance in favour of the fundamental rules of morality, that with each successive generation the results both of heredity and environment tend more and more to make them instinctive. The lines which Tennyson, the great poet of modern thought, puts into the lips of his Goddess of Wisdom—

> "And because right is right, to follow right
> Were wisdom in the scorn of consequence"—

are becoming more and more every day the instinct, not of higher minds only, but of the mass of the community.

Such a foundation for morals is clearly both more certain and more comprehensive than one based on doubtful revelations. It is more certain, for it does not depend on evidence which, with the progress of science, is fast becoming incredible. The command not to murder is not weakened by proof that the book of unknown origin and date which contains it, gives a totally erroneous account of the creation, and is therefore not inspired; nor does adultery cease to be a crime because the narrative of Noah's deluge is shown to be fabulous. It is also more comprehensive, for no hard-and-fast written code can long conform to the conditions of an ever-varying society. It will err both by enjoining things which have become obsolete, and by omitting others which have become imperative. Thus the Mosaic code classes sculptors with murderers and thieves, and makes Canova and Thorwaldsen as great offenders against Divine commands as the last criminal who was convicted at the Old Bailey. On the other hand, there is no injunction against slavery or polygamy, but, on the contrary, an implied sanction of them, from the example of the patriarchs who are held up as patterns of holiness. The feeling against slavery is a conspicuous instance of the development of a moral instinct in quite recent times. It is the result of advancing civilization leading to more humane ideas, and to a clearer recognition of the intrinsic sacredness and dignity of every human soul.

In like manner, a multitude of moral ideas have come to be part of our mental furniture which had no place in the early code of the Jews, or even in the more advanced period of early Christianity. The Christian ideal, to a great extent, ignored courage, hardihood, self-reliance, foresight, providence, and all the sterner and harder qualities that make the man, for the softer and more feminine virtues of love, patience, and resignation. The æsthetic side of life also, the recognition and love of all that is beautiful in art and nature, was not only ignored, but to a great extent condemned by it, owing to an exaggerated and one-sided antithesis between the flesh and the spirit.

Among the modern ideas which are fast becoming moral instincts is that of the duty of following truth for its own sake. Doubt is no longer regarded as a crime, but as a duty, when there are real grounds for doubting. We may parody the words of the poet and say—

> "And because truth is truth, to follow truth
> Were wisdom in the scorn of consequence."

And this allegiance to truth carries with it the virtue of sincerity. A man must not palter with his convictions, and profess to hold one set of opinions because they are expedient, while he holds others because they are true. If it be a *fact* that the human race has risen by evolution through long ages from palæolithic savagery, he has no right to admit the fact and at the same time profess to believe that he is a fallen creature descended from the Biblical Adam. His duty is to use his reason to ascertain which statement is true, and, having done so, to the best of his ability and without bias or prejudice, to cleave with his whole heart to the truth, and not remain a miserable, half-hearted Mr. Facing-both-ways.

So far, therefore, as morality is concerned, we need not much concern ourselves about the future of religion. Morality can take care of itself, and, with or without theological creeds, it will go on strengthening, widening, and purifying its instinctive holds on the character and conduct of civilized communities. As regards conduct, which is, after all, the practical test of the goodness or badness of theoretical opinions, a system which can produce a life like that of Darwin is good enough for anything. Conduct is, fortunately, not dependent on creeds, and good men and women can be found plentifully among all classes of belief, from Orthodoxy to Agnosticism. But it cannot, I think, be denied that the leaders of scientific thought, such as Darwin, Herbert Spencer, Lyell, Huxley, and other honoured names, have led, on the whole, simple, noble lives, and present characters worthy of imitation. Nor is there any reason to believe that the vast and increasing number of the rank-and-file, who have more or less adopted the views of these great leaders, are in any respect below the average type, or lead worse lives than those who walk in the narrower paths of pre-scientific traditions.

Thus far the Religion of the Future has been comparatively plain sailing. Intellectually, it is clear that evolution has become the mould of thought, and that the lines of Agnostic Christianity and of Agnosticism pure and simple, but recognizing Christianity as one of the forces of evolution, have converged so closely that the difference between them is almost reduced to a name. What Herbert Spencer calls the infinite, eternal energy, which underlies all phenomena, and of whose existence we feel certain, though we can never know or define it, Bishop Temple calls "God." Accurate thinkers may prefer the former definition, for the term "God" has come to be associated with a number of anthropomorphic and other ideas, which imply knowledge of the unknowable; but practically the bishop and the philosopher mean much the same thing, and the converging lines of science and religion approach so nearly that they may be said to coincide. Morally, it is equally clear that there is nothing to fear from such a view of religion, and that the moral instincts are based on something much more permanent and certain than intellectual conceptions or antiquated traditions. But when we come to practical religion there is a great deal comprised in the word which it is not so easy to dispose of.

In the recent controversy between Herbert Spencer and Frederic Harrison the latter reproached the former with offering to the world the mere ghost of a religion. Religion, he says, must be something positive; it must have a "creed, doctrines, temples, priests, teachers, rites, morality, beauty, hope, consolation;" and these, he adds, can be found only in a religion which is intensely anthropomorphic. "You can have no religion without kinship, sympathy, relation of some human kind between the believer, worshipper, servants, and the object of his belief, veneration, and service."

As Mr. Harrison not only admits, but asserts strongly, that science has upset all existing anthropomorphic creeds and theories, his logical conclusion apparently ought to be that there can be no more any religion. But he escapes from his dilemma by offering us a new religion—Positivism, or the religion according to Comte. For the dethroned Deity of the Christians, who has been, by the confession of his own theologians, "defecated to a pure transparency," we are to substitute 'Humanity," the symbol of the new Divinity being a woman of the age of thirty, with her son in her arms; and Christian worship is to be replaced by an elaborate series of rites and ceremonies, evolved from the inner consciousness of the French philosopher, and which, to the apprehension of an ordinary observer, are for the most part puerile and ridiculous. Thus among the

Positivist saints, who are to be canonized in order of merit, Gall, who, in conjunction with Spurzheim, wrote an obsolete book on phrenology, gets a week, while Kepler gets only a day; Tasso is assumed to be a seven times greater poet than Goethe, and Mozart a seven times greater musician than Beethoven; while in politics Louis XI., the crafty and sinister French king, depicted by Walter Scott in *Quentin Durward*, is to be worshipped as a seven times greater saint than Washington. Of the only two new forms of positive religion which have been started in my recollection, Positivism and Mormonism, I may be excused if, barring the plurality of wives, I give the preference to the latter, which has, at any rate, proved its vitality by laying hold, not without a certain amount of success, of colonization, temperance, and other problems of practical life. Herbert Spencer had little difficulty in answering this attack. He showed that his definition of the "Unknowable" was very different from the mere negation, or algebraical symbol, which Harrison assumed it to be, and that it was distinctly the assertion of something positive and actually existing, though beyond our faculties. In fact, it is very much the same as Wordsworth's

> "Sense sublime,
> Of something far more deeply interfused,
> Whose dwelling is the light of setting suns,
> And the round earth, and in the mind of man."

And if such a feeling can inspire noble poetry, why not a noble religion? The retort was obvious, that, if the Unknowable were too refined an idea on which to base a religion, at any rate it was better than Humanity; for the first is based on a fact, while the second has no foundation but a phrase.

It is an undoubted fact that, when we trace phenomena back to their source, we arrive at a substratum, or first cause, which we cannot understand, or even form any conception of. But what is Humanity? It is but a convenient expression, like gravity or electricity, by which we sum up a number of separate, individual facts, which have certain attributes in common. The only thing real about gravity is, that individual bodies attract one another directly as the mass and inversely as the square of the distance. Annihilate the individual masses, and you cannot anthropomorphize the law of gravity; for instance, following the example of Comte, under the symbol of a heavy woman with a fat child. No more can you individualize and anthropomorphize "Humanity," apart from the individual human beings, good, bad, and indifferent, of whom the aggregate has been, is, and will be composed. "*Parturiunt Montes*"—the mountains labour to produce a new religion; and the result of Positivism is to make a fetish of a phrase.

At the same time it must be admitted that, while Positivism is no more likely than Mormonism to become the world's religion of the future, the new creed to which we are tending, whether we call it Agnostic Christianity or Christian Agnosticism, places in jeopardy a great deal of what has hitherto been included under the word religion. Mr. Harrison's definition is not an unfair one, that the term includes "creed, doctrines, temples, priests, teachers, rites, morality, beauty, hope, consolation." Of these, the four last may be called spiritual, and the six first practical elements of religion. As regards the spiritual elements, they will remain unaffected, and, in some cases, will be strengthened. Morality, as we have seen, depends on rules of conduct, which have, to a great extent, become instinctive; and it would be strengthened, rather than impaired, by getting rid of the Calvinistic conceptions of a cruel and capricious Deity, condemning untold millions to eternal punishment for the offence of a remote ancestor, and only partially appeased by the sacrifice of his only son. Beauty, again, would certainly gain by getting rid of the idea that all pleasant things are of the domain of the flesh and the devil, and substituting an enlightened æstheticism for a narrow and sordid asceticism. Hope would, as at present, find its field in the possibilities which lie behind the veil, and time, the one great consoler of human sorrows, would still exert its beneficent influence to assuage the poignancy of recent afflictions.

But what will become of the "creed, doctrines, temples, priests, teachers, and rites," which constitute what may be called the machinery or practical side of existing religions? Is the creed the keystone of the fabric, and will it crumble to pieces if this creed ceases to be credible? In other words, if the creeds of Christian Churches, instead of being definite doctrines, as embodied in the Thirty-nine Articles, or the dicta of infallible Popes and Councils, are sublimated into such vague and remote conceptions as enable Huxley to say that the three bishops have conceded all he asks, and Mivart to remain a good Catholic while admitting all the most advanced conclusions of Darwinian science and of Biblical criticisms, can sincere men become Christian priests and officiate in Christian churches?

I judge no one, and can appreciate the reasons which may induce enlightened and excellent men to cleave to old creeds and remain in positions when they feel that they are doing good, as long as it is possible for them to allegorize or explain away accepted doctrines, without feeling that they are consciously insincere. But I confess that it is not easy to understand how this can go even the length it has, and, still more, how it can go further and become general, without degenerating into hypocrisy and insincerity. Take, for instance, the Apostles' Creed, which, I suppose, contains the *minimum* of doctrine that is generally considered consistent with a profession of Christianity. I can understand how, by an allowable latitude of construction, a Broad Church divine may adopt the first Article and confess a belief in God. But when we come to the subsequent, more precise and definite Articles, which profess a belief in the miraculous conception, birth, and resurrection of Jesus, the carpenter's son of Nazareth, I fail to see how any one can subscribe to them who believes in the permanence of Natural Law and the Darwinian theory of Evolution. Even in the form of Bishop Temple's theory of original impress, as opposed to special acts of supernatural interference, it must be admitted that miracles, if not impossible, are in the highest degree improbable, and that it would require an immense amount of the clearest possible evidence to admit occurrences which are so entirely opposed to all we know of the real facts of the universe, and which, in so many cases, have been shown to be mere delusions of the imaginations. And the slightest acquaintance with Biblical criticism is sufficient to show how weak the evidence really is, and how utterly unfounded the claims of the various books of the Old and New Testament to anything like Divine inspiration. But, if the creeds go, what become of the priests, and, without priests, where are the churches, rites, and ceremonies? And, if these disappear, what an immense gap does it make in the whole framework of existing society! Consider the priests, including in the word all ministers of all denominations. It is easy to denounce priestcraft, and to show by a thousand examples that wherever priests have had power they have done infinite mischief. They have too often been cruel persecutors and narrow-minded bigots; and, even at the best, have been opposed to freedom of thought and progress. But, for all this, the question has another side, and there is a good deal to be said for the existence of a special class, set aside from the ordinary pursuits of life, for spiritual instruction and works of mercy and charity.

In countries like England, where priests have long since ceased to possess any temporal power, and where they live—more and more every day—in an atmosphere of free and liberal thought, there can be no doubt that they are, as a class, much better than they were in former ages. Few exercise an influence actively injurious, many are respectable and harmless, and a considerable number set a good example of virtuous lives, and devote themselves to the promotion of works of charity and benevolence. They have, no doubt, to a considerable extent, lost touch with the masses of population in large towns and industrial centres; and where they have preserved it, chiefly among dissenting congregations, it is too often exerted towards narrowness of views and sectarian prejudices. Still, on the whole, it is exerted for good; and in many rural parishes and poor districts, like the East-end of London, the priest is a powerful factor in organizing charities, visiting the sick, rescuing the fallen, and giving consolation to the suffering. To take an extreme case, what would a poor parish in the West of Ireland be without its priest? He is the sole centre of civilization in a district of, perhaps, twenty square miles; he is not only the spiritual guide of his flock, but, to a great extent, their Education Board and Poor-Law Guardian; he is their friend and adviser in all their difficulties, and, in case of need, their "Village Hampden," who fights their battles with tyrannical landlords, and negotiates the compromises by which they are enabled to retain their humble roofs over their heads. He is worth all the magistrates and policemen put together in repressing crime and preventing outrages. It will be long before a population like that of rural Ireland can dispense with priests.

Again, priests and churches go together; and although church services have to a great extent become a repetition of formulas, and sermons an anachronism, still there is a good deal in institutions which bring people together on one day in the week, cleanly in dress and decorous in behaviour, to join in services and listen to discourses which appeal, however faintly and drearily, to higher things than those of ordinary prosaic life. Especially to the female half of the population attendance at church or chapel is, in many cases, a great pleasure, and, if it were only to see and be seen and criticize one another's bonnets, it is a relief from the monotony of life, gives them topics of interest, and promotes a feeling of decency and respectability. Those, therefore, who hold larger views, and feel that they cannot without insincerity subscribe to creeds which to them have become incredible, would do well to be liberal and tolerant towards traditional opinions and traditional practices, and trust with cheerful faith to evolution to bring about gradually such changes of form as may be required to embody changes of spirit.

In the meantime, the course of those who worship Truth above all other considerations is plain. There are abundance of duties clear enough for men of all creeds: the difficulty is to live up to them. But for those who hold the larger views the first duty is to be doubly careful as to conduct. It would be too great a scandal if the larger creed were made the excuse for a looser life. Those who are Darwinians in theory ought to try to be like Darwin in practice: like him, high-minded, modest, gentle, patient, honourable in all relations of life, loving and beloved by friends and family. This, at least, is within the reach of every one, high or low, rich or poor, if not to attain to, at any rate to aim at, as an ideal. Nor do I think that Freethinkers will be wanting in this passive side of conduct. On the contrary, as far as my experience has gone, while more liberal and large-minded, they lead lives quite as good, on the average, as those which are more directly under the traditional influences of religion. But what the Agnostic must beware of is, not to be content with the passive side of virtue, but to cultivate also its active side, and not let himself be surpassed in works of charity and benevolence by those whose intellectual creeds are narrower than his own. There is no doubt that the evangelical faith in Jesus has been and is a powerful incentive with men like Lord Shaftesbury, General Gordon, Dr. Barnardo, and thousands of other devoted men and women who fight in the foremost ranks against sin and misery. With such as these all men can sympathize; and a more intellectual creed ought to be no obstacle in giving aid and co-operation, but rather an incentive to show that a belief in the truths of science is not inconsistent with active charity and benevolence.

Another point which Agnostics would do well to attend to is to cultivate a love of Nature and Art, so as to keep alive the imaginative and emotional faculties which might wither in the too exclusive atmosphere of pure reason. A prosaic life is a dwarfed and stunted life, which has been more than half a failure; and, as old dogmatic religions fail to supply the spiritual stimulus, it is the more necessary to find it in the wonders of the universe, the beauties of nature, and in communion with great minds through music, painting, and books. These are now brought to a great extent within the reach of every one, and there is no more hopeful symptom of the times than to find that really good books by great authors, when brought out in cheap editions, circulate by the millions. Shilling and even sixpenny editions of Shakespeare, Scott, Carlyle, and other standard authors, are continually brought out, and must be sold in tens of thousands to make them a paying speculation. Who buys them? Certainly not the upper classes, who, in former days, were the only buyers of books. They must circulate widely among the masses, and especially among the more thoughtful members of the working-classes, and the rising generation of all classes who are earnestly seeking to improve their minds and widen their range of sympathies and culture. To read good books rather than silly novels is a practical measure within the reach of every one, and it is supplying, more and more every day, a larger and more liberal education than was ever afforded by theological controversies and conventional sermons.

Another hopeful symptom is to see the growing

demand among the working-classes for schools, libraries, museums, music-halls, excursion trains, and all manner of clubs and societies for mutual help, instruction, and amusement. These are the plastic cells multiplying and forming new combinations, out of which, in due time, will be evolved the "priests and temples, the rites and ceremonies," and other institutions requisite to give life and form to the demonstrated truth of the "great Unknowable," and leave the magnificent conception of Darwin and Herbert Spencer no longer the ghost of a religion, but the foundation of a rational, lovable, and, on the whole, happy existence, useful and honourable while its little span of life lasts, and looking forward with hope and manly fortitude to whatever may await it behind that veil which no mortal hand has ever lifted.

CHAPTER VIII. (*continued*).

PART III.

Practical Philosophy—Zoroastrian Theory—Emerson on Compensation—Good and Evil—Leads to Toleration and Charity—Matthew Arnold and Philistinism—Salvation Army—Conflict of Theology and Science—Creed of Nineteenth Century.

THE philosophy which I have found work best, both in reconciling intellectual difficulties and as a guide in practical life, is that which I have described elsewhere[1] at some length as "Zoroastrianism," or "Polarity." It amounts to this, that the infinite, eternal, and inconceivable essence of all phenomena, which theologians call God, and philosophers the Unknowable, manifests itself to human apprehension under conditions or categories which are equally certain and equally incomprehensible. We know that it is so, or so appears to us; but we do not know why. Thus Space and Time are fundamental moulds of thought, or, to use the phraseology of Kant, imperative categories. Another of such categories is that of Polarity: no action without reaction, no positive without a negative, no good without evil. In the physical world this is a demonstrated fact. Matter is made of molecules; molecules are made of atoms; atoms are little magnets which link themselves together

[1] *A Modern Zoroastrian.*

and form all the complex creations of an ordered cosmos, by virtue of the attractive and repulsive forces which are the results of polarity. Ordered and regular motion also—whether it be of planets round suns, of an oscillating pendulum, or of waves of water, air, or ether, vibrating in rhythmic succession—is a result of the conflict between energy of motion and energy of position.

As Emerson well says in his Essay on Compensation: "Polarity, or action and reaction, we meet in every part of nature: in darkness and light; in heat and cold; in the ebb and flow of waters; in male and female; in the inspiration and expiration of plants and animals; in the undulations of fluids and of sound; in the centrifugal and centripetal gravity; in electricity, galvanism, and chemical affinity. Superinduce magnetism at one end of a needle, the opposite magnetism takes place at the other end. If the South attracts, the North repels. To empty here you must condense there. An inevitable dualism besets nature, so that each thing is a half, and suggests another to make it whole: as spirit, matter; man, woman; odd, even; subjective, objective; in, out; upper, under; motion, rest; yea, nay." This principle, applied to the higher problems of religion and philosophy, leads to results singularly like those which, if we may believe the sacred books of the Parsees, were taught 3000 years ago by the ancient Bactrian sage, Zoroaster. His religion was one of pure reason. He disclaimed all pretension to found it on miracles, or to define the indefinable by dogmas; but, taking natural laws and human knowledge as his basis, he asserted, in the identical words used by Emerson thirty centuries later, that an "inevitable dualism besets nature," and embodied the two conflicting principles under the names of Ormuzd and Ahriman. To Ormuzd belong all things that are bright, beautiful, pure, lovely, and of good repute, both in the material and moral universe; to Ahriman all that is foul, ugly, and evil. Apart from certain archaisms of expression and ritual observances which have become obsolete, the Zendavesta might have been compiled to-day from the writings of Herbert Spencer and Huxley. This conception of the universe has this enormous advantage over all those which rest on the idea of an anthropomorphic Creator—that it does not make religion a means of perverting the fundamental instincts of morality, by making an Omnipotent Creator the conscious author of evil. This is a dilemma from which no anthropomorphic form of religion can escape: either its God is not omnipotent, or he is not benevolent. Sin and suffering are *facts*, as much as virtue and happiness; and if the good half of creation argues for a good Creator, it is an irresistible inference that the bad half argues for one who is evil.

Theologians, in attempting to escape from this dilemma, have been only too apt to confuse the instincts of morality, by arguing that actions which would be cruel, unjust, and even devilish, in the case of a human despot, become merciful and righteous if done by an Almighty Ruler in Heaven. Such a dogma is, to all intents and purposes, devil-worship, and degrades man into a slave crouching under the lash of a harsh master. How infinitely superior was the ideal of the old Roman poet of the "*justum et tenacem propositi virum*"; the upright and firm-minded man, whom no threats of a frenzied mob or raging tyrant could shake from his purpose, or induce to palter with his convictions; nay, not even though the earth and sky fell in ruins about his head, could the convulsion of nature daunt his steadfast soul.

"*Victrix causa Deis placuit sed victa Catoni.*"

But, with a Polar theory of existence, the difficulty is relegated to the realm of the unknown, and, instead of sinking with Cowper into the despairing depths of religious madness, we may hold with Wordsworth—

"The cheerful faith that all which we behold
Is full of blessings."

A serene and cheerful faith is, of itself, one of the greatest blessings, and it is specially needed in an age in which so many gospels of pessimism are abroad, and so many failures in the struggle for existence tell us that society is a sham, civilization an imposture, and life a mistake.

Another advantage of this Polar theory of the universe is that it teaches us to take a large and tolerant view of men and of events. The true charity which "suffereth long and is kind" is scarcely compatible with a bigoted and one-sided adherence to a particular set of opinions. Whether in politics or in religion, if we believe that all those who differ from us have a double dose of original sin, we can scarcely comprehend or love them. Good natures may pity them, bad natures hate them, conscientious natures feel it a duty to stamp them out; but we can never really feel towards them as brothers and sisters, who have gone "a kenning wrang," and been drawn a little too far by the attraction of the opposite polarity to that under the influence of which we ourselves live and have our being. Thus, in politics, the cosmos of an ordered society can only be maintained, as in the orbit of a planet, by a due balance between the centripetal and centrifugal forces. If we were all Conservatives, society would condense into a sluggish and inert mass; if all Radicals, it would be apt to fly off into space. Evolution will surely bring about in their appropriate time the results which are fittest to survive. Why quarrel, then, and entertain hard and bitter thoughts, because our own individual atom is acting in one direction, while that of our neighbour is acting in another? Act strenuously in that direction which, after conscientious inquiry, seems to be the best; do the duty which lies most nearly and plainly to our hands; and trust to what religious men call Providence, and scientific men Evolution, for the result.

A large-minded and large-hearted creed is the more needful, as the weak part in the otherwise admirable British nature is a tendency to that peculiar form of narrowness which is commonly called Philistinism. Why the Philistine, or dweller in the land of palms on the border of the Mediterranean, should have been taken as the type of straight-laced and narrow-minded conventionality, is hard to see. But the fact is there, and the word expresses it; and it is beyond doubt that there is a great deal of truth in Matthew Arnold's indignant diatribes, and that the average well-meaning and respectable citizen is apt to be an awful Philistine. It is not confined to classes; in fact, there is probably more of it in the upper and middle classes than among workmen. But whether it be the cut of a coat, or of a creed, and whether going to a court or to a chapel, the essence of the thing is the same—viz. that some class or coterie fences itself in behind some narrow conventionality, and ignores the great outer world. If the pale be one of fashion, those not within it are outsiders, cads, commoners; if of religion, they are sons of perdition. To the narrow-minded Tory all Irish are dynamiters, all Radicals rebels, and Gladstone is Antichrist. To the narrow-minded Radical all landlords are robbers and all parsons hypocrites. Socialists seek to regenerate society by abolishing capital; capitalists to save it by ignoring that property has duties as well as rights. It is all Philistinism, and incapacity to see that there are two sides to every question, and that one thing only is certain, that falsehood lies in extremes. Half the difficulties which perplex us would disappear if we could enlarge our minds, so as, in the words of Burns—

"To see ourselves as others see us";

and to act on the precept of the wise old Rabbi Hillel, now 1900 years old: "Never to judge another man till you have stood in his shoes."

Another advantage of this Polar philosophy is that it enables us more readily to assimilate with those who hold different forms of belief. What matters it whether the Parsee embodies his good principle in an Ormuzd, the Christian in a Jesus, the Stoic in a Marcus Aurelius, or the philosopher finds no need for any personification at all? The essential thing is that they are all soldiers fighting together in the cause of goodness and light, against evil and darkness. Practically, a great many modern Christians are Zoroastrians, with Jesus for their Ormuzd. They care little for dogmas, except as exalting the character of the object of their veneration, and giving expression to their transcendental love and adoration for his person and character. Listen to the simple preaching of the Salvation Army, and you will find how exclusively it turns upon the one element of the love of Jesus. You would never discover that Christianity had been identified with mysterious dogmas and metaphysical puzzles, and that salvation depended on holding the Catholic faith as defined by St. Athanasius. But sinners are exhorted to give up drink and evil ways for the love of the dear Redeemer who died for them; and if this touches simple natures, and if calling themselves soldiers, marching in ranks, and beating drums, aid in the work, why should any one object to it? We are nearer to these simple souls than we are to the divines who beat the drum ecclesiastic, and tell us from pulpits, that, unless we believe all the articles of the Catholic faith without doubt we shall perish everlastingly.

To sum up, the duty of a man of the nineteenth century is clear. He has to follow truth at all hazards. Questions of the highest importance have been raised, which he cannot shirk without narrowing his whole nature, and shutting himself up in an ever-contracting circle of ignorance and prejudice. There are two theories of the universe and two of man, which are in direct conflict. Of the universe, one, the theological, that it was created and is upheld by miracles—that is, by a succession of secondary supernatural interferences by a Being who is a magnified man, acting from motives and with an intelligence which, however transcendental, are essentially human; the other, the scientific, that it is the result of original impress, or of evolution acting by natural laws on a basis of the Unknowable. In like manner, of man, one theory, the theological, is that he is descended from the Biblical Adam, created quite recently in a state of high moral perfection, from which he fell by an act of disobedience, entailing on his descendants the curse of sin and death, from which a portion were redeemed by the sacrifice of the Creator's own son, incarnate in Jesus of Nazareth; the other, the scientific theory, that man is a product of Evolution from palæolithic ancestors, who lived for innumerable ages in a state of savagery, but always gradually progressing upwards in arts and civilization.

Both theories cannot be true; they are in direct contradiction upon fundamental facts, which are a question of evidence. The evidence for the theological theory is based entirely on the assumption that the Bible is an inspired record of Divine truth, attested by miracles. The scientific theory rests on the evidence of a vast and ever-accumulating mass of facts, which admit of no doubt or contradiction. It seems to me that an unlearned man need not go farther than to contrast the theories of man's descent. Let him go to the British Museum and look at the implements of flint and bone which have been found in conjunction with remains of extinct animals, in caves and river gravels of immense antiquity. How can the theological theory hold water, unless it could be proved that these, and the hundreds of thousands of similar human remains, including skulls and skeletons, which have been discovered in similar deposits over the four quarters of the earth, were placed there by a conspiracy of scientific men, who wished to discredit the Bible? Even the Duke of Argyll, who has conspiracy on the brain, would hardly contend for such a conclusion, or maintain that the narrative of Noah's deluge gives a *true* account of the manner in which animal life has been diffused over the different zoological provinces in which it is actually divided.

The more he extends his researches and enlarges his knowledge, the more will every honest and conscientious inquirer find that the scientific theory is victorious along the whole line. If he is a lover of truth, therefore, he will find himself constrained to adopt the larger creed. But, in doing so, let him show that it is not merely a speculative creed or intellectual deduction; but that the larger creed leads to a larger life; that it makes him more liberal and tolerant, more pure and upright, more loving and unselfish, more strenuous, as becomes a soldier fighting in the foremost ranks in the campaign against sin and misery; so that, when the last day comes which comes to all, it may be recorded of him that his individual atom of existence left the world, on the whole, a little better, rather than a little worse, than he found it.

CHAPTER IX.

THE HISTORICAL ELEMENT IN THE GOSPELS.

Huxley and Dr. Wace—Sermon on the Mount, and Lord's Prayer—English and German Biblical Criticism—Papias—His Account of Origin of the Gospels—Confirmed by Internal Evidence—Common-sense Conclusions—Miracles a Question of Faith—Evidence Required—The Ascension—Early Christian and Mediæval Miracles—St. Thomas-à-Becket—Faith—Historical Element—Virgin Mary—Guiding Principles of Historical Inquiry—Minimum of Miracles—Admissions which tell against—Jesus an Historical Person—Born at Nazareth—Legends of Nativity—St. John the Baptist—Kingdom of God—Socialistic Spirit—Pure Morality—Nucleus of Fact in Miracles—Precepts and Parables—Disputes with Scribes and Pharisees—Jesus a Jew—Messiahship—Dying Words—Passion and Crucifixion—Improbabilities—Pilate—Resurrection—Contradictions—Growth of Legend—Probable Nucleus of Fact—Riot in the Temple—Return of Disciples to Galilee—Conflicting Accounts of Resurrection—Return of Apostles to Jerusalem and Foundation of Christian Church.

Professor Huxley in a recent article in the *Nineteenth Century* refers to the great difficulty he has felt in his efforts to define "the grand figure of Jesus as it lies in the primary strata of Christian literature. What did he really say and do? and how much that is attributed to him in speech and action is the embroidery of the various parties into which his followers tended to split themselves within twenty years after his death, when even the threefold tradition was only nascent?"

I have felt the same difficulty myself, and after reading a mass of critical literature, both English and German, I must confess to having found myself more than ever perplexed. In English Biblical criticism the tone is almost invariably that of advocates rather than of judges. The opponents of Orthodoxy insist too much on finding arguments against inspiration in every text, while its supporters are almost invariably guilty of the fallacy which is known to logicians as the *petitio principii*, and begin by assuming the very points which they profess to prove. Thus Dr. Wace, in his reply to Huxley, starts with the assumption that the Sermon on the Mount and the Lord's Prayer prove the divinity of Jesus and the inspiration of the Gospels; and this being proved, it follows that we must believe everything we find recorded in these Gospels as true, down even to the miracle of the Gadarene swine, under pain of making Jesus out to be a liar. Of course we must, if we admit the theory of divine inspiration, but this is the very point to be proved. How does Dr. Wace attempt to prove it? By lengthened arguments to show that the omission of all mention of the Sermon on the Mount and Lord's Prayer by Mark is not a fatal objection; that the Synoptic Gospels, or parts of them, were probably written not later than from 70 to 75 A.D., and other doubtful points of really very little importance. But he totally ignores what is the real difficulty in the way of accepting his fundamental axiom that the Sermon on the Mount and Lord's Prayer compel us to admit inspiration. The difficulty is this, that their precepts, admirable as they are, are not original. There is scarcely one which is not to be found, identical in substance and often almost in the exact words, in the older writings of earlier religions and philosophies. Thus the cardinal precepts, such as to "love your neighbour as yourself," to "do as you would be done by," to "return good for evil," &c., are found in the old Egyptian ritual, the Vedic literature, the maxims of Confucius, and still more conspicuously in the oldest writings of the Buddhist and Zoroastrian religions.

And what is even more important, the Talmudic or Rabbinical literature of the age immediately preceding that of Jesus is full of them; the writings of Jesus the son of Sirach, of Hillel, and of Philo, contain many of the same precepts, almost verbatim, and they were the common possession of the Jewish world at the time when the Sermon on the Mount is supposed to have been preached.

These facts are undeniable, and it is equally undeniable that, if so, the bottom is knocked out of Dr. Wace's assumption; for if these precepts and this code of morality could be evolved in other ages and countries by natural means, why should they require the miracle of Divine Inspiration to account for them in the New Testament? The Sermon no doubt has its value in bringing to a focus a number of excellent precepts, and helping to form the ideal of Jesus and his teaching, which has become the fundamental fact of Christianity, but as anything like reasonable proof of miraculous inspiration it is worthless. Nor is there anything in the Lord's Prayer which might not have been the prayer of any pious Jew of the time, or, for the matter of that, of any pious Gentile, for "Our Father which art in heaven" is a iteral translation of Jupiter, or Dyaus-piter, the father of gods and men identified with the vault of the sky. And it cannot be reasonably denied that the omission of all mention of it in Mark tells strongly against its authenticity, for, if really taught by Jesus, it would have been the very thing to be committed to memory, and taught to all converts by his immediate disciples.

I refer to this argument of Dr. Wace's to illustrate what I find to be the great fault of English theologians, viz. that they shirk the obvious difficulties which present themselves to the minds of ordinary men using their reasoning faculties, and either refuse to reason and appeal to faith, or battle about minor points which hardly touch the real objections.

When I turned to German criticism I found it less obscured by theological, but more by theoretical prepossessions. Every professor had his own theory to establish, and that of his predecessors to demolish, and in doing so applied an enormous amount of erudition to points which, for the most part, seemed to me to remain doubtful, or to be of minor importance. The effect produced on my mind by critics such as Strauss, Baur, Volckmer, and Reuss was to leave a sort of blurred and hazy image, as of a landscape in which the essential features are lost in the multitude of details.

For instance, it seemed to me that the enormous mass of literature which has been written to assign the precise date of each Gospel, their respective priorities, how many successive editions they went through, and how far each copied from the others or from older manuscripts, might have been greatly abridged if the learned authors had been content to take the simple, straightforward evidence of the earliest Christian writer who gives any account of their origin, viz. Papias.

Papias was Bishop of Hieropolis, one of the churches in Asia Minor, which was reputed to have been founded by St. John, and who suffered martyrdom for his faith when an aged man, about 160 A.D. He was certainly in a position to know what was accepted as of authority by the early Christian Church of his period. He had been in close personal communication with Polycarp and others of the generation preceding his own, who had been themselves disciples of the apostles, and his information was therefore only removed by one degree from being that of a contemporary and eye-witness. His work is unfortunately lost, but Eusebius, who was a great collector of information respecting the Gospels in the fourth century, happily preserves the most important part of it in a long quotation.

What does Papias say? Practically this—that he preferred oral tradition to written documents, of which he expresses a somewhat contemptuous opinion, assigning as a reason that there were only two written records which possessed any real authority: one a collection of anecdotes or reminiscences, taken down without method or order from the mouth of St. Peter by Mark, his interpreter; the other a collection of logia, or sayings of Jesus, written by St. Matthew in Hebrew, and badly translated into Greek by various writers.

This statement of Papias, if correct, proves several things:—

1. The Gospel of St. John could not have been known to Papias, or he, a bishop of a church reputed to have been founded by that apostle and a friend of Polycarp and others who had known him personally, could never have expressed an almost contemptuous preference for oral tradition over any written records, and made no mention of what has been always considered the most important and spiritual of all the

Gospels, proceeding direct from the Apostle whom Jesus loved.

2. The same remark applies to the Gospel and Acts of St. Luke, which contain by far the most precise details of the crowning miracles of the Resurrection and Ascension.

3. It is equally clear that he could not have known the Gospels of Mark and Matthew as they now exist, for they are connected biographies of the life and teachings of Jesus, and not fragmentary anecdotes and sayings such as Papias describes.

4. It is evident, however, that two written records —one attributed to Mark and the other to Matthew— were known in the time of Papias, and received as of sufficient authority to make him refer to them in his general depreciation of written as compared with oral testimony.

This is a perfectly clear and intelligible statement, made apparently in good faith, without any dogmatic or other prepossession; and it is confirmed by all the evidence we possess of this obscure period—whether it be the external evidence that the Gospels in their present form are not quoted or referred to as an authority by any Christian writer earlier than the second century, or the internal evidence derived from the Gospels themselves. That of Mark has exactly the appearance of having been compiled into a biography from a series of such reminiscences as Papias describes. It is full of little life-like touches which have no special significance, but seem to have come from the recollection of an eye-witness. For instance, that the throng was so great to hear Jesus that not only the room but the doorway were crowded, and that the hurry and bustle were such that they had not time even to eat.

It is true that such touches are not conclusive, and may have been added to give local colour and a life-like character to the narrative, a remarkable instance of which is afforded by the episode of the woman taken in adultery, in St. John, which is not found in the oldest manuscripts, and is doubtless an interpolation. This episode has every appearance of being taken from the life; the abstracted air, the writing with the finger on the sand, the exact words spoken, all give it an air of reality, and yet it must have been interpolated at a comparatively late date after several manuscripts of the Gospel were already in existence. Such an instance may make us hesitate in judging of similar passages from internal evidence, but it hardly applies to Mark, whose characteristic traits are much shorter and simpler, and whose level of culture and literary ability is much lower than that of the compiler—whoever he may have been—of the Gospel according to St. John.

The Gospel of Matthew, again, has exactly the appearance of having been compiled from such a collection of logia as Papias describes, woven into a biography by the aid of the original Mark and other early traditions, and embellished by the addition of much mythical matter intended to show the fulfilment of Messianic prophecies, and to meet objections.

It has always seemed to me, therefore, that all theories as to the date and origin of the Canonical Gospels were comparatively worthless which did not take into account the fundamental fact of this statement of Papias. It is either true or false. If true it is worth a hundred theories evolved, like the ideal camel, from the inner consciousness of German professors, and is conclusive of the fact that the Gospels in their present form were not known, or not accepted as an authority, by the early Christian Churches of the East in the first half of the second century, though this is quite consistent with their containing passages and traditions which may date back to the siege of Jerusalem, or even to a much earlier period. If, on the other hand, Papias is to be rejected, let us know the reason why, and give us some sort of an intelligible explanation of how such a passage came to be quoted from his work by Eusebius.[1]

[1] The difference to which I have referred between the conclusions of common-sense, and those of erudite ingenuity acting under the influence of theological prepossession, is well illustrated by the attempt of Bishop Lightfoot, in his *Essays on Supernatural Religion*, to answer the obvious inference from this passage of Papias. Common sense says, if the Canonical Gospels, and especially that of St. John, had been extant in their present form and accepted as an authority by the early Christian Church, Papias must have known them. If he had known them he could not have referred in such contemptuous terms to written records as inferior to oral tradition, and could not have mentioned the disconnected anecdotes of Mark and the Hebrew logia of Matthew as the only records of importance. Nor could Eusebius have quoted this passage alone from Papias, which obviously tells against his own views, without quoting other passages which refer to the Canonical Gospels, if any such had existed in other portions of the work of Papias. The Bishop replies—

1. That the design of Eusebius may have been to quote only references to the Apocryphal writings, and in the case of the Canonical Gospels anything which threw light on their origin; and therefore that the silence of Eusebius is no proof that there may not have been references to and quotations from these Gospels in the writings of Papias.

But this, which is in itself a very far-fetched supposition, is contradicted by the words of Eusebius himself, who says, "As my history proceeds, I will take care to indicate what Church writers from time to time have made use of any of the disputed books, and what has been said by them concerning the *Canonical and acknowledged Scriptures.*"

2. That when Papias says, "I thought I could not derive so much advantage from *books* as from the living and abiding oral tradition," he meant books which were not Gospels, but commentaries on Gospels.

I give this as an illustration of the way in which the more I studied these professional works of Biblical criticism, the more confusion became worse confounded. At length, after having abandoned the subject for a time, I resolved, almost in despair, to see what conclusion I could form for myself by the application of common sense and the ordinary rules of evidence. I succeeded thus in forming a tolerably clear and consistent view of what might be the real historical element in the origin of Christianity, and the personality of its Founder. I do not pretend to impose on others my own solution of this extremely difficult and obscure question, but I think it may perhaps aid some sincere inquirers in giving clearness and precision to their ideas, and defining the boundaries between what may be accepted by the ordinary rules of reason, and that which lies outside the province of reason, and can only be accepted as an article of faith.

To begin with, I believe that miracles lie entirely within the domain of faith. I mean real miracles, for

---

Here again this far-fetched supposition is contradicted by Papias himself, who says "books" without any qualification, and refers to written records, viz. the notes of Mark and the logia of Matthew, which assuredly were not commentaries or interpretations of existing Gospels, but historical records of the sayings and doings of the Founder of the religion as much as the Canonical Gospels themselves; or rather they were the primary matter and first forms of the Synoptic Gospels, and could not have been so referred to, if the Gospels, in their more complete and elaborate form, and especially that according to St. John, had been known to Papias and received as authorities.

The closer the connection is drawn between Papias and the Apostle John through Polycarp, and the Bishop insists greatly on this in his *Essays*, the more impossible does it become that, if Papias had known of such a Gospel as is attributed to John, he could have written such a sentence as is quoted from his lost works by Eusebius, saying that he could get "little profit from books," and have referred, as he does, to Matthew and Mark, without saying a word of John, or of the Gospel which is pre-eminently the foundation-stone of Christian theology.

a large number of those narrated by the Gospels may well be natural occurrences described in the language of the day. For instance, casting out devils, faith-healing, or curing paralytic affections of the nerve or will by a strong impulse; and the effects of religious excitement, the sympathy of crowds, dreams, visions and hallucinations, are all well-known causes at the present day, of effects which in former ages would undoubtedly have been considered as miraculous. These may very well have actually occurred, and be as historical as any other part of the narrative.

But when we come to such miracles as raising the dead, or permanently curing organic diseases, they require a special supernatural interference with the laws of nature. Now what does reason say to such miracles? It tells us that in thousands of such cases of alleged miracles, both in Pagan, early Christian, and mediæval ages, once firmly believed in and attested by what seems strong contemporary evidence, not one now holds the field and is seriously accepted, with the possible exception of some half-dozen, which are accepted solely on the authority of the New Testament.

Take, as an illustration, the statement that one who was really dead returned to life. There are some thousand millions of people living in the world who are renewed by death and birth at least three times in every century, and this has been going on for some fifty centuries. That makes some 15,000,000,000 human beings who have died, and of whom it may be said with certainty that not one has ever returned in the body to life. You wish to establish some five or six exceptions to this rule, or rather one, for if the return to life of Jesus cannot be proved, few would be disposed to rest their faith in miracles on any other of the alleged cases of resurrection. And the historical truth of the appearances of a living and tangible Jesus after death hinges mainly on the account of the Ascension given by St. Luke in the Acts of the Apostles. This is the crowning miracle of all, the appropriate conclusion of his mission on earth, and strongest proof of his Divine nature; and it is described in the fullest detail as having occurred in the presence of a large number of witnesses. St. Paul says of this, or of some other appearance not recorded in any of the Gospels, five hundred witnesses, many of whom remained alive till his day, and in a definite and well-known locality close to the large city of Jerusalem. If the evidence for this miracle fails us, how can we believe in others more obscure and less well authenticated?

Surely the evidence for an event which is a solitary exception to 15,000,000,000 experiences, requires to be proved by testimony far stronger than would be required to prove an ordinary occurrence. But how stands the evidence for the miracle of the Ascension? Of the four witnesses called into court, one, Mark, the oldest of all, and probably deriving his information direct from St. Peter, makes no mention whatever (if we omit the last verses, which are an obvious addendum, and, as the authors of the revised edition tell us, are not found in the oldest manuscripts) of the Ascension, or of any other supernatural event connected with the Resurrection. Matthew says distinctly that the message sent by Jesus to his apostles was to "depart into Galilee," and that they went there accordingly, where they saw him, but "some doubted," and makes no reference to any Ascension. John describes certain miracles occurring at Jerusalem, but places the concluding scene of the Resurrection, when Jesus took his final farewell of his disciples, in Galilee, and, like Mark and Matthew, makes no mention of any Ascension.

Observe that Luke says distinctly that Jesus charged the apostles "not to depart from Jerusalem," and that all the miraculous appearances, including the Ascension, occurred there. There cannot be a more flagrant contradiction than that between Matthew and Luke. Consider now what would be the chance of establishing, not a stupendous miracle, but such a commonplace event as the signature of a will, if the first witness called was a solicitor who said that the testator in his last illness asked him to remain in London to draw and attest his will, which he did, while the second witness was another solicitor, who swore that the testator told him he was going down to his place in Yorkshire on the chance that the air of the country might revive him, and asked the witness to follow him there by the next day's train, in order to complete his will, which instructions he accordingly carried out. And let any candid and dispassionate person say how, if tried by the ordinary rules of reason, this differs from the direct contradiction between Matthew and Luke.

With this conclusive proof of the impossibility of establishing the greatest of all miracles by the ordinary rules of evidence, it is almost superfluous to refer to the many other circumstances which, on the showing of the Gospels themselves, lead to the same result. For instance, the next greatest miracle to those of the Resurrection, the raising of Lazarus, is related only in one Gospel, and that the latest and least authentic; while if it really occurred, it must have been known to and recorded by the three other evangelists. Or what can be said of the admission that even the minor miracles of casting out devils and faith-healing depended on faith, and could not be performed in the sceptical atmosphere of Nazareth, where Jesus and his family and surroundings were well known; or of the refusal of Jesus to comply with the perfectly reasonable request of the Pharisees to prove his Messiahship by a sign from heaven, a refusal which, if he possessed the power, was unfair to men who, if narrow and fanatical, were doubtless many of them sincere and zealous for their country and religion.

I do not see how it can be doubted that the evidence for many early Christian and mediæval miracles, which no one any longer believes, is much stronger than for those of the Gospels. St. Augustine, a perfectly historical and leading personage of his day, testifies that in his own time, and in his own bishopric of Hippo, upwards of seventy miracles had been wrought by the relics of St. Stephen. The friend and biographer of St. Ambrose relates numerous miracles, one a resurrection from the dead, which had been notoriously wrought at Milan by the saint during his lifetime. Eginhard, the secretary of Charlemagne, who was a well-known historical character, relates, as from his own experience, a number of miracles wrought by the relics of two Christian martyrs which an emissary of his had purloined from Rome, and which he was transporting to Heiligenstadt. To come to later times, St. Thomas-à-Becket was as well known an historical character as King Henry, and no miracles were attributed to him in his lifetime; but after his murder, under circumstances causing universal horror and excitement, a whole crop of miracles sprung up about his shrine at Canterbury. Any one who will consult the authorities cited by Freeman will be astonished to find how very precise and circumstantial is the evidence for many of these miracles. One instance is that of the attestation of the mayor and several burgesses of a Northern borough, to the fact that a fellow-townsman of theirs, blind from his youth, had gone to the shrine and returned with perfect sight. There is nothing in the account of any miracle in the New Testament at all approaching this in what constitutes the force of evidence, precision of date, place, persons, and circumstance. And yet for millions who believe on the weaker evidence, there is scarcely one who retains any belief in such miracles as those related of St. Thomas-à-Becket.

The reason is obvious; miracles are in a totally distinct province, that of faith. What is faith? St. Paul tells us it is "the assurance of things hoped for, the proving of things not seen." Hardly of "things not seen," for in that case, mathematicians and chemists who believe in atoms and molecules would, of all men, have the largest faith. But say of "things not proven," and it is a very accurate definition. There can be no doubt that there are men, often of great piety and excellence, who have, or fancy they have, a sort of sixth sense, or as Cardinal Newman calls it, an "illative sense," by which they see by intuition, and arrive at a fervid conviction of the truth of things unprovable or disprovable by ordinary reason. The existence of a personal God, the divinity of Christ, the inspiration of the Bible, and consequent reality of miracles, appear to them to be fundamental and necessary truths beyond the scope of reason. They feel that if their belief in these were shaken their whole life would be shattered, and they would lose what Wordsworth says Nature was to him—

> "The anchor of my purest thoughts, the nurse,
> The guide, the guardian of my heart, and soul
> Of all my moral being."

With such men I have no quarrel. Let them hold to their faith, and leave reason to poor ordinary mortals, who, like myself, have no such transcendental intuitions. Only do not let them confound the two provinces, and try to ride on two horses at the same time. Faith is either a delusion, or something which is above and beyond reason. If the latter, they only weaken it by seeking to prop it up by weak and sophistical arguments. If, for instance, a man tells me that he believes in the miracle of the Ascension by faith, I have no more to say; but if he proceeds to back up his assertion by arguing that there is no contradiction between Luke's account of it and that of the other evangelists, I say, "This man is either insincere or illogical." His motto is, "Believe if you can; if you can't, Cant."

I do not, therefore, so much deny the truth of the Christian miracles, as affirm that they are altogether outside of the province of reason, and have no place in such an historical *resumé* as I am attempting to give in this essay.

Another reservation I have to make, that if the historical element in the life of Jesus may seem to be reduced to very slender proportions, this does not necessarily affect the vital truth of the Christian religion. This religion has always been to a considerable extent, and is becoming more and more every day, not so much a question of external evidence or of dogma, as of a sincere love and reverence for the ideal which has come to be associated with the name of Jesus. This ideal is a *fact*, and has long been, and will continue to be, an important factor in the progress of human evolution from lower to higher things. How the ideal grew up and came to be established is of far less importance than what it is. Love, charity, purity, compassion, self-sacrifice are not the less virtues because the ideas and emotions of so many good men and women, for nineteen centuries, have taken form and crystallized about a comparatively small nucleus of historical fact.

My meaning will be best explained by an illustration. In Catholic countries there is a figure which competes with, if indeed it do not often supersede, that of Jesus—the figure of the Virgin Mary. Now here we can trace the historical nucleus down to a *minimum*. What do we really know of the mother of Jesus as an historical fact? That she was a Jewish matron, the wife of a mechanic in a small provincial town, the mother of a large family, for four brothers of Jesus are mentioned as well as sisters. Apart from the legends of the Nativity, which are obviously mythical, nothing else is known of her, except that she was probably one of the sceptical friends and kindred at Nazareth whose want of faith prevented the working of miracles there, and whose impression seems to have been that Jesus was not altogether in his right mind. Her relations with her Son do not appear to have been very cordial, from his refusal to go out to her when she came to the door asking to see him, and his emphatic assertion that those who believed in him were dearer to him than his blood relations.

The only other mention of Mary by St. John, who describes her as sitting at the foot of the Cross, is apocryphal, being directly contradicted by the very precise statement in the three other Gospels, that the Mary who was present on that occasion was a different woman, the mother of Salomé. The motive of this introduction of Mary, the mother of Jesus, by the author of the fourth Gospel is obvious, viz. to exalt the character of St. John, as is apparent throughout this Gospel, in which the "Boanerges," the violent and narrow-minded John of the other Gospels, is converted into the gentle and amiable apostle whom Jesus loved.

What is the sort of figure which, if we relied on historical evidence only, we should draw from these scanty records? That of a plain, motherly Jewish woman, who did her own scrubbing and washing, and was probably too much oppressed by household cares, and those of a large family, to know or care much for the spiritual aspirations and prophetical pretensions of her eldest son.

And yet from this homely figure what a world of beautiful ideas and associations have flowered into life. The Madonna has become an embodiment of all female virtues carried to a point where they become divine. Love, purity, innocence, maternal affection, human suffering, have all found their highest ideal in the "Mother of God," the "mild and merciful Madonna," the "Blessed Virgin." Do you tell me this is not a fact because it is not based on historical evidence? I tell you it is a *fact*, far more certain and more important than nine-tenths of the events related in history. If you doubt it, look at Raffaelle's *Madonna di San Sisto*, or Murillo's *Immaculate Conception;* or listen to Mozart's *Ave Maria*, or Rossini's *Stabat Mater*, and you will see that this ideal worship of the carpenter's wife of Nazareth has produced works which will remain for ever as high-water marks which have been reached in the evolution of modern art. You will say with Byron—

"Ave Maria, oh, that face so fair,
Those downcast eyes beneath the Almighty dove.
Ave Maria, may our spirits dare
Soar up to thee and to thy son above."

And so of Jesus; the historical figure, though a good deal more certain and definite than that of his mother, is but a small matter compared with the ideal which has grown up in the course of ages about it. It is but as the fragment which, dropping into a saturated solution, attracts molecule after molecule, until it grows into a large and lovely crystal which all eyes admire.

With these reservations, which may go some way to mitigate the scruples of orthodox readers, if I should happen to have any, viz. that miracles are a question of faith, and that the historical element does not materially affect the vital truth of Christianity, I fall back on my own humble province of reason, and attempt to show what can be gathered by it from the earliest records as to the personality and teaching of Jesus.

I begin by stating the two principles by which I have been mainly guided in the research. The first is what I may call the "Minimum of Miracle." Of different biographies of the same person, that which contains the fewest miraculous legends is almost certain to be the earliest and most authentic. It is far more likely that such legends should be added or invented than that, if they actually occurred, or were generally accredited, they should be designedly omitted. As an illustration of what I mean by this, take the case already referred to of St. Thomas-à-Becket. If newspapers had existed in his time which published a biography of eminent men on the day after their death, such a biography would have contained no miracles; one written a few weeks later would have doubtless contained some reference to the miraculous vision of the monk who watched by his remains, and some of the miracles said to have occurred at his shrine; while still later accounts would have multiplied the miracles into scores and hundreds. There can be no doubt here that the succession in point of time would have been, no miracles, few miracles, many miracles. And the same holds good of all biographies of eminent men, saints, and martyrs. The outlines of their historical figures are almost lost in the accumulation of myths and legends, which in uncritical times have grown up about them.

The second even more important principle is, that admissions of events and sayings which tell against the point of view of the writer, are far more likely to be historical, than those which have the appearance of being introduced to show the fulfilment of prophecies, to answer objections, or to support dogmatic views. Thus if Jesus is described as being born and bred at Nazareth, the son of a carpenter whose family and surroundings were well known there, the statement is far more likely to be true than one which describes him as having been born at Bethlehem, and attributes to him a whole series of marvellous and miraculous incidents.

Tried by both these tests, the Gospel of Mark has every appearance of being the earliest and most authentic record, and when this is confirmed by the clear and explicit statement of Papias, I have no hesitation in assuming it to be the surest basis of our historical knowledge, and in all probability mainly derived from the reminiscences of Peter himself, or of other contemporary witnesses of the events described.

Starting from this basis, I assume, as beyond all

doubt, that Jesus was an historical personage. There is nothing in Mark which would lead to the supposition that any considerable portion of his Gospel was legend or myth. The time is too modern, and the narrative too precise, to allow us to suppose that the whole story had been elaborated by later theologians from Oriental myths and Messianic prophecies. The age was long past when religions could originate in solar myths and misunderstood personifications of natural phenomena. Every great religious movement which comes fairly within the historical period, from Buddha and Zoroaster down to Mahomet, had some real personality as its starting-point, about whom myths and dogmas accumulated, until almost obscuring the historical nucleus. So also was doubtless the case with Jesus.

The next point I consider to be quite certain is, that he was born of humble parents at the little town of Nazareth in Galilee. The legends of the Nativity and Infancy may all be dismissed as purely mythical. The two accounts and genealogies in Matthew and Luke do not agree, and are each hopelessly inconsistent with the evidence of the other Gospels. It is plain that during his life and afterwards Jesus was supposed to have been born at Nazareth, that this was cast in his teeth as being irreconcilable with any claim to be the Messiah, and that neither he nor his apostles ever attempted to deny it, or made any claim to his having been born at Bethlehem. If such a series of startling events as are described by Matthew had really occurred, the inhabitants of Nazareth could hardly have ignored his claims as a prophet on the ground that he was a mere ordinary fellow townsman, "the Son of the carpenter, whose brothers and sisters are with us every day."

The accounts of the Nativity, Infancy, and early Manhood of Jesus may be dismissed as purely legendary. I do not say so merely because they contain so many miracles, but on the ordinary grounds of historical criticism. In the first place, the two accounts of Matthew and Luke are contradictory. The second admits that Nazareth was the abode of Joseph and Mary, and accounts for the birth of Jesus at Bethlehem by the supposed necessity of Joseph's going there to be taxed, as being of the family of David; while the first assumes that Bethlehem was the abode of the parents, and says that they only went to Nazareth some years later from fear of Archelaus, who had succeeded to his father Herod. Matthew describes the Massacre of the Innocents at Bethlehem, and says that Jesus escaped it by flying into Egypt; while Luke omits all mention of the massacre, the miraculous star, and the wise men of the East, and says that the parents took the babe straight to Jerusalem. In both cases all the events are described as happening in fulfilment of prophecies. The other two evangelists, Mark and John, make no mention of any such occurrences, and begin their biographies with the visit of Jesus, when a grown-up man, to John the Baptist.

But the most conclusive fact is that these legends are identical, both in their general tenor and in many minute details, with similar legends of earlier religions. Thus the miraculous birth from a virgin is related of Horus, of Krishna, of Buddha, and of many of the celebrated heroes and gods of antiquity, and is almost certainly derived from a solar myth of the sun rising in the constellation of Virgo. The story of the massacre of the innocents is related of Krishna, and if we accept the narrative of Matthew, we have to suppose that there were two wicked kings, one in India and another in Judæa, separated by an interval of many centuries, who both adopted the same expedient of a massacre of all male children under two years of age, to destroy a Divine Incarnation who was born in one of their cities. The escape by flight, owing to a miraculous warning and other particulars, are almost word for word the same in the two legends, and we may fairly assume that both are alike unhistorical. We know that a whole crop of such legends grow up in early Christian tradition, for we have the Gospel of the Infancy, which is full of the most childish and absurd magical tricks, supposed to have been performed during the boyhood of the Messiah.

The first firm historical ground is afforded by the Gospel of St. Mark, who begins with the visit of Jesus to John the Baptist. This is very likely to be true, for we know from Josephus that the time was one of great religious and political excitement, and that there were several such preachers or prophets as John the Baptist is described to have been, who went about holding what may be called camp-meetings, and in some cases causing local insurrections, which had to be repressed by the Roman soldiery. Nothing is more likely than that a young man of original genius and strong religious sentiment, should go to one of such meetings, not far from his home, to hear a celebrated preacher. That such a young man was not altogether satisfied with the narrow and fierce denunciations of a rude ascetic, and did not enroll himself as one of his disciples, was also very probable; but that John really did make a considerable impression on him is evident from the fact that he left his home immediately afterwards, assumed the character of a wandering missionary, and began to preach identically the same gospel as that of John—"Repent ye, for the kingdom of heaven is at hand."

Let us pause for a moment to consider what was meant by the kingdom of heaven being at hand. It did not mean such a millennium as certain enthusiasts may now suppose, after nineteen centuries of unfulfilled expectation, that is, the advent of an era of purer morals and better laws, but the literal end of the world and last judgment, to take place within the lifetime of some of the existing generation. "The sun was to be darkened, the moon not to give her light, and the stars fall from heaven." And then they were to see the "Son of Man coming in clouds with great power and glory," and his angels to gather all mankind from the four winds of heaven before the judgment-seat, where the tares are to be separated from the wheat, the goats from the sheep, the good rewarded and the wicked cast into everlasting fire. Nothing can be more explicit than the assurance that this event would come to pass in the lifetime of the present generation. "Verily I say unto you, This generation shall not pass away until all these things are accomplished."

Such was evidently the current opinion among the apostles and early Christians; and even the cultured and educated Paul, some twenty years later, repeats it with the fullest conviction, and describes how "the Lord shall descend from heaven with a shout, with the voice of an archangel, and with the trump of God;" and how "the dead shall rise first; then we that are alive, that are left, shall together with them be caught up in the clouds, to meet the Lord in the air."

It is clear that, according to all rules of ordinary

reason, predictions thus confidently made and conclusively refuted, are an irresistible argument against the possession of any inspiration or special foresight on the part of the prophets, and that prophecies, like miracles, must be relegated to the province of faith. But, on the other hand, they bring us nearer to the human and historical element in the New Testament. They supply a motive power which may explain the early conversions and the rapid spread of the new religion. Evidently the hope of a large and immediate reward was present in the minds of the apostles. These humble peasants and fishermen were "to sit on twelve thrones judging the twelve tribes of Israel," and "every one who has left houses, or brethren, or sisters, or children, or lands, for My Name's sake, shall receive a hundred-fold." And this not in a remote future, but in the lifetime of the existing generation. It is conceivable also that many educated Jews, who despaired of an armed resistance to the overwhelming power of Rome, might be inclined to view with favour the idea of a spiritual Messiah who should bring about the advent of an end of the world and last judgment, in which the elect children of God should be rewarded and the heathen punished.

Another element which must have contributed largely towards the reception of the Gospel by the poorer classes, is the extreme socialistic spirit which is uniformly displayed. For "rich" write "capital," and for "poor" "wages," and the preaching of Jesus is almost identical with that of modern socialists. The poor are to be rewarded and the rich punished in the kingdom of God, irrespective of any merit or demerit. Thus, "blessed are ye poor," "woe unto you that are rich." Even the rich young man, who had kept all the Commandments, is told that he cannot be saved unless he "sells all his possessions, and gives to the poor;" and the remark of Jesus is, that it is "easier for a camel to go through a needle's eye, than for a rich man to enter into the kingdom of God." For anything that appears to the contrary, Lazarus may have been a loafing vagabond, who had brought poverty and disease upon himself by his own misconduct, and Dives a man who, having inherited a large estate, spent it hospitably in entertaining his neighbours; but no moral is inculcated. It is enough that Lazarus is poor and Dives rich, to place one in Abraham's bosom and the other in eternal fire.

It is evidently neither in these falsified prophecies nor in this exaggerated socialism that we are to find the fascination which the ideal of Jesus has exercised over so many minds for so many centuries. It is rather in the interpretation which he gave to the first words of the Baptist's formula, "Repent ye, for the kingdom of God is at hand." Repentance, as taught by Jesus, meant not merely an outward obedience to formal laws and abstinence from direct breaches of moral commandments, but such a spiritual conversion as embraced the whole sphere of human life, and made the very idea of sin insupportable. Men were to be good, pure, merciful, compassionate and charitable, because the principle of "loving God, and thy neighbour as thyself," was so wrought into the soul that it became a second nature. The law was to be observed, but in a liberal, tolerant, and comprehensive spirit, and the intention was to be looked to rather than the outward act. The widow's mite was of more value than the rich man's offering, and the publican's remorseful prayer was more acceptable than the formal and lengthened devotions of the strait-laced Pharisee.

It is remarkable, when we come to consider it, how much more the ideal of Jesus, which is the central fact of Christianity, is founded on the precepts and parables by which this spiritual religion is taught, and by the human incidents of his life which illustrate it, than it is on the alleged miracles. The Sermon on the Mount, the Parable of the good Samaritan, the tenderness to children, the affectionate and "sweetly reasonable" intercourse with his humble followers, these and such as these are the traits which build up the ideal character which draws all hearts.

The miracles, on the other hand, are at best but capricious instances of a supernatural power, healing one and leaving thousands unhealed, and failing when most required as evidences, as in the case of the incredulous Nazarenes and the Pharisees who asked for a sign; while at the worst, some of them are wholly inconsistent with the historical character of the just and gentle Jesus. Thus the miracle of the Gadarene swine, if true, obviously detracts from this character. It is an act of cruelty to animals, for what had the poor swine done to deserve death, and it is a wanton destruction of property cruel to the owners. Doubtless these swine had owners, some perhaps poor Galilæan peasants, who, like those of Donegal or Galway, depended on the pig to pay their rent and save them from eviction. It was a wanton and a cruel act to send their humble property to destruction in order to please a pack of devils. Again, the miracle of the fig-tree reads rather like the hasty curse of a passionate fool, than the act of a gentle, long-suffering, and sweetly reasonable man.

But to return to the historical narrative, I find no difficulty in believing that the accounts of the commencement of the mission of Jesus, of his comings and goings among the small towns of Galilee, of his camp-meetings, and of most of his preachings, parables, and sayings, are substantially accurate. There is nothing improbable in them, except in some of the miracles taken literally, and these may readily be explained, or indeed were inevitable, in such a medium of excited crowds of poor and ignorant men, where every one believed in miracles as events of daily occurrence, and where many natural acts of faith-healing and casual coincidences had given a popular prophet the reputation of being a worker of mighty works.

Indeed many of the miracles appear as if they had a nucleus of historical fact, which became expanded into legend. Thus, the legends of Jesus and Peter walking on the sea appear to be based on the first simple narrative, how a sudden squall having overtaken the boat in which they were crossing at night, they awoke Jesus, who was asleep, and the squall passed over.

Those, again, of the "loaves and fishes" may have readily arisen from the recollection of some occasion when a scanty supply of food had lasted out longer than was expected, owing very probably to many of those who attended the camp-meeting having brought their own provisions, a conjecture which is confirmed by the abundance of baskets in which to collect the fragments, and which could not have been required to carry seven or five loaves.

These, however, are mere conjectures, and not to be taken as facts, and I only mention them to show that a good many of the miraculous legends need not neces sarily detract from the general historical value of Mark's simple narrative of this early part of the career of Jesus in Galilee.

And I think the sayings and parables may generally be taken as authentic. It is true that many of both may be found in the literature of the Talmud and of older religions, but this does not negative the probability that Jesus may have used them in his popular addresses, and at any rate they afford a view of what his doctrine and style of preaching really were; and many of the parables and shorter sayings are just such things as would be readily retained in the memory and transmitted by oral tradition. Many of the details also of the incidents and wanderings to and fro of this Galilean period are very like what might be expected from the reminiscences in old age of an apostle like Peter, who had accompanied Jesus from the first, though we must always recollect that the author who worked up these reminiscences, as described by Papias, into a connected biography, may have added a good deal from other sources.

I am inclined also to accept as authentic a good many of the controversies between Jesus and the Scribes and Pharisees. They are exactly in the style of the verbal conflicts which were so common in the East, and which survived down to the scholastic tournaments of the Middle Ages. An opponent makes a desperate thrust by a puzzling question, it is parried by an adroit answer, both leaving the root of the matter untouched. Thus the celebrated answer, "Render unto Cæsar the things that are Cæsar's, and unto God the things that are God's," is clever, but no answer to the real question whether a conscientious servant of Jehovah could voluntarily pay taxes to a heathen power which had usurped his place. The position was precisely that of a conscientious Dissenter in our own days, who was in doubt whether to pay Church rates, or let his chattels be seized. He would have got little enlightenment from being told to pay Queen Victoria the things that were hers, and render to God what was God's. The question was, what things were Cæsar's and what God's.

Again, the puzzle of the Sadducees, whose wife she would be in heaven who had been married successively to seven brothers, remains a puzzle to this day. It is no question of marrying in the kingdom of heaven, but of marriages which have taken place on earth. Shall we preserve our personal identity after death, so that two souls which have been united by the holiest and closest ties while living, shall be united in a future life? Shall we know and recognize those whom we have loved and lost:

> "See every face we feared to see no more;"

or is Arthur's last wish, that Guinevere should cling to him and not to Launcelot, when they meet before "the fair father Christ," a vain dream? If it be not, who can answer the Sadducees' question, or say more than our greatest poet—

> "Behind the veil, behind the veil"?

What Jesus might have said, but did not, is, The rule is an abominable one; it degrades the sanctity of marriage, and reduces woman to a mere chattel, who is to be handed over like an ox or an ass—they to bear burdens, she to bear children—for their master Man.

Up to this point, therefore, I see no difficulty in

accepting the Synoptic narrative, best told in the earliest and simplest Gospel of Mark, as being in the main historical. And if so, the best picture I can form of it is something very like the Salvation Army of the present day. The movement had evidently no political significance, and attracted little notice, or Josephus must have mentioned it, and there is no trace of any interference with it, in the earlier stages, on the part of the authorities. In fact, the modern Salvationists have suffered more from provincial Bumbles and Justice Shallows than Jesus and his disciples seem to have done while they remained in Galilee. But, like the Salvation Army, there was a loose organization of a general, twelve principal officers, and a body of disciples or professed adherents, who went about holding camp-meetings, and preaching the advent of the kingdom of God and a new and better life to excited crowds, who listened eagerly and on the whole sympathized with them. The only difference was in the superior genius, eloquence, and attractiveness of the personality at the head of the movement, and the purity, spirituality, and general excellence of his doctrine.

There are one or two points in this doctrine which it is interesting to consider. Did Jesus consider himself as a Jewish reformer, or as the founder of a new religion? Decidedly the former. The declarations are quite explicit: "Think not that I come to destroy the law or the prophets, but to fulfil;" "Till heaven and earth pass away, one jot or one tittle shall in no wise pass away from the law;" "I was not sent but unto the lost sheep of the house of Israel." He was as far as possible from Paul's doctrine, that he was sent to liberate the Jews from the bondage of the law, and to introduce a new and universal religion for Jews and Gentiles alike. But in a few exceptional cases he healed Gentiles who had shown extraordinary faith, and his interpretation of the law was a large and liberal one, looking to the spirit rather than the letter of the Mosaic commandments, and rejecting the trifling and vexatious rules which the Scribes and Pharisees had introduced in later times. Thus, he strolled through the fields on a Sunday afternoon with his disciples, plucking ears of corn, and declared that "the Sabbath was made for man, not man for the Sabbath," a saying in respect of which our modern Pharisees have generally sided with those of old rather than with the liberal-minded and tolerant Jesus.

What did Jesus believe respecting his own Messiahship? This is a very perplexing question, aggravated by the tendency, after the doctrine was firmly established, to invent or adopt traditions showing that he had fulfilled the conditions attached to such a character by the prophecies of the Old Testament, and by the prevailing expectations.

But it is tolerably clear that in the early part of his career he advanced no such pretension. The Gospels all agree in describing the remarkable persistency with which he endeavoured to suppress all evidence which tended to support such a claim. The evil spirits who recognize him, the patients whom he miraculously cures, Peter when he calls him the Christ, are all enjoined to "tell no man anything." When the little damsel is supposed to have been raised from the dead, his first care is to "charge them much that no man should know this." In any ordinary case the inference would be that he did not wish miracles, which passed muster with ignorant disciples, to be investigated by impartial and educated critics. If this explanation be negatived as inconsistent with his pure and holy character, the only other that can be suggested is, that he did not wish it to be supposed that he was a supernatural being attested by miracles, believing miracles to be vulgar things of which even false prophets might be capable, but that he preferred to rely on the excellence of his doctrine and his own powers of eloquence and persuasion.

It would seem, however, that later in his career the conviction began to dawn on him that he might be the Messiah of the prophecies, and that he stood in some peculiar relation to God, and would be His vicegerent in inaugurating His kingdom and holding the assizes of the last judgment.

The most distinct assertion of this is found after he had gone to Jerusalem, in his reputed reply to the adjuration of the high priest to say whether he was "the Christ, the Son of the Blessed," to which he replied, according to one version, "I am," and to another, "Thou sayest."

It is evident, however, that he never thought of equalling himself to God, or representing himself in the literal sense as being "of one substance with the Father," and he would probably have torn his clothes and shouted "blasphemy" if he had heard the articles of the Athanasian Creed. To the last he uses the term "Son of man" in speaking of himself, even in his answer to the high priest, and he never adopts the language of the evil spirits who address him as "Jesus, thou Son of the Most High God," or as "the Holy One of God." He never doubts that "my Father is greater than I," or that God alone knows things which he does not know.

The best clue to his conception of himself is, to my mind, afforded by the pathetic dying words, "Eloi, Eloi, lama sabachthani?" These, if any, must be historical, for they tell against the orthodox view, and could never have been invented, while they are just the sort of thing which would impress itself, in the actual words spoken, on the memory of his affectionate disciples. But if these words were really spoken, they show that he really believed himself to be the promised Messiah, and trusted up to the last in some signal miraculous act of deliverance, such as the advent of the last day, or the descent from heaven of "more than twelve legions of angels."

It is worthy of remark that the author of Luke seems to have felt the force of this objection, for he transforms the expression into "My God, into Thy hands I commend my spirit," and inserts "Forgive them, for they know not what they do," which words are not found in any other record. It is evident that if Luke's version had represented the words really spoken, they could never have been altered by eye-witnesses or by early tradition, into words conveying such a totally different impression as "My God, my God, why hast Thou forsaken me?"

We come now to the concluding scene at Jerusalem, when it becomes more than ever difficult to distinguish between fact and legend. The narrative of the three Synoptic Gospels are fairly consistent up to the Crucifixion, when they become hopelessly discordant. That of John is apparently founded on the same tradition, though, after the fashion of the author, dealt with in a very free-hand way, altered, transposed so as to make it the ground-work for several dogmatical speeches and visits to Jerusalem, and embellished by various amendments and details. But the primitive narrative is clear enough. Jesus and his apostles go up to Jerusalem to keep the Passover; they are received there with a triumphal procession; Jesus clears the Temple of the money-changers; the authorities become alarmed, but are afraid to arrest him openly, as the people are in his favour; one of the apostles betrays his hiding-place, and he is arrested at night; he is tried and condemned by the Sanhedrim and by the Roman Governor; Pilate believes him to be innocent, and tries to save him, but the Jews clamour for his blood; Pilate yields, and he is crucified.

Thus far the story is consistent, and it involves nothing that is impossible. But it is full of the gravest improbabilities. Why should the Jews, who one day are so much in his favour that the authorities are afraid to arrest him, be converted in a single day into a furious crowd clamouring for his execution? Why should an appeal to Pilate be necessary for a religious offence against the Mosaic law, when Stephen, under precisely similar circumstances, was publicly stoned to death, and Paul made havoc of Christians without any Roman mandate? Why should false witnesses, whose testimony was inconsistent, be required to prove an offence which Jesus avowed in open court?

But the portion of the narrative which relates to Pilate is that which is open to the gravest suspicion. It is opposed alike to human nature and to Roman practice, that a high functionary should first publicly proclaim his belief in the innocence of a prisoner whom he was trying, and go through the solemn act of washing his hands to show that he would not be guilty of his blood, and immediately afterwards condemn him to a cruel and ignominious death. Nor is it conceivable that such a Governor, if forced to yield by the threat of being reported to Cæsar for disloyalty, should insist, against the remonstrances of the Jewish rulers, in placing an inscription on the cross which proclaimed Jesus to be "the king of the Jews."

In fact, the whole episode of Pilate has very much the air of being an interpolation of much later date, when the feeling of hatred between Christians and Jews had become intense. The object evidently is to show that this hatred was justified by the Jews having imprecated the blood of Jesus on their own heads and those of their sons, and to represent the heathens as having been better than the Jews, inasmuch as Pilate tried to save Jesus, and to a certain extent believed in him. It is difficult to believe that such a narrative could have come from men like Peter, John, and James, who remained devout Jews, zealous for their faith and country.

Nor, again, is it easy to see how, if the events had really assumed the publicity and importance assigned to them, there should be no mention of them by Josephus, or any contemporary writer, especially if there was, as the Gospels say, a miraculous darkness over the land, an earthquake, the veil of the Temple rent, and ghosts walked about the streets. The Gospel narratives also, though consistent in the main outlines, contain a number of discrepancies in details which show that they were not derived from any one written document or from any fixed tradition. Thus, Judas' death is differently described. Herod is introduced by Luke and not mentioned by the others. Jesus carried his own cross in one account, while Simon of Cyrene bore it in another.

Jesus gave no answer to Pilate, says Matthew; he explained that "his kingdom was not of this world," says John. Mary his mother sat at the foot of the cross, according to John; it was not his mother, but another Mary, the mother of Salome, who "beheld from afar," according to Mark and Matthew. There was a guard set to watch the tomb, says Matthew; there is no mention of one by the others.

These, however, are minor discrepancies which are only important as showing that there was no clearly fixed historical tradition, except of the general outline of the course of events, when the different Gospels were compiled, and subsequent to the Crucifixion there is, as we have already seen, a hopeless discordance.

In some cases it is almost possible to trace, step by step, how the legends grew with each successive repetition. Thus, according to Mark, two women went to the tomb, found the stone rolled away and the tomb empty, and saw a young man clothed in white who gave them a message to Peter and the disciples, that Jesus had risen and gone before them to Galilee, where they would see him—a message which they never delivered, being afraid. In Matthew the young man has become an angel who rolled the stone away and sat on it, delivering the same message to go to Galilee, where his disciples would see him, which they ran and delivered. In Luke there are the same two Marys, with another woman named Joanna, and several others, and they saw, not one but two men in dazzling apparel; "go to Galilee" is changed into "as he spoke unto you while yet in Galilee," which in the Acts is enlarged into a positive injunction "not to depart from Jerusalem;" and Peter is introduced as running to the tomb and finding it empty. In John there are two angels; John runs along with Peter to the tomb; and Mary Magdalene has a miraculous vision of Jesus, whom she at first mistakes for the gardener. No one who reads these narratives by the ordinary light of reason can doubt that the simple story of Mark is nearest to the original fact or tradition, and that the successive amplifications of one into two, men into angels, the introduction of Peter, and finally of Peter and John, and the miraculous vision of Mary Magdalene, have grown up about it. If the facts had really happened as described by Luke and John, no one could have subsequently cut them down into the bald statement of Mark, while the opposite process is what we know to be historically true in the case of so many early Christian martyrs and mediæval saints. It is the clearest possible case of the application of the principle of the "Minimum of Miracle."

I may here remark, however, that, as I said before, the historical nucleus is of minor importance compared with the fact that the belief in the Resurrection did somehow come to be entertained, and became the chief agent in the establishment and evolution of the new religion, and that there is no reason to doubt that it was honestly entertained by sincere men, who, if they did not see it with their bodily eyes, saw it with the eyes of faith, and to whom visions, dreams, hallucinations, and subjective impressions, were as much facts as objective realities.

In trying to disentangle the historical nucleus from these legends, the best ray of light I can discover is afforded by the account of the riot in the Temple, and assault on the traders who changed money and who sold doves and other objects of sacrifice. This is found in all the Gospels, and could hardly be an invention, while if true it must have been followed by immediate consequences. Prompt and stern repression must have been exercised both by the Jewish and the Roman authorities.

We must recollect that their point of view would not be that of later Christians, when the faith in the Divine character of Jesus had been established for centuries, but that of contemporaries who knew nothing of him but as the provincial prophet of an obscure sect. To recur to the simile of the Salvation Army, it would be as if a body of Salvationists, who had preached without interruption in some remote province of Russia, came to Moscow, and in a fit of religious enthusiasm invaded the cathedral, and broke the windows of the shopkeepers in its vicinity who exhibited Ikons and other sacred objects of the Greek ritual. Undoubtedly the Metropolitan would complain to the Governor, and the leader of the rioters would be summarily arrested, and if not crucified, sent to Siberia.

Supposing this narrative to be true, it affords a natural explanation of many of the incidents recorded. A disciple might well be bribed to disclose the hiding-place of his master; the arrest might be made under the circumstances described; the disciples might disperse in alarm, and Peter deny his connection with them; Jesus might be taken before the high priest, and by him referred to the Roman Governor. The incredible legends about his trial and Pontius Pilate might resolve themselves into the fact that Jesus had no defence to make, and was condemned, not on theological grounds, or on the charge of having proclaimed himself as a temporal king of the Jews, but on the simple charge of having been the ringleader in a serious riot. Crucifixion would, as we know from numerous instances in Josephus, have been a common Roman method of dealing with such leaders, and its various incidents, such as the brutality of the soldiers and the procession to Golgotha, are only what might be expected. The historical part of the narrative can hardly be carried farther than that Jesus came up to Jerusalem with a body of his followers, that a riot took place in the Temple, and that he was arrested, tried, and executed by the Roman Governor at the request of the Jewish authorities. His entombment and the finding of the tomb empty rest, according to Mark, who is the best authority, on the testimony of two women, Mary Magdalene and Mary the mother of James, who are alone mentioned as seeing where the body was laid, and of these two women and Salome, who found the tomb empty, but, being afraid, said nothing at the time to any one.

The next historical question is one of great importance. Did the apostles, as directly asserted by Matthew, and indirectly by Mark, return immediately to Galilee, where the belief in the Resurrection took form; or did they, as asserted with equal positiveness by Luke, remain at Jerusalem, where a series of startling miraculous appearances took place?

There can be little doubt in considering the Galilean tradition to be the true one. Independently of the great weight of authority for considering the narrative of Mark, which is substantially the same as that of Matthew, to be the earliest and most authentic, it is inconceivable that, if events had really occurred as described by Luke, any author or compiler of any other Gospel should have ignored them and transferred the scene to Galilee. However simple-minded such an author may have been, he could not but have seen that he was weakening immensely the evidence for the cardinal fact of the Resurrection, if, instead of referring to such precise and definite statements of miracles, including the Ascension, occurring in or near the capital city Jerusalem, in the presence of numerous witnesses, many of whom survived to attest their truth twenty or more years afterwards, he either omitted all mention of such occurrences like Mark, or like Matthew transferred the scene to a remote province and to a select few of his own disciples, and whittled down the evidence to the vague statement that these went into the "mountain where Jesus had appointed them," where "some worshipped him and some doubted."

Such a perversion of Luke's narrative might well have come from an enemy of the new faith, but hardly from an apostle. On the other hand, at a subsequent period, when the eye-witnesses were dead, and the original records and traditions were obscured by time, and when the dogmas of the Resurrection and Divine nature of Jesus were firmly established, nothing is more likely than that the birthplace of the new religion should be transferred to Jerusalem, and the vague statements of occurrences in Galilee should be transformed into a series of stupendous miracles occurring at the sacred city in the presence of a large number of witnesses.

The probabilities of the case also are all in favour of the return to Galilee. The disciples had come to Jerusalem on a special pilgrimage, to keep the Passover there, which was over; there was no intimation of any intention to remain, nor could they well have brought with them any sufficient resources for a long stay. They were in mortal fear of the Jews, and several of them had wives and families at home, to whom they would hasten to return. If we could believe John, they not only returned, but resumed their original occupation as fishermen; but I lay little stress on this, as the author of John, whoever he was, was evidently a man of considerable literary attainments and dramatic genius, which he displayed in writing a Gospel, great parts of which may be most aptly described as a theological romance.

But it is useless to dwell on details, as the conclusive argument is, that Mark and Matthew could by no possibility have written as they did if the course of events immediately after the death of Jesus had really been, or even had been generally supposed to be, as described by Luke.

With the return of the disciples to Galilee the curtain falls on what may be fairly called the historical drama of the life of Jesus, and we enter on a region where all is conjecture and uncertainty. The belief in the Resurrection evidently grew up in Galilee. It probably originated with the women, for they are mentioned in all the accounts as the first to have seen the risen Jesus, or to have brought a message from him or from angels, and this is hardly likely to have been invented.

If at first they were afraid to tell any one, nothing is more natural than that, when they found themselves in their own country, and among friends, their tongues would have been loosened, and they would begin to talk of the wonderful things they had seen, or fancied they had seen, at Jerusalem.

The only thing certain is, that the belief in the Resurrection, once started, grew rapidly, but that the various accounts of how it grew are so vague and contradictory that it is hopeless to attempt to draw any certain conclusion respecting them. This will be apparent if we simply place in juxtaposition the five different records which have come down to us in the New Testament.

The most certain and authentic record is that related by St. Paul in his Epistle to the Corinthians. It is true that Paul was not an eye-witness, or at all likely to have examined the evidence critically, and he places the appearance to himself, which, whether supernatural or not, was obviously in the nature of a vision, on precisely the same footing as the others. Still it is good evidence that, some twenty years after the event, the appearances he mentions were currently believed by the early Christian community at Jerusalem.

They are six in number, and presumably, though he does not mention the place, all at Jerusalem, except that to himself on the road to Damascus. Viz.:

1. To Peter.
2. To the twelve.
3. To above 500 brethren at once.
4. To James.
5. To all the apostles.
6. To himself.

Compare this with the other accounts, beginning with that of Mark, which probably came direct from St. Peter.

In the genuine Mark of the oldest manuscripts—

Miraculous appearances.

None. Only a message from a young man in white delivered to the two Marys and Salome.

In the addition to Mark, introduced later than the date of the oldest manuscripts—

Three. 1. To Mary Magdalene.
2. To the two walking from Emmaus.
3. To the eleven.

1 and 2 being distinctly stated not to have been believed by those to whom they were told, at the time of their alleged occurrence.

According to Matthew—

Miraculous appearances. Two.

1. To Mary Magdalene and the other Mary at Jerusalem.
2. To the eleven on a mountain in Galilee, when some worshipped and "some doubted."

According to Luke—

Miraculous appearances. Four—all at Jerusalem.

1. Messages of two men in dazzling apparel, probably angels, to the two Marys, Joanna, and other women.
2. To the two disciples walking from Emmaus, who at first did not recognize him.
3. To the eleven, when he eat the broiled fish.
4. The Ascension, when he was bodily taken up in a cloud to heaven in the presence of the eleven.

According to John—

Miraculous appearances. Four—first three at Jerusalem, fourth in Galilee.

1. To Mary Magdalene alone, who at first took him for the gardener.
2. To the disciples sitting in a room with closed doors.
3. A second time to the disciples, to remove Thomas' doubts.
4. By the sea of Galilee, when Peter and six other disciples caught the miraculous draught of fishes, when at first none of them recognized him.

And John expressly states that this last was the third appearance to the disciples after Jesus had risen from the dead, thus excluding all others except 1, 2, and 3.

It will be remarked, that of the five miraculous appearances recorded by St. Paul as being the current belief at Jerusalem twenty years after the event, three, those to Peter, James, and above 500 brethren at once, are not even mentioned in any other account. The latter can hardly be the same as Luke's Ascension, which comes in its natural place as the concluding scene of the great drama of the life and resurrection of Jesus, and the spectators are confined to the eleven apostles.

Paul's No. 5, or second appearance to all the apostles, may refer either to that described by John to convince Thomas, or to Luke's Ascension; but Paul makes no mention either of Thomas or of the Ascension, which would be very strange if the bodily Ascension to heaven was a cardinal article of faith when Paul visited Jerusalem, which it must have been if it really happened as described by Luke. There remains therefore only the vague tradition that Jesus had appeared to the twelve, as to which the enumeration by Paul of five miraculous appearances receives the slightest confirmation from any of the Gospels.

The Gospel accounts, again, vary so much that there is not a single case in which any one is confirmed by any of the others. The nearest approach to it is in the appearances to the women; but here John says distinctly it was to Mary Magdalene alone; while Matthew says it was to the two Marys; Luke, that the vision was to the two Marys, Joanna, and other women, and was one of angels, and not of Jesus; Mark, that the message was given to the two Marys and Salome by a young man. Evidently the tradition as to the women was very vague.

Again, the Ascension at Jerusalem, the greatest of all the miracles, rests on Luke alone, and is negatived by the testimony of Matthew and John, that the apostles returned to Galilee, and that the final scene, whatever it may have been, took place there; and still more significantly by their silence and that of Mark, respecting an event which, if it took place as described by Luke, must have been known and mentioned.

The appearance to the two disciples returning from Emmaus rests also on the sole authority of Luke, and that to convince Thomas on that of John. The miraculous draught of fishes is mentioned by John, and by John alone. The appearance to the eleven is the only event mentioned by three of the Evangelists, but of these, two place it in a room at Jerusalem, while one places it on a mountain in Galilee.

It is evident that it would be futile to attempt to form any historical estimate from such accounts as these; they must be left, with miracles generally, to the province of faith rather than that of reason. All we can rationally infer is, that, as in the case of St. Thomas à Becket and so many other saints and martyrs, the growth of miraculous myths was very rapid, and that probably those records which contain the fewest of them must date back very closely to the original events, and to the actors who took a principal part in them. I have never been able to see any explanation of the silence of the Gospel according to St. Mark respecting any miraculous appearances after the Resurrection, and the brief and vague reference to them in St. Matthew, except in the supposition that the account given by Papias is true, and that they are really based on written notes taken down by Mark from Peter, whose authority was sufficient to prevent later compilers and editors from adding to them legends and traditions which were floating about in the early Christian world, unsupported by any direct apostolic authority.

Here then the curtain falls on any attempt to realize the historical element in what Huxley so appropriately terms "the grand figure of Jesus as it lies embedded in the primary strata of Christian literature." We see him crucified at Jerusalem, his disciples returning to Galilee, and the faith in his Resurrection growing up there, and soon becoming an assured conviction, though with no agreement as to the facts on which it was founded, and rapidly becoming surrounded with an atmosphere of myths and miracles.

The next stage is even more obscure. We have no information as to when and how the apostles returned from Galilee to Jerusalem, and became, as we find them twenty years later, pillars of the Church there, and leaders of a great religious movement. The Acts of the Apostles may contain some authentic records of their proceedings at a later period, after they had established themselves at Jerusalem, and exchanged the profession of fishermen for that of missionaries of the new religion; but Luke's account is discredited by the obvious fact that his earlier narrative of what occurred during the first period of the Crucifixion is unhistorical. It is clear that some time must have elapsed, and considerable changes taken place at Jerusalem, during the interval between the departure of the disciples for Galilee in mortal fear of the Jews, and their return to the capital, where they seem to have preached publicly, and made numerous converts, without any serious interference by the populace or the authorities.

The narrative of this early period in the Acts, up to the date of Paul's appearance on the scene, is full of improbabilities. The miracles attributed to Peter, his deliverance from prison by angels, the gift of tongues by the Holy Ghost, which did not enable Peter to dispense with an interpreter, these and many other incidents have rather the air of legends than of genuine history. They stand in marked contrast with the naïve and natural incidents recorded by Mark, how the crowd overflowed into the street, how the bustle was such that they had no time to eat, how Jesus slept through a night-squall which endangered the boat. I can find no solid historical ground until Paul met the pillars of the Church at Jerusalem, except the general fact that the apostles returned there from Galilee, preached publicly, made numerous converts, and that Peter probably played a leading part. But with the death of Jesus and the flight of his disciples to Galilee the first chapter ends, and the second opens with the history of the early Christian Church, when the preoccupations of the principal actors were doctrinal rather than historical, and we enter on a new and wider phase of religious controversies and metaphysical speculations. It requires all the erudition of the most learned divines and professors to find any clue through this labyrinth, and takes us far from that which is the sole object of this essay—to endeavour to form some conception of what may be the historical element in the records of the life and death of the founder of the religion.

CHAPTER X.

SCEPTICISM AND PESSIMISM.

Carlyle—Causes of Pessimism—Decay of Faith—A Prosaic Future—Denial of these Charges—Definition of Scepticism—Demonology—Treatment of Lunatics—Witchcraft—Heresy—Religious Wars—Nationality has superseded Religion—Wars more Humane—Originality of Modern Events and Characters—Louis Napoleon—Bismarck—Gladstone—Parnell—Abraham Lincoln—Lord Beaconsfield—Darwin—Huxley—Poetry—Fiction—Painting—A Happier World.

CARLYLE was a great genius, but he was a dreadful croaker. Barren, brainless, soulless, faithless were the epithets he commonly applied to the age in which he lived, and his favourite simile for his contemporaries was that of apes chattering on the shores of the Dead Sea. In the case of Carlyle, the cause of this pessimism is not far to seek. He suffered from chronic dyspepsia. If with the many other excellent qualities of his peasant progenitors, he had inherited some share of the "dura messorum ilia," and been able to eat his three square meals a day, and feel all the better for it, his views of the age and of his contemporaries would have been materially altered. He would have seen an age which is one of the most marked chapters in the history of human evolution; an age of great events and marvellous progress, progress not material only, but fully to an equal extent social, political, moral, and intellectual. The shores of the Dead Sea would have blossomed with verdure, and instead of chattering apes he would have seen human faces, "men my brothers, men the workers," with a great deal of human nature in them, good and bad, weak and strong, joyous and sad, healthy and suffering, but on the whole working up to a level which, if not necessarily happier, is at any rate higher.

For such dyspeptic pessimists there is an excuse. Pessimism is probably as inevitably their creed as optimism is for the more fortunate mortals who enjoy the "mens sana in corpore sano." But there are a large number of our modern pessimists for whom no such excuse can be pleaded.

There are the would-be superior persons, who think their claim to superiority is best established by affecting a lofty air of superfine disdain for the rude realities of real life; the critics who, as Lord Beaconsfield wittily says, are the failures; the minor poets, painters, and writers who, in their own opinion, would have been shining lights if their tapers had burned in a more congenial atmosphere; the prejudiced politicians and aristocratic classes who feel that knowledge, and with it political power, is passing over to the masses. And above all there are the orthodox divines, and good but narrow-minded religious public, whose one idea of religion is that it consists of adherence to traditional dogmas, and an unbroken belief in the truth of every word of the Bible as the inspired word of God, and the *ne plus ultra* of human knowledge.

With prejudices such as these it would be a waste of time to attempt argument; but there are a certain number of earnest and thoughtful men who hold what are substantially the same views upon different grounds, which deserve more careful consideration. They are not confined to social swells, would-be superior persons and orthodox theologians, but even a man of light and leading like Mr. Frederic Harrison can see no salvation except in the exceedingly improbable contingency of the world adopting the cult of humanity as evolved by the inner consciousness of M. Auguste Comte. What they say is substantially this, Science is killing faith; scepticism and democracy are advancing on old creeds and old institutions, like the lion of the desert, who in Tennyson's splendid simile—

> "Drawing nigher,
> Glares at one who nods and winks behind a slowly-dying fire."

Religion, they say, is becoming extinct, not only in the simple, old-fashioned sense of belief in creeds and catechisms, but in the higher sense of doubting the truth of the essential principles on which the Christian scheme of theology, and ultimately all spiritual faith and all religions, depend. A God who, according to one eminent Anglican divine, has been "defecated to a pure transparency," and, according to another, removed behind the primeval atoms and energies into an "original impress" acting by unvarying laws, is, they tell us, practically equivalent to no God at all, and instead of agnostics we ought to call ourselves atheists. Without a lively faith in such a personal, ever-present Deity, who listens to our prayers, modifies the course of events, records our actions, and finally rewards or punishes us after death according to our deserts, there can be, they say, no real religion; and they hold, and I think rightly hold, that the only support for such a religion is to be found in the assumed inspiration of the Bible and the divinity of Christ.

Destroy these, and they think the world will become vulgar and materialized, losing not only the surest sanction of morals, but what is even more important, the spiritual aspirations and tendencies which lift us above the sordid realities of daily existence, and give poetry to the prose of life. The Muses will take their flight with their sister Theology to happier spheres; imagination, idealism, heroism, and originality will disappear, leaving the world to a barren and prosaic sort of Chinese civilization. In short, their forecast of human existence is very similar to that which astronomers make of the planet upon which the human race live, viz. that as its inner heat radiates away in the course of ages, it will become, like its satellite the moon, a barren and burnt-up cinder.

To these gloomy forebodings I venture to return a positive and categorical denial; and to assert, on the contrary, that scepticism has been the great sweetener of modern life, has not only given us truer and juster views of the realities of the universe, but has made us more liberal-minded, tolerant, merciful, charitable, than in the hard, cruel days of mediæval superstition, and, in a word, that almost in exact proportion as we have drifted away from the letter, we have approached nearer to the spirit of true Christianity.

This, I am aware, will appear to many a strong assertion, and I must be prepared to justify it by specific instances, which I proceed to do. But first let me define what I mean by the term "scepticism." In a general way it means allegiance to truth; the habit of mind which makes a man, like a conscientious juryman, require evidence before he delivers his verdict, and if the evidence is insufficient makes him return one of "not proven." Doubt of doubtful things is to such a one as sacred a duty as affirmation of what is true and denial of what is false. His cardinal maxim is that of Dr. Johnson, "Clear your mind of cant." Don't say you believe when you really disbelieve, or only half believe, and try to hide your misgivings from yourself and from the world by loudness of asseveration or bitterness of denunciation.

But to this general meaning of the word scepticism a more limited and precise significance has come to be attached, and it is commonly used to denote disbelief in the inspiration of the Bible and the dogmas of theological Christianity. In this sense I accept it, and proceed to join issue with those who deny my assertion, that the world is a better place to live in on account of scepticism.

I will begin by taking a specific instance—the treatment of lunatics. Ever since the establishment of Christianity there has been a controversy between doctors and theologians. Theologians, and the public generally, relying on texts of Scripture, held that lunacy, with its kindred diseases of epilepsy and nervous affections, were caused by demons, or evil and unclean spirits, taking bodily possession of the unfortunate patients. Doctors, who for a long time alone represented the cause of science, relying on fact and experiment, and the teachings of great physicians of pre-Christian times, such as Hippocrates and Galen, held that such diseases were simply cases of pressure on the brain and over-wrought nervous systems. This was held to be so contrary to the truths of revealed religion, that doctors were looked upon as infidels of the worst sort, and the saying became general, "*Ubi tres medici duo Athei;*" atheist being the polite appellation with which every one was pelted who dared to appeal from Scripture to reason, and think for himself.

This radical divergence of view respecting the cause of lunacy led naturally to a corresponding difference in the mode of treatment. From the orthodox point of view the lunatic was a loathsome and repulsive object, whose body, probably for sins of his own or of his ancestors, had been taken possession of by an evil spirit. The only hope of cure was, so to speak, to bully the demon out of him by portentous exorcisms in ecclesiastical latin, and, worse still, by ill-treatment amounting often to the most horrible torture. Bedlam, with its row of raving madmen, chained like wild beasts to the wall, was a type of the usual mode of treatment.

Even such a great and good man as Sir Thomas More ordered acknowledged lunatics to be publicly flogged; and throughout rural England there were many what were called bowsening-places, for curing of madmen, consisting of deep walled cisterns full of water, into which, as Carew describes it in his *Survey of Cornwall*, "the lunatic was suddenly plunged by a blow on his breast, tumbling him headlong into the pond, where a strong fellow, kept for the purpose, dragged him about till he was quite exhausted;" when he was taken to church, masses said over him, and if he did not recover he was "bowsened again and again, while there remained any hope of life in him."

This simple picture of what was going on every day in remote country parishes of England enables us to realize the practical consequences of the theory of demoniacal possession, better perhaps than an enumeration of the Papal bulls and sermons of eminent divines, which urged the civil to unite with the ecclesiastical authorities and the Inquisition, in rooting out the bond-servants of Satan.

The medical men, on the other hand, of whom two out of every three were reputed to be atheists, took the opposite view, that madness was nothing but a form of brain disease, that its victims were rather objects for compassion than for aversion, and that gentle treatment was far more likely to effect cures than exorcisms and tortures.

Here, then, was a distinct issue joined between the Doctors of Divinity and the Doctors of Medicine, between the "theologici" and the "athei." If the question were to be decided by texts, the "theologici" had it all their own way, and the "athei" were nowhere. Nothing can be clearer than that Jesus over and over again asserted the theory of demoniacal possession. The demons knew him, he knew them, they conversed together; and he was so well acquainted with their ways that he could tell what sort could only be ejected by prayer and fasting. In the famous instance of the Gadarene swine, a raging madman was cured by evicting a legion of devils, and instead of leaving them homeless on the roadside, as if they had been Irish peasants, allowing them to occupy as caretakers the bodies of more than two thousand unfortunate pigs.

Nothing can be more explicit. Orthodox Christians were quite right in struggling to the last against a theory of lunacy which was in such direct contradiction with the express words of Scripture and of Jesus himself. We cannot wonder at Bossuet preaching his two great sermons, "Sur les Demons"; and John Wesley insisting that "most lunatics are really demoniacs," and that "to give up witchcraft is to give up the Bible, and to take ground against the fundamental truths of theology."

There cannot be a clearer illustration of the logical strength of Dr. Wace's formula, that if you believe in the inspiration of the Bible and in the Divine nature of Jesus, you must believe these things, or make him out to be a liar—I may add, a liar of the worst description, for if he were Divine and Omniscient, he must have known not only that he was fostering a delusion, but that this delusion would be in future ages the cause of misery and torture to thousands of the most helpless of the human race. But I reply, not without some little tone of indignation, "It is you, not I, who make Jesus out to be a liar; it is your assumption of Divine inspiration and Divine nature which defaces the pure and noble image of the Man Jesus, and places us in the alternative of either believing incredible things, or making him out to be an utterer of falsehoods. As a man no taint of falsehood or insincerity attaches to him in admitting that he used the language and shared the mistakes of his age and country. But as a God there is; and a God who teaches theories which are demonstrably false, and which lead to barbarous and revolting practices, is an incarnation not of goodness, but of evil."

For the theory of demoniacal possession is demonstrably false. If, instead of appealing to texts, the appeal is made to facts, the verdict is reversed; it is the "athei" who hold the field, and the "theologici" who are nowhere.

Which cure or alleviate the larger number of cases of lunacy—exorcisms and tortures, or gentle treatment? Which is most in harmony with the best instincts of human nature—love, charity, mercy, and compassion, Hanwell, with its harmless and happy inmates, or Bedlam, with its row of chained wild beasts? If a Doctor of Divinity says of a lunatic that he is possessed by a devil, while a Doctor of Medicine says he is suffering from a lesion of the brain; if the lunatic dies, and his brain is dissected, which do you find, the devil or the lesion? Nay, has not medical science gone so far that you can often predict the exact spot where the pressure on the brain is taking place, and by an operation remove the tumour, and restore the patient to reason?

If these things are true, and if the modern treatment of madness is really an improvement on the old one, it is quite clear that we are indebted for the change to scepticism, for it was impossible as long as the authority of Scripture was held to be the supreme tribunal, superior to fact and reason, and whose dicta it was impious to dispute. Montaigne, Hume, Voltaire, and a host of what used to be called infidel writers were the precursors of Pinet and Tuke; and with Galileo, Newton, and the triumphs of modern science, created the purer sceptical and scientific atmosphere of the present age, in which the monsters of mediæval theology simply die out like the Saurians of the secondary period, leaving a few fossil remains and degenerate descendants.

Witchcraft affords another test case in which the humanizing influence of scepticism is most apparent. Down to a comparatively recent period the belief in witchcraft was universal, and whole hecatombs of miserable victims were sacrificed to a superstition which is no less barbarous and degrading than that which exists to the present day in Dahomey, and among the cannibals of Central Africa. Why? Because the texts of what was supposed to be the inspired Word of God explicitly asserted the reality of witchcraft, and contained the command—"Ye shall not suffer a witch to live."

The case is the same as that of the belief in demoniacal possession as the cause of lunacy, except that the treatment of witches was even more cruel than that of lunatics, being founded more on texts of the Old Testament, dating back to a barbarous age. It was a form of cruelty also for which Protestants were even more responsible than Catholics, its worst excesses occurring in Protestant countries after the Reformation. In Germany alone, it is estimated that in the great age of witch-burning which followed that event, more than 100,000 persons perished by an excruciating death in the course of a single century.

On a smaller scale, one of the worst and latest outbreaks of the witch-burning epidemic occurred in Puritan Massachusetts at the close of the seventeenth century, incited and fanned into a flame by the efforts of the Mathers and other leading Calvinistic divines. Hundreds of innocent men and women of good characters were tortured into confessions, or convicted on the testimony of private enemies and professional witch-hunters, and perished in the flames, as was clearly proved when the epidemic subsided, and reason began to resume its sway, though divines like Cotton Mather held out to the last, and groaned over the evil spirit of unbelief which had thwarted the glorious work of freeing New England from demons.

Nobody now believes in witchcraft, and foolish old women and hysterical young ones may talk as much nonsense as they like without fear of being burned alive. Surely the world is the better for this; but how has it been brought about? Not that the texts have become more ambiguous, but that people have ceased practically to believe in them. I say *practically*, for there are a good many who still retain a sort of half-belief, and who would be shocked either to confess that the Bible is not inspired, or to say, with John Wesley, that "to give up witchcraft is to give up the Bible," but as the Ichthyosauri died out, and left harmless lizards as their successors in the purer air of the Tertiary era; so this, with other barbarous superstitions, has lost all real hold on the minds and consciences of those who, happily for themselves, live in the atmosphere of a scientific and sceptical age.

If the idolatry of Scriptural texts has caused so much human misery in the case of lunacy and witchcraft, the same idolatry, expanded from texts into dogmatical creeds and confessions, has been even more destructive in the case of heresy. Heresy, or the holding of different beliefs from those of the Church, is either a harmless and necessary incident in the use of human reason, or it is an act of pernicious and contagious wickedness which it is the duty of the State to aid the Church in stamping out. This depends on whether we do or do not believe the Creeds. If we believe the Athanasian Creed, which contains the fullest summary of the articles of the Catholic faith, and which is still retained in the Anglican ritual, all men will "without doubt perish everlastingly" who do not believe in every single article of that remarkable Creed. What right have we to rail against Torquemada, or blame Calvin for burning Servetus, if we really believe this to be true? They were simply carrying out, conscientiously and logically, the principles to which all orthodox Christians profess to adhere. Surely if it is right to stamp out the cattle plague, it must be still more right to stamp out a moral cattle plague which is eminently contagious, and which beyond all doubt causes those who contract the disease "to perish everlastingly." There is no possible answer to this, except that we do not believe the Creeds; that we feel the burning of men for differences of opinion to be cruel, and the suppression of freedom of thought to be mischievous. In short, that our attitude has become that of the poet who says—

> "There is more life in honest doubt,
> Believe me, than in half the Creeds."

If this is not "scepticism," I do not know what the meaning of the word is.

We live, fortunately, in an age when scepticism has so effectually killed the class of ideas which led to persecutions for heresy, that we have almost forgotten what the Inquisition and the fires of Smithfield really were. From first to last, hundreds of thousands of victims perished in horrible tortures for the crime of thinking for themselves. There is hardly a man of light and leading of the present century who would not have been sent to the stake if Spain had conquered England, and the integrity of the Catholic faith had been enforced by the civil power, or if Calvin had ruled in England as he did in Geneva. Darwin, Huxley, and Herbert Spencer would certainly have been burned; Carlyle, George Eliot, Byron, and Shelley would have shared the same fate; and Dean Stanley, Bishop Temple, and the whole Broad Church would have been in imminent peril. Spain, where the Inquisition so long reigned supreme, is an instance not only of the devilish cruelty which a misplaced religious earnestness can inspire, but of the inevitable political and social decrepitude which follow from successful attempts to stamp out freedom of thought.

Religious wars were only an outcome on a larger scale of the ideas which inspired religious persecutions. At bottom it was a firm conviction by those who held one set of opinions that those who held different ones were miscreants, enemies of the human race, who ought to be forcibly converted or exterminated. Given the conviction, the persecutions and wars followed as a matter of course, or rather of conscience. Destroy it, and the persecutions and wars cease. We no longer persecute and go to war in the name of religion. Why? Because the age has become too liberal, enlightened, tolerant, and humane. And why has it become so? Because scepticism has triumphed over orthodoxy. That the age has become more sceptical, and that faith in the old hard-and-fast lines of orthodox religion has declined, are facts which all acknowledge, though some deplore. It is evident, moreover, that these two facts are not merely concurrent, but stand to one another in the relation of cause and effect. It is a case not merely of *post hoc*, but of *propter hoc*. Voltaire, who may be taken as the representative of the literary scepticism of the last century, was inspired in his attacks on orthodoxy by his indignation at one of the last "autos-de-fé," or acts of faith, in the burning of a heretic. His shafts of ridicule wounded the monster to death more effectually perhaps than could have been done by solid arguments. The name of Darwin, again, may be taken as the representative of the scientific scepticism which has effected the greatest revolution of thought in the history of the human race, and substituted the idea of original impress, acting by unvarying law, for that of secondary supernatural interferences with the course of Nature. No educated man any longer believes in the sense in which our forefathers believed the Bible, and in which Mahometans still believe in the Koran. The assured faith in the Bible, as an ultimate and exhaustive record written by God's finger, has vanished never to return, and has quite lost its power as a practical factor in the life of nations. We retain our affection and reverence for it, from old associations, and as containing many beautiful and excellent things, but we no longer make it an idol. We criticize it freely, and find it to be a collection of various writings of various ages, by unknown or doubtful authors, and containing, with much that is of the highest truth and highest interest, much that bears evident traces of the ignorance, superstition, ferocity, and immorality of the rude and barbarous ages over which its traditions extend. No one now would think of appealing to every single text of Scripture as an ultimate tribunal from which there was no appeal, or, like the Caliph Omar, burning all the other books in the world because, if they agreed with the Bible they were superfluous, and if they disagreed with it, mischievous.

A better proof cannot be afforded of the extent to which ecclesiastical religion has ceased to be a motive power in human affairs, than by a reference to the great wars of the last half century. By an irony of fate, the first great exhibition in Hyde Park, which was thought to have inaugurated an era of peace, has been, like opening the temple of Janus, the signal for a series of the greatest wars recorded in history; wars great not only in the magnitude of the scale on which they were waged, but in the momentous importance of the issues involved. In all these wars the element of religion was entirely absent, and its place was supplied by the new element of Nationality. The net result of these wars has been the consolidation of a great Germany, a great Italy, and a great United States. Everywhere people of the same race, speaking the same language, and having a common literature and common interests, however broken up and divided into fragments by internal dissensions or foreign foes, have tended with irresistible force to consolidate themselves into great nations. Even the weaker races—the Greeks, Roumanians, Servians, and Bulgarians—have felt the same impulse, and the half-satisfied aspirations of the Eastern Christians constitute the peril of Europe, and threaten us with the impending shadow of another war. Nearer home, Irish nationality is the root of our Irish difficulty. We have taught the Irish people to read and write, we have given them a free Press and Parliamentary institutions, and the result is that they claim an increase of self-government and recognition of their separate nationality which we hesitate to concede, because we fear that it would destroy the old system of English ascendancy, and subvert many of the settled principles of English law, especially as regards the tenure of land and the rights and duties of landlords. If we have saved our colonial empire, it is only by conceding with the freest hand to Canada, Australia, New Zealand, and South

Africa all that we once contended for, and giving them the fullest scope to work out their destinies as independent communities, attached to the mother country by ties of common interests and affections, rather than by the hard-and-fast lines of superior force.

Now in all these great movements it is remarkable that ecclesiastical religion has not only not been an appreciable factor, but that in many cases they have gone on in the teeth of whatever influence it might be supposed to have remaining. In Italy, the head-quarters of ecclesiastical authority, the Pope, though still tho venerated head of millions of Catholics, has been utterly powerless when opposed to the idea of Italian nationality. The Catholics of South Germany fought as stoutly at Gravelotte and Sedan, shoulder to shoulder with the Protestants of the North, to make a great Germany, as their ancestors did under Tilly and Wallenstein against the ancestors of the same Protestants to secure the ascendancy of their respective Creeds. Austria has to forget the traditions of the Thirty Years' and the Seven Years' wars, and ally herself to heretic Prussia. France has for more than a century been intensely national, and very little religious. Even in Spain a dominant ecclesiasticism died out with the embers of the Carlist insurrections, and Spanish colonies in far-off Mexico, Buenos Ayres, and Chili are entering on a career of progress and prosperity almost exactly as they have emancipated themselves from the rule of priests and adopted modern ideas.

Has this change from religious to national wars been on the whole beneficial? One thing is certain, that war among civilized states has become infinitely more humane. Compare the picture by a military correspondent, of the advance of the Crown Prince's army through France, with the details of the Thirty Years' War, as given in Schiller's history. In the one case you see French peasant girls standing at the doors of their cottages to see the brilliant staff ride by, and exchanging nods and smiles with the German soldiers; in the other you have Tilly's Pappenheimers tossing heretic babies on the points of their pikes at the sack of Magdeburg.

The most signal instance, perhaps, of the humanizing influence of modern ideas is afforded by the action of the United States after the close of the great Civil War. A war of unexampled magnitude, costing tens of thousands of lives and millions of money, had been fought out with unexampled determination. The vanquished had begun the war, and in the view of the victors were rebels, but not a single hair of their heads was touched after the contest was over, not a single political prisoner was brought to trial. Jeff Davis was not hanged on a sour apple-tree, and the leading generals and politicians on either side for the most part returned quietly to civil occupations. I sometimes wonder what an historian writing a century hence will think of this record, compared with our English one of twenty-five members of Parliament imprisoned as common felons for political offences. To pursue this further would, however, lead me too far towards the burning region of contemporary politics, and I content myself by drawing this conclusion. If the spirit of the age be really sceptical and democratic, as all admit and many deplore, then scepticism and democracy must be included among those "ingenuas artes" of which the Roman poet says—

"Emollit mores nec sinit esse feros."

Nor is it in war only that milder manners and a more humane and charitable spirit have accompanied, if they have not been created by, the development of these two great principles of modern society. The air is full of projects, visionary or otherwise, which are all based on the spirit, if not on the letter, of true Christianity, of assisting the poor and suffering, and sweetening the conditions of life. Bismarck and the German Emperor adopt large schemes of State socialism, and aim at a universal insurance of workmen against poverty and old age. Trades Unions, Provident Societies, and Savings Banks do the same on an ever-widening scale in English-speaking communities. The old harsh principles of English law, which always sided with the strong against the weak, with man against woman, with landlord against tenant, with capital against labour, are being broken down in all directions. The rigid conclusions of political economy are no longer accepted as axioms. The duties of property, so long ignored, are coming into formidable antagonism with its rights.

So far from impairing the sanctions of morality, moral considerations are coming more and more to the front in this age of material progress. Slavery, long sanctioned by Bible texts and immemorial usage, offends the public conscience and disappears. We began by burning heretics, burning softened into boycotting, and finally this last vestige of intolerance has disappeared, and we live in an England where,

> "Girt by friends or foes,
> A man may speak the thing he will."

That world-old though newly-named institution, the "boycott," is no longer applied to differences of opinion, but confined to conspicuous offenders against the unwritten laws of a nation's conscience; to respondents in divorce courts, exceptionally bad landlords, and heartless profligates. The poor are always with us, but we no longer pass them by on the other side like the Pharisee, muttering our ecclesiastical texts and economical formulas. We feel for them, our consciences are touched, a daily diminishing number ignore them, and an increasing number try, in their respective spheres, to assist them by active effort, or sympathize with those who do.

The truth is, that morals are built on a far surer foundation than that of Creeds, which are here to-day and gone to-morrow. They are built on the solid rock of experiences and of the "survival of the fittest," which, in the long evolution of the human race from primeval savages, have by "natural selection" and "heredity" become almost instinctive. Every day of civilized society, working in an atmosphere of free discussion and free thought, tends to make the primary rules of morality more and more instinctive, and to extend and widen their application.

The other charge against the spirit of the age is still more easily refuted. It is said that scepticism has killed spiritualism, and stripped life of its poetry and higher aspirations, while democracy has reduced everything to a dead level of prosaic mediocrity. Those who say so see the reflection of their own souls. The man must be indeed hopelessly commonplace and prosaic, who fails to recognize the grandeur, splendour, and dramatic interest of the events of the age in which we live, and the striking originality of its principal characters. Was there ever in classic or mediæval times such a tragic drama of human life as is afforded by the career of Louis Napoleon. See him in his early years a dreamy youth, dabbling in obscure conspiracies, and musing over vague ideas and destinies connected with the name he bore. Then comes the attempt at Strasburg; the life in London, half Bohemian, half on the outskirts of fashionable society; the ludicrous fiasco at Boulogne; the romantic escape from the prison at Ham. The curtain falls on the first act, and when it rises we find the obscure adventurer clearing the streets of Paris with grape-shot, imprisoning all that is noblest and most respectable in the public life of France, and finally firmly seated on the Imperial throne. He proclaims the Empire to be peace, and he plunges France into four great wars, the Crimean, the Italian, the Mexican, and the Franco-German, all alike senseless in the view of any possible French interest. He inaugurates the system of armed peace and excessive armaments, and for quarter of a century is the disturbing element in European politics. The attitude of all other nations is, to use the expression of the witty Frenchman, that of spaniels watching the eye of their master at the Tuileries. Then comes the collapse, and in the closing scene we see a wretched creature driving out in a hack carriage from Sedan to give up his sword to the German Emperor, and sitting on a wooden chair with Bismarck, in front of a little wayside cabaret, to discuss the terms of the surrender as prisoners of war of his last army of 120,000 men. What must have been the emotions on that fatal day, hid under the mask of an imperturbable countenance and an eternal cigar. And all the time the man was essentially the same. Kind-hearted, easy-going, utterly unprincipled, vague, moony, idealistic; easily influenced by those about him, and twisted round his finger by a strong and practical nature like that of Bismarck. As his best counsellor and most intimate friend, the shrewd, cynical, polished, and worldly De Morny, once said to me, when the Emperor was in the height of his power, "The world will some day discover that the man has a better heart and a worse head than it gives him credit for."

I have mentioned Bismarck. There is a man indeed, a man such as Europe has not produced since Luther and Cromwell. Think of his career from a wild student, a provincial Tory squire, training himself by degrees to be first a diplomatist, and then a statesman; startling the starched representatives of the German Confederation at Frankfort by lighting his cigar without the permission of the Austrian Envoy, with the same cool courage and happy audacity which led him to Sadowa and Sedan, and now the founder of the German Empire, the great Chancellor, the arbiter of the peace of Europe. What made him what he is? His solid strength of character, his sagacious sincerity, his keen insight, glancing through the outward show of things into their real essence, and above all, his indomitable courage, which never quailed before hostile parliaments or vacillating emperors, and led him to stake his head on the success of the Prussian needle-gun and Prussian discipline, against the veteran legions of Austria and the showy *prestige* of imperial France.

At the opposite pole from Bismarck is our own "Grand Old Man." Opinions may differ as to Mr. Gladstone's policy, and whether his powerful personality is an element for good or for evil in English history; but no one who is not a purblind political partisan, can deny that, whether for good or evil, he is a grand and striking figure. Where will you find a man of such universal attainments, wide sympathies, and persuasive eloquence? Where look for an intellect which combines such scholastic subtlety with such argumentative power, such a grasp of details, such juvenile energy, and such a fervid white heat of passionate conviction. What a rich and complex nature must it be, which has in it the evolution from the ecclesiastically-minded Oxford student who was the rising hope of the Tories, to the great financier of Free Trade, the disestablisher of the Irish Church, the statesman who is at the head of all Liberal movements, the man whose eager sympathies side with liberty and with the masses "of our own flesh and blood," from Ireland to Italy. His mind is like the steam-hammer, which can either crack nuts, or mould masses of stubborn iron, and even in extreme old age there are no signs that his natural vigour has abated.

There is another striking personality of our times, whom, at the risk of offending political prejudices, I should like to mention, the "uncrowned king of Ireland"—Parnell. I am accustomed to call him the Irish Bismarck, for in many of his essential traits he resembles the iron Chancellor. Here again I pass no judgment as to his aims and policy, but look simply at the man and his career. What a career it has been! A young man with no special gifts of position or fortune, little likely as a Protestant and a landlord to enlist the sympathies of the Irish race, gifted with no showy qualities of oratory, the very antipodes of the former great Irish leader, O'Connell, silent, self-restrained, reserved, I may almost say, unsocial. I recollect this young man when I first knew him in the House of Commons, an obscure member even of his own Home Rule party; one of a little knot of five or six Irish members, who thought Isaac Butt's leadership too tame, and whose ruling idea was to force the attention of the House to Irish grievances by organizing obstruction. They succeeded, and soon became very conspicuous, and intensely obnoxious. Step by step Parnell came to the front, and first rivalled and then displaced Shaw in the leadership of the Irish party left vacant by the death of Butt. Like Carnot he organized victory, and even more than Bismarck, forged his own weapons as the strife went on. For Bismarck had his sturdy emperor, his admirable Prussian army, and his great strategist, Von Moltke, made to his hand; Parnell had nothing but what he made himself. His strength of character, practical sagacity, and far-seeing insight, by degrees gave him an ascendancy which secured him the support of the great majority of the Irish race at home and abroad, enabled him to wean them from impossible dreams of rebellion and revenge, to the practical policy of constitutional agitation; and finally has placed the return of some eighty-five out of one hundred and five members for Ireland in the hollow of his hand, and what was apparently more hopeless, has silenced the conflicting jealousies and interests which, in former days, marred all Irish movements, and drilled these eighty-five members into a compact body, acting as one man, under the control and advice of their leader. He has thus, almost single-handed, advanced Home Rule from being a dream as wild as the restoration of the Heptarchy, to be the burning question of practical politics. He has got four-fifths of Ireland, two-thirds of Scotland and Wales, and the bulk of the Liberal party in England on his side, and few dispassionate observers can doubt that, whether for good or evil, the realization of the main features of his policy has become a question of more or less, and of sooner or later, rather than of absolute and permanent rejection.

This is a good deal for an undergraduate of Magdalen to have done before he has passed the meridian of middle life, and to have done it for a hopeless minority, an unpopular cause, and a down-trodden race, by sheer force of individual character. Of the epithets which their contemporary age has attached to these three leading personalities, the "Great Chancellor," the "Grand Old Man," and the "Uncrowned Irish King," I think there is little doubt that the Macaulay of a future century will find them to have been justly applied, and that without reference to the success or failure of their work which is in the womb of the future.

It would not be right to close the list of the great political personalities of the day without saying one word of Abraham Lincoln, one of the greatest, as he is certainly one of the most original and interesting of modern statesmen. Wise, far-seeing, steadfast, simple, and noble, as Washington, he had a fund of genial humour, and a touch of the quaintness and eccentricity of the old Illinois rail-splitter, which endears his memory to the affectionate respect of all classes of English-speaking men, and makes him a bright example for all time of the height of heroism to which a self-taught working-man of the new democracy may attain.

If we turn from what may be called the epic of modern history to its romance, what figure can be more original and interesting than that of Lord Beaconsfield. What a career, from a second-rate novelist and dandy about town, seeking notoriety by resplendent small-clothes, to become the minister of a great country, the favourite of sovereigns, the superior of dukes, the champion and hero of a proud aristocracy and of a great historical party. And yet, as the novel of his last years shows, essentially the same man throughout. Brilliant, audacious, a master of phrases, and believing in them as stronger than facts. A sort of glorified Gil Blas, or hero of a Spanish comedy; and yet with qualities which endeared him to friends, captivated the popular imagination, and enabled him to play his part to perfection in all the varied vicissitudes of his extraordinary career. Infinite cleverness, infinite courage, infinite self-possession, and at bottom a genial and artistic temperament, which made him always, whatever else he might be, a finished gentleman. No one ever heard of him, whether as leader of a Government, or as leader of an Opposition, doing a coarse, vulgar, or ungentleman-like thing. He never lost his temper; he fought, like a courtly duellist of one of Dumas' romances, with the keen rapier of polished sarcasm and pungent epigram, but he fought fairly and left the coarser work, the flouts and jeers, to titled subordinates. His ideas, if vague and visionary, were always grandiose, and according to his lights, imperial and patriotic. He had no prejudices, and although the leader of bucolic squires and favoured guest of ducal drawing-rooms, he was fully convinced that Toryism could only survive by becoming democratic. Here surely was a product of the age as piquant and original as any to be met with in the romance of history.

I turn gladly to the serener regions of science and art. Here also, while we find everywhere the influence of the spirit of the age, we find everywhere genius and originality of character. It is the age of science; its marvellous triumphs have given man an undreamt-of command over the forces of nature, and revolutionized his ideas both of the material and of the spiritual universe. But what I wish principally to remark for the present purpose, these triumphs have been achieved, not by a mechanical process of second-rate specialists working each in his separate groove like wheels and pulleys in the mill of progress, but by a succession of great men, worthy leaders of great events. Take Darwin, the greatest of all. Who in the school-boy scolded by his master for wasting the time which should have been devoted to hexameters in trying rude chemical experiments and collecting beetles, could have foreseen the great philosopher who was to revolutionize the whole course of modern thought? At college he was, like many another careless student, thinking more of partridge-shooting than of books, and looking forward to taking orders, and becoming a college don, or vicar of a country parish. But his beetle-hunting saved him, it brought him into connection with men of science at the University like Henslow, and the merest accident led to his being appointed as naturalist to accompany Captain Fitzroy in the exploring voyage of the *Beagle*.

He saw new lands and new races of men, and his mind, rapidly expanding, acquired a storehouse of new facts and ideas which were the germ of his future greatness. See him next a martyr to ill-health in his quiet cottage in a secluded Kentish village, thinking out his ideas, trying simple experiments, clipping out extracts, and patiently collecting information, until one day he woke to find himself famous, and to have his name associated with the greatest revolution ever known in man's conception of the universe. In less than forty years "Darwinism," that is, evolution by unvarying law, superseded "Supernaturalism," or the theory of a world created and maintained by a succession of secondary interferences, as completely as the Copernican theory superseded that of Ptolemy.

Before he died he could see all educated thought, all men of light and leading in all countries, converts, if not to all the details, to the leading ideas and facts of his world-wide theory. And what a simple, noble character he was. Patient, candid, magnanimous, modest, loving and beloved in all intercourse with family and surroundings down even to his little dog, faithful friend, single-minded worshipper of truth, one might say that, apart from his fame, here was a model man of the nineteenth century, and if scepticism can give us more like him, we may well be content to take what the outcome of a sceptical age has in store for us without much apprehension.

And if Darwin was the Napoleon of science, what a brilliant array of marshals marched under him at the head of its various divisions—men not of one idea and cramped intellects, but large-minded men of genius and originality, men such as Lyell, Huxley, Herbert Spencer, and a host of others.

Take Huxley as a typical instance. If he had never made a discovery in science, he would go down to posterity as the greatest master of style and best writer of English prose in the whole range of modern literature. To a wit keen as that of Voltaire he adds a far greater range of accurate knowledge and force of pungent logic; his grave irony and undercurrent of genuine humour are delicious, and every sentence goes straight to the mark like a rifle-bullet. In controversy he is like a sun-god shooting his arrows of light through the thickest cuirass of ignorance and prejudice. Given something to say on a theme of science or philosophy, and I know of no writer, past or present, who can say it as well as Huxley.

Of all these, and of the hundred other names which might easily be added to the list of generals and captains of the army of modern science, it may safely be said, that as a rule they lived true, simple, and noble lives, giving no cause of scandal or offence to the world, and showing that the high-priests of truth need not fear a comparison as regards character and conduct with those of any stereotyped and formalized religious creed or caste.

The remaining complaint of the pessimists, that the world is becoming uninteresting and prosaic, is easily disposed of. I reserve for another essay what I have to say as to the creeds of the great poets, but for the present it is enough to ask whether Byron and Shelley were believers or sceptics, and whether their poems show any falling-off in the poetic faculty? Swinburne, whatever we may think of him otherwise, has the gift of word-music and of brilliant imagination in an eminent degree; and Victor Hugo, though too turgid and rhetorical for an English taste, strikes a powerful lyre whose chords resound loudly in the souls of his sceptical countrymen. Above all, Tennyson, the great poet of modern thought, attains a height of inspiration which has been seldom if ever equalled. I care not what his creed may be, but he is thoroughly the man of his age, imbued with its science, from which many of his noblest similes are drawn, and a sharer in its strength and weakness, its hopes and fears, its grandest aspirations, and its blankest misgivings. The stanzas in *In Memoriam*, which conclude with the solemn words, "Behind the Veil," are the profoundest expression of the deepest thoughts of the most earnest minds of the nineteenth century.

In fiction, we have a hundred writers and a thousand readers, of works of a fairly high standard of excellence, for one of former centuries. Nothing gives me more hope for the future of that inevitable democracy which is advancing on us with such rapid steps, than the multitude of standard works which are circulated in cheap editions. Shakespeare, Walter Scott, Dickens, Thackeray, George Eliot, as well as works on history, philosophy, and art, like those of Macaulay, Carlyle, and Ruskin, are published in ever increasing numbers and at ever lower prices. Who reads them? They must be bought by hundreds of thousands, or it would not pay to publish them, even in pirated editions like those of America. They must be read by millions who never read before, but who now read with intelligent interest for education and self-culture.

If we turn to painting we find the same phenomenon. It is becoming more popular and more democratic. Prints and chromo-lithographs hang on the walls of every cottage; illustrations, often admirable, like those of the modern school of wood-cut, adorn the pages of pictorial newspapers and magazines, and have become almost a necessary accompaniment of every work of wide circulation. And how has this affected the higher class of painting? Has it become more prosaic? Distinctly the reverse, it is far more poetical; that is to say it aims far more at expressing the real essence and typical spirit of the varying moods, whether of external or of human nature. The contrast between the modern French school and that of conventional classicism affords the best instance for my present purpose, for France is *par excellence* the country whose scepticism and democracy may be supposed to have killed poetry. Compare a landscape of Corot's with a landscape of Poussin, which is most poetical? Or take Millet, who has caught for all time the type of the true French peasant, with his simple or even sordid surroundings, his narrow horizon as he bends with an almost ferocious intensity of labour over his paternal clods, yet illumined by gleams of humble poetry, as in the *Angelus*, or of pure domestic affection, as in *Teaching the Baby to Walk*. Surely this is real poetry, and worth a thousand of the academic pictures of the school of David.

In the English school of art, the same tendency is manifest. All the great modern masters aim at representing types and ideas rather than traditional conventionalities or prosaic realities. Thus Millais' *North-West Passage* and *Boyhood of Raleigh* give us the essence of that spirit of maritime adventure which has made Britannia rule the waves; Faed's pictures of humble Scottish life are as tender and true as if they were poems of Burns transferred to canvas; Peter Graham, Brett, and Hook paint the sea as it never was before painted, in all its moods of strength, repose, and of the joyous freshness of its rising flood. And so of a host of others. They aim at and often succeed in painting pictures which are really poems, true and touching phases of human characters, types of nature which speak to the varying emotions of the human soul, and their masterpieces find a ready response in the hearts of millions.

All this does not look like the advent of a drab-coloured age of prosaic mediocrity; or as if the fresh bracing breeze of modern science and free thought, sweeping through the confined air of mediæval cloisters, were going to do otherwise than sweeten and purify the atmosphere, and make the blue of heaven more blue, the grass greener, and the earth, on the whole, a better and more genial place for man to live in. Blow, brave North-Wester! sweeping over the free and boundless ocean of Truth, chilling to worn-out creeds and decrepit superstitions, but filling the lungs with ozone, bracing the nerves and brightening the eye.

"Who loves not knowledge, who shall rail
Against her beauty? may she mix
With men and prosper, who shall fix
Her pillars; may her cause prevail."

CHAPTER XI.

CREEDS OF GREAT POETS.

What is a Great Poet—Ancient and Modern Poets—Byron, Shelley, Swinburne, Browning, Pope, Dryden, Coleridge, Spenser—Chaucer—Wordsworth—Nature-Worship—*Ode on Immortality*—Byron and Shelley—Burns—Gospel of Practical Life—Shakespeare—Self recorded in Hamlet and Prospero—The Sonnets—Views of Death—Behind the Veil—Prospero—Views identical with Goethe's *Faust*—And with the Maya or Musiar of Buddhism—Pantheism—Ignoring of Religion—Patriotism and Loyalty his ruling Motives—Practical Influence of Religion Exaggerated—Religious Poets—Dante—Milton—Contrast between Greek Tragedy and Modern Poetry—Tennyson—Poet of Modern Thought—*In Memoriam*—Practical Conclusions.

WHAT is a poet, and what is a great poet? A poet I take to be one whose nature is exceptionally susceptible to impressions from the surrounding universe, especially those of a character which comes within the domain of art, and who unites with this a certain musical faculty and command of language, which enables him to translate these impressions into apt and harmonious verse. The poet's brain may be compared to a photographic plate which is extremely sensitive and retentive of images which flash across it; or to a delicate Æolian harp which vibrates responsive to harmonies of nature, unheard, or only half-heard, by the coarser fibres of ordinary mortals.

This of itself, where it exists in an exceptional degree, may make a pleasing or even a considerable poet, but to make a great poet something more is required. To this fine susceptibility and musical nature must be added a great intellect; an intellect capable of casting flashes of insight into the varying phases of human character, and the deepest problems of man's relations to the universe; an intellect so imbued with the spirit of the age and abreast of the knowledge of the day as to be able to sum them up in a few glowing lines which embody their inmost essence. Such poets are extremely rare. Of the ancient world, Homer, Æschylus, Sophocles, and Euripides of the Greeks, Lucretius and Virgil of the Romans, still shine as stars of the first magnitude among the "stars of mortal night," though dimmed by distance and seen under greatly altered conditions. Of moderns, I hardly know that the very first class can be assigned to other names than those of Shakespeare, Dante, Milton, Goethe, Burns, Wordsworth, and Tennyson. Many come near it from exceptional excellence in some of the qualities which are most essential to true poetry. Shelley, for instance, is equal to the very greatest in the exquisite susceptibility to all that is beautiful in nature, and the faculty of reproducing it in the loveliest and most musical of lyrics. His *Skylark* and *Cloud* may well stand as the high-water mark to which lyrical poetry has ever attained. But he was cut off at an early age, before his intellect had got over the stage of youthful effervescence, and settled down into the sober and serene wisdom requisite to reflect truly the spirit of an age, and guide a world towards better and higher things. He and Keats have given us "things of beauty" which are "joys for ever," but scarcely wise counsels and consoling words, to enable us better to live our lives and face our destinies. The same may be said of Byron, the vigour of whose verse and vividness of feeling and description are unsurpassed, but whose ideal of life and character, be it real or be it affected, is about the last any one would do well to follow.

Of living poets Tennyson alone comes up to the highest standard. Others approach it on different sides, but on special sides only, and fail as conspicuously in many of the attributes of the highest poetry as they excel in others. Swinburne, for instance, almost equals Shelley in the exquisite musical susceptibility of rhythm and language, but the ideas behind the words are for the most part rhetorical and exaggerated, like those of his prototype, Victor Hugo. Browning again has intellect and insight, but his style is so rugged and obscure that to read his poetry is almost like trying to solve chess-problems. He is to Shelley or Tennyson what Wagner is to Rossini or Beethoven; caviare to the multitude, and almost outside the range of the true art which is based essentially on the beautiful.

Of other well-known poets, Pope is a great master of the art of weaving appropriate words into harmonious verse, and his ideas are for the most part clear and sensible. But they are not profound, and in his chief philosophical work, the *Essay on Man*, he rather reflects, with point and precision, the somewhat conventional and commonplace views of the average intellect of his age than gives flashes of insight drawn from his own inward struggles and experiences. The same may be said of Dryden, who had a singular gift of terse and vigorous expression, which has made so many of his lines survive in the form of standard quotations. But he was hardly a deep and original thinker, and however much we may admire his poetry we learn little from it.

Coleridge I hardly mention as a poet, for his principal work, as a religious philosopher influencing to a certain extent the spirit of his age, was done in prose and in conversation. His *Aids to Reflection* was long the text-book of the advanced thinkers of Anglican theology, but his *Christabel*, *Kubla Khan*, and *Ancient Mariner*, admirable as they are, are little more than the dreams of a gorgeous imagination. They might be the visions of an "English Opium-Eater," in the earlier stages of the seductive drug as described by De Quincey.

Of the early English poets, the names of Chaucer and Spenser stand out pre-eminent. Spenser, indeed, has perhaps as large a share as any other, even of the greatest poets, of that which is the substratum or first requisite of all true poetry; the exquisite susceptibility to all that is beautiful in the surrounding universe. But his philosophy does not go much beyond an allegorical representation of vices and virtues as they appear in the abstract, rather than in the concrete form of living individuals. Compare Una, who is his most distinct and lovable character, with Imogen, and you feel at once that Shakespeare gives you a living woman, in contact with an actual world; while Spenser's embodiment of nearly the same ideal is shadowy and mystic, half woman and half allegory, living in a world of impossible giants and monsters.

Chaucer, on the other hand, stands on solid earth, and deals with real characters. In the dramatic faculty of depicting actual living men and women he has no rival except Shakespeare, and is inferior to him rather in the narrower width of his canvas, and in the complexity and variety of the characters depicted, than in the truth and vividness of the portraits themselves. In his *Canterbury Tales* we have the real England of the reign of Edward III. brought before us as distinctly as if we had been one of the company assembled at the Tabard, and had ridden on the Dover road to the shrine of St. Thomas, with the worthy knight, the dainty and soft-hearted abbess, the jolly wife of Bath, and the other typical representatives of the various classes who made up what was the framework of English society in the fourteenth century. How like they are to us, how completely we feel that they are our own flesh and blood, and that five centuries have made but little change either in human nature itself, or in the special form of it which may be called English nature.

In reading Chaucer I am also struck by the wonderful anticipations of the most advanced modern thought, which occasionally crop up in the most unlikely places, and which only require to be translated into modern language to be at once recognized. For instance, I came across a passage the other day which, if expressed in the terminology which would now be used to convey the same ideas, would read as follows—

"The inscrutable First Cause of the universe knew well what He was about when He established the fair chain of love or of mutual attraction. For with this chain He bound the elements, fire, air, water and land together in definite forms so as not to fly asunder into primeval chaos.

"In like manner He established certain periods and durations for all creation beyond which nothing could pass. This needs no authority to confirm it, for it is proved by universal experience. Men, therefore, by this order of the universe may easily discern that the laws of nature are fixed and eternal. And any one who is not a fool can understand that as every part is derived from a whole, nature cannot have originated from any part or parcel of a thing, but from something that is perfect and stable, passing by evolution from the homogeneous into the heterogeneous, until it becomes subject to change and corruption. The Creator of the universe has therefore in His wise Providence so established its order, that definite pieces and progressions of things shall not be eternal, but come into existence and pass away in due succession.

"Thus the oak which grows so slowly and has so long a life, at last wastes away and dies. Even the hard rock in time wasteth away; broad rivers run dry; great cities decay and disappear; and all things have an end. So also of the human race. All die; some in youth, others in old age; kings as well as commoners; some in their beds, some in the deep sea, some in battle-fields.

"There is no help; all go the same way; all die. What causeth this but the Ruler and First Cause of all things, who draws back into His own essence all that was derived from it, against which decree it availeth no living creature to strive. Therefore it seems to me to be wise to make a virtue of necessity and make the best of that which we cannot prevent; and that a man is a fool who grumbles at that which is the universal fate, and rebels against the law to which he is indebted for his own existence."

If any one came across this passage without knowing

its origin, he would be apt to attribute it to some writer who was conversant with the works of Herbert Spencer, Darwin, and Lyell, and about the last guess he would make would be, that it came from the father of English poetry writing in the fourteenth century. And yet if he would turn to the speech of Duke Theseus in the *Knight's Tale*, he would find that it is a literal though modernized version of what Chaucer puts into the mouth of his representative of perfect manhood and mature wisdom. Religions and philosophies have changed, knowledge has increased, but these lines of Chaucer remain as a summary of the best and truest attitude in which a man can face the insoluble mysteries of the universe.

This passage alone should be sufficient to justify Chaucer's claim to rank among the great poets.

My object, however, is not so much to review poetry generally, or to assign to each poet his proper place in the hierarchy of Art, as to ascertain what have been the real creeds, or inmost convictions, of those who, by universal consent, are ranked among the highest. And when I talk of creeds, I do not mean the outward professions, which, with poets as with other men, may be mainly affairs of time and circumstance; but the deeper insight with which they "see into the life of things," and find with Wordsworth—

> "The anchor of the purest thoughts, the nurse,
> The guide, the guardian of the heart, and soul
> Of all the moral being."

In Wordsworth's case the answer is easy; he gives it himself. He finds it in nature. Not in a dead or mechanical nature, or one limited to seas and skies, mountains and rivers; but one which includes

> "The still sad music of humanity."

And which lives with

> "A presence which disturbs me with the joy
> Of elevated thoughts; a sense sublime
> Of something far more deeply interfuse
> Whose dwelling is the light of setting suns,
> And the round ocean and the living air,
> And the blue sky, and in the mind of man;
> A motion and a spirit that impels
> All thinking things, all objects of all thought,
> And rolls through all things."

This is very nearly pure Pantheism, and it is remarkable how closely he approximates in other respects to the Oriental philosophy which finds its expression in the religions of Brahma and of Buddha, and which tinged the speculations of Plato. In the *Intimations of Immortality*, he adopts, to a considerable extent, the doctrine of the transmigration of souls, or to express it in modern language, the "Conservation of Energy," applied to the immaterial soul as a distinct and indestructible essence.

The problem of immortality hinges on two questions: life before birth, life after death. They hang very much together, for if from nothing we came—*i. e.* nothing in the sense of no conscious personal identity, it is more than probable that to nothing we shall return. Wordsworth, in common with Brahmins, Buddhists, and Platonists, solves this problem by postulating pre-existence.

> "Our birth is but a sleep and a forgetting;
> The soul that rises with us, our life's star,
> Hath had elsewhere its setting,
> And cometh from afar."

It is remarkable that this pantheistic view of the universe is essentially that of other great modern poets, who, in other respects, differ most widely from the calm and self-contained character and serene wisdom of Wordsworth. Byron, in his moments of best and truest inspiration, expresses in still more passionate and vigorous language, the same feeling for one great living whole, comprising nature, humanity, and himself.

> "All heaven and earth are still—though not in sleep,
> But breathless, as we grow when feeling most;
> And silent, as we stand in thoughts too deep—
> All heaven and earth are still; from the high host
> Of stars to the lulled lake and mountain-coast,
> All is concentred in a life intense,
> Where not a beam, nor air, nor leaf is lost,
> But hath a part of being, and a sense
> Of that which is of all Creator and defence.
> Then stirs the feeling infinite, so felt
> In solitude when we are least alone:"

And again in the rush of the midnight storm he wishes to be

> "A sharer in thy fierce and far delight
> A portion of the tempest and of thee!"

Shelley, again, was essentially the poet of Pantheism, and derived all his best inspiration from

> "Earth, ocean, air, beloved brotherhood!"

The song of the skylark, the fleeting cloud, the forest at noonday, the

> "Waste and solitary places, where we taste
> The pleasure of believing what we see
> Is boundless, as we wish our souls to be,"

spoke to him and he to them as living beings, vibrating in unison with the most delicate harmonies.

Of Death he speaks as

> "The boundless realm of unending change,"

where

> "All that we feel, and know, and see
> Shall pass like an unreal mystery."

In other words, his glance of insight into the mysteries of the universe is essentially Pantheistic and Agnostic.

In sharp contrast with the ethereal poetry of Shelley, Burns, while equal to him or any other poet in the exquisite delicacy of his lyrics, stands on solid earth, and teaches what may be called a gospel of practical life. He may not always have acted up to it, but his poetry is pre-eminent in laying down sound and sensible maxims of conduct, and investing common things and ordinary life with a halo of tenderness and dignity drawn from the inspiration of the highest feelings of human nature. Thus, when he says—

> "To make a happy household clime
> For weans and wife
> Is the true pathos and sublime
> Of human life,"

he presents an ideal universal in its application, within reach of all, common to all sorts and conditions of men; and he presents it in a way which lifts the fundamental fact of the family tie from the region of prose into that of poetry. The poorest man who lives even approximately, up to these lines, may feel that he has not lived in vain. By industry, prudence, self-restraint, good temper and kindness, he has made his humble home a shrine of affection and happiness, and has made good his title to rank as one of Nature's gentlemen. Goethe means much the same thing when he says that "no man carries it farther than to perpetuate the species, beget children, and nourish them as well as he can."

But how cold and ironical does this sound when contrasted with Burns. One is prose, the other poetry; one a criticism on life, the other an incentive to purify and exalt it.

No one equals Burns in the keenness of insight with which he looks through the outer husks and habiliments of things to their real essence. Carlyle's clothes philosophy in *Sartor Resartus* is but a sermon on the text—

> "The rank is but the guinea stamp,
> The man's the gold for a' that."

A manly independence, based on the qualities which Tennyson attributes to the Goddess of Wisdom,

> "Self-reverence, self-knowledge, self-control,"

is to Burns, as it is to every one, the solid basis of all the manly virtues. It is a basis which is more readily provided to those who live by work, whether of the hand or head, than to those who are born with a silver spoon in their mouths, and are cradled in comfort and luxury. A man never knows what is really in him until he has measured himself with his fellows in real honest work. I have known many a man who fancied himself one of the *crême de la crême*, and looked down on the rest of the world as "cads" and "outsiders," who was not honestly worth twenty shillings a week of any man's money. He could ride, but not well enough to be a whipper-in; shoot, but did not know enough of wood-craft or rearing pheasants to be a gamekeeper; dance, sing, or draw perhaps, but nothing well enough to earn a penny by it. Strip him of his cotton-wool wrappings of wealth and rank, and land him at Sydney or Melbourne without a sixpence in his pocket, and what could he do to earn a living? Possibly drive a cab, or be a waiter at an eating-house. How can such a man feel the same manly independence as one who knows that, wherever he goes, he has muscles or brains to sell which are honestly worth their price in the world's market.

No one sets forth so forcibly as Burns the dignity of labour, and the compensations which go so far to equalize the lot of the rich and poor. If I wanted to convert to sounder views some narrow-minded social democrat, whose one idea was envy of the rich, I would make him read Burns' *Twa Dogs*, where the relative advantages and disadvantages of different stations of life are set forth with so much force and humour. Against the hardships and privations of the working masses, alternating with the enjoyments of the evening rest, the healthy appetite, and the sound sleep, he would read of the non-working classes, how

> "Gentlemen, and ladies worst,
> With even-down want of work are curst,"

and learn

> "It's no in riches or in rank,
> It's no in wealth like London Bank,
> To bring content and rest.
>
> If happiness has no its seat
> And centre in the breast,
> We may be rich, or wise, or great,
> But never can be blest."

He may learn also from the *Cotter's Saturday Night* how peasant life may rise to the level of patriarchal dignity; and from *Highland Mary* or *Bonnie Jean* how the romance of love may be as true and tender by the "banks and braes o' bonnie Doon" as in Belgravian drawing-rooms. Nor will the lesson be wanting from

Willie brewed a peck o' maut and *Auld Lang Syne*, that frank joviality and hearty friendship are not the exclusive appanage of any class or condition of mortal men.

From Burns to Shakespeare is a long stretch, but any attempt to ascertain the creeds of great poets would be incomplete without some analysis of what seems to be the inmost and truest attitude of the greatest of all poets towards the deepest problems of life. In the case of Shakespeare this is not easy to discover, for his genius is so essentially dramatic that his characters speak and act their own lives, and are not mere masks behind which the author discourses to the public. Thus Childe Harold, Conrad, Lara, and Manfred are only Byron himself posing in different attitudes, while Othello and Macbeth, Falstaff and Dogberry, are types of themselves reflecting Nature, and not Shakespeare. All we can say from them of Shakespeare's individuality is, that it must have been wide enough and rich enough to realize, with a certain amount of sympathy, all the varied range of human passions and emotions, strength and weakness, wisdom and folly. Even the humorous drolleries, and rogueries, and sheer imbecilities of human nature are noted and reproduced with a genial smile.

We cannot say that Shakespeare had any resemblance to Falstaff, but we may be sure that he had noted some one like him; some humorous ton of flesh, unblushing compound of braggart, coward, liar, and glutton, yet who half redeemed these evil qualities by his ready wit and unfailing good-humour, and left us almost sorry for him when he died babbling of green fields in Mistress Quickly's hostelry.

It is only in one or two of his characters that we can discover something of the real Shakespeare himself, projected from within outwards, and fashioned in some mood of his own image. This is the case mainly with Hamlet and Prospero. Of Hamlet I think we may say with some certainty, that no one could have conceived such a character who had not a Hamlet in him. He must have felt the irresolution, the despondency, the metaphysical thought sicklying over the "native hue of resolution," the burden of life almost too heavy to be borne, which made a noble nature and high intelligence drift the sport of circumstances, rather than "take arms against a sea of troubles," and incur the pain of coming to a definite decision.

The Sonnets, in which Shakespeare speaks in his own person, reveal a good deal of this frame of mind. The general tone is that of thought rather than of action, with an under-current of despondency and gentle melancholy. Thus, if the 29th Sonnet be really Shakespeare's, what a sermon is it on the vanity of human things, to find the supreme artist of the world, the man who had apparently led the most prosperous life, who had risen from a poor country lad to be the admired friend of the highest nobles and best intellects of his day, and who had in a few years achieved fame and competence, writing such lines as these—

> "When in disgrace with fortune and men's eyes,
> I all alone beweep my outcast state,
> And trouble deaf Heaven with my bootless cries,
> And look upon myself, and *curse my fate.*"

Or think of such a man, when recalling his past life to the "sessions of sweet silent thought," thus summing it up—

> "I sigh the lack of many a thing I sought,
> And with old woes new wail my dear time's waste;

Then can I drown an eye, unused to flow,
For precious friends hid in death's dateless night,
And weep afresh love's long-since cancelled woe,
And moan the expense of many a vanished sight.
Then can I grieve at grievances foregone,
And heavily from woe to woe tell o'er
The sad account of fore-bemoaned moan."

No one can mistake the analogy between these Sonnets and the melancholy musings of the Prince of Denmark.

Again, the 66th Sonnet is almost identical with the enumeration of the ills of life which make death desirable in Hamlet's famous soliloquy—

"Tired with all these, for restful death I cry,—
As, to behold desert a beggar born,
And needy nothing trimmed in jollity,
And purest faith unhappily forsworn,
And gilded honour shamefully misplaced,
And maiden virtue rudely strumpeted,
And right perfection wrongfully disgraced,
And strength by limping sway disabled,
And art made tongue-tied by authority,
And folly, doctor-like, controlling skill,
And simple truth miscalled simplicity,
And captive good attending captain ill:
Tired with all these, from these would I be gone."

The evidence of this identity between Shakespeare and Hamlet is strengthened if we examine in detail the enumeration of the "whips and scorns of time" which might almost compel a man to suicide. As a general rule Shakespeare's characters speak with an admirable dramatic propriety of place and circumstance. They say nothing but what such characters in such conditions might have said. But in this soliloquy there are things which Hamlet hardly could have said, and which must be Shakespeare speaking of his own experiences. Thus, the "law's delay" would hardly be included among the serious ills of life justifying suicide by any one who had not known it by personal experience. We can hardly suppose the high-born and accomplished heir to the Danish throne to have been a party to a Chancery suit, or to have trod for years, like Peter Peebles, the corridors of a Copenhagen Court of Session. Nor was he likely to have suffered from

> "The insolence of office, and the spurns
> That patient merit of the unworthy takes."

If then Hamlet's soliloquy expresses the real sentiments of Shakespeare, we have his judgment on the great questions of death and immortality summed up almost in the identical words of Tennyson—

> "Behind the veil, behind the veil."

To die is "to sleep—to sleep! perchance to dream." Death is "the undiscovered country from whose bourne no traveller returns." There is no assurance, absolutely none! He cannot say with the Materialist, we shall certainly perish, or with the Christian, we shall certainly live.

The character of Prospero affords even a better test than that of Hamlet for ascertaining what were Shakespeare's mature views on these subjects. There can be little doubt that in Prospero Shakespeare has an eye to himself, retiring in the plenitude of his powers from London and the stage, to spend the autumn of his days in a round of domestic duties in his native town. The magic which Prospero abjures can hardly be other than the poet's imagination, and the staff which he breaks and book which he drowns,

> "Deeper than did ever plummet sound,"

the poet's pen, which had bodied forth so many of these airy nothings, and given them

"A local habitation and a name."

It is well worthy of remark how nearly this practical solution of the problem of life coincides with that of another of the world's greatest geniuses, Goethe.

The drama of *Faust* concludes by showing how the hero is delivered from the power of evil, and how the sins and miseries of his career while commanding the powers of magic are condoned, by devoting himself to the practical work of real life—reclaiming a waste tract from the sea, colonizing it, and making it the abode of healthy human industry.

The moral is precisely the same in the two cases, that man's true life is in the natural and not in the supernatural, or, as Goethe expresses it elsewhere, that "here is your America,"—not in visionary continents across unmeasured oceans, but in doing, as Carlyle phrases it, "the duty that lies nearest to your hand, as the best guide to further duties."

But Shakespeare, speaking through Prospero, in his farewell address to the world goes beyond the sphere of practical life, and gives us his views of the highest problems of the universe in the well-known lines—

"And like the baseless fabric of this vision,
The cloud-capped towers, the gorgeous palaces,
The solemn temples, the great globe itself,
Yea, all which it inherit, shall dissolve,
And like this unsubstantial pageant faded,
Leave not a rack behind. We are such stuff
As dreams are made of, and our little life
Is rounded with a sleep."

If, in the case of Wordsworth, I had to remark on

the singular approximation of modern poetry to the Pantheistic views of Oriental religions and philosophies, this passage of Shakespeare carries the comparison still closer. It is the pure doctrine of Maya or illusion, which plays such a great part in the systems of Brahma and Buddha. There is no reality but the great Unknowable; all the manifestations of the universe are illusive dreams, rising and falling like mists from the Ocean of the Infinite. Individual existence is but one of these illusions, destined to disappear like others when its "little life is rounded with a sleep."

Observe that in this latest utterance Shakespeare has gone beyond the phase of thought which dictated the soliloquy of Hamlet. There, death was a sleep indeed, but a sleep in which there might be dreams, an undiscovered bourne where there might be anything. But here there is not merely Agnosticism, but the positive assertion that sleep is all, and that the individual life is absorbed, like everything else, in the great Ocean from which it came, of the Infinite and Absolute.

Goethe's theory of the universe is very similar to that of Shakespeare, but he approximates to the Oriental philosophy rather on its positive or Pantheistic side than on the metaphysical side of Illusion. Thus, in the famous reply of Faust to the simple inquiry of Margaret whether he believes in God, "Wer darf ihn nennen?" he says—

"Who dares to name Him?
Who to say of Him, I believe?
Who is there ever
With a soul to dare,
  To utter, I believe Him not?
The All-encompasser, the All-upholder,
  Enfolds, sustains He not
Thee, me, Himself?"

And he goes on to say how the over-arching sky, the solid earth, the everlasting stars, the depths of human emotion, are but manifestations of the eternal essence, call it what name you will.

> "Words are but mist and smoke
> Obscuring Heaven's glow."

This is almost identical with Wordsworth's—

> "Sense sublime
> Of something far more deeply interfused."

In a word it is pure Pantheism. So also is the hymn of the Earth Spirit, who sits weaving the varied shows of the universe—

> "And at Time's humming loom prepares
> The garment which the Eternal Spirit wears."

It has often been observed to what a little extent religion, that is, the formal religion of theological creeds, appears in Shakespeare's plays. Love, ambition, jealousy, all the various motives which practically influence human conduct and character, are depicted to the life; but religious belief is as completely ignored as if it had no existence. One would have thought that in an age which had witnessed the martyrdoms of Latimer and Cranmer, the destruction of the Spanish Armada, and the innumerable wars and conspiracies of the reign of Elizabeth, almost every one must have been a keen partisan either of the Protestant or of the Catholic persuasion. And yet such is Shakespeare's indifference or impartiality that it is impossible to say to which side he inclined. The only conjecture that has been hazarded is, that he leant towards the old faith, because his friars, especially Father Lawrence in *Romeo and Juliet*, are depicted in a favourable light. But this can hardly be carried further than to show that he was not one of those bigoted Protestants to whom everything connected with Rome was an abomination. On the other hand, we find no trace of it, where it might have been most expected, in ridicule or abuse of the Puritans.

The Puritans were already a considerable sect, and from their bitter hostility to the stage must have appeared to Shakespeare almost in the light of personal enemies. His observant eye could not have failed to notice many of the traits which, as in Butler's *Hudibras*, laid them open to ridicule. Many of his characters, as for instance that of Malvolio, would have enabled him with perfect dramatic propriety to sharpen the shafts of his satire by introducing an element of Puritanism. But he entirely abstains from doing so by a single word or insinuation. Malvolio is a prig, but not a Puritan.

The fact is that patriotism and loyalty seem to have been such ruling motives in Shakespeare's breast as to have left no room for political or theological differences. The dithyrambic and almost Jingoist praises of England which he puts in the mouth of John o' Gaunt and other characters are evidently written *con amore*, and express his real sentiments; and so also are the glowing eulogiums on the "imperial votaress throned in the West." Had he lived a generation later, we may conjecture that he would have been a Cavalier, and charged with Rupert rather than with Cromwell; but at the first threat of foreign interference he would have been for England, whether under a king, a Protector, or a Parliament.

Perhaps Shakespeare is right, and after all religion plays a less part in the real life of individuals and of nations, than we are apt to assign to it. It becomes important when it happens to coincide with great currents of feeling or opinion which are setting in the same direction, but it has little effect when it runs counter to them. Thus at the present day, we see that the feeling of nationality is vastly more powerful than any differences of religious denomination. Frenchmen, Italians, and Germans are for national independence and greatness alike, whether they are Catholics, Protestants, or Freethinkers, just as English Catholics were Englishmen first and Catholics afterwards at the time of the Armada. Catholic Ireland bows the Pope's rescript respectfully out of Court when it comes in conflict with National feeling, and follows the lead of an "uncrowned king" who is a Protestant. In private life nothing can be clearer than that the Christian theory is, that it is better to be poor than rich; while the Christian practice is, that it is better to be rich than poor. The example of Lazarus and Dives does not prevent the immense majority of mankind from striving to be better fed, better clothed, better lodged, and more independent; and the precept to "take no thought for the morrow" is nowhere in competition with Burns's ideal of life—

> "To make a happy household clime
> For weans and wife."

An ideal which, under existing conditions, is only to be realized by the constant exercise of providence and foresight. So also nine-tenths of the very men who preach and who repeat the command, "Thou and thy servant shall do no work on the Sabbath," go home to a hot dinner, which compels their cook to do the same work on the seventh as on the other days of the week.

The fact is, that these remote and metaphysical speculations, whether of theology or philosophy, exert wonderfully little influence on practical life. The spiritualist who holds with Berkeley that matter has no real existence, walks on solid earth exactly as does the materialist who believes in nothing but matter. The determinist, who holds that everything is the result of pre-established harmony, or of mechanical necessity, when it comes to practical action differs in no perceptible degree from the believer in free-will, who holds with Tennyson that

> "Man is man, and master of his fate."

In either case the practical incentive is that

> "Because right is right, to follow right
> Were wisdom in the scorn of consequence."

In other words, that the rules of right and wrong, which have become almost instinctive by the operation of heredity, education, and environment, influence conduct far more than any theoretical considerations as to the origin of morals, and practical life is made up mainly of the conflict between these instincts and the lower inducements of selfishness, sensuality, and passion, which tempt us to disregard them.

Of great poets who may be considered to have drawn their inspiration from theology there are two—Dante and Milton. In the case of Dante, however, it is doubtful whether the phantasmagoria of mediæval horrors in the *Inferno* can be considered as anything more than the canvas on which he has painted his immortal pictures. He is a great poet, from the passionate insight with which he has described contemporary events and characters, his knowledge of universal human nature, his vivid power of description, and the occasional gleams of pity and tenderness which lighten up his gloomy landscape. His inspiration is to a great extent political and personal rather than theological. He loves and hates with the intense vehemence of an exile whose life has been marred by the struggles of contending factions, and who has known the misery of eating the bread of charity, and mounting the cold stairs of haughty patrons. He takes the regions of Tartarus, the tortures of the damned, and the malignity of devils, as he finds them ready to his hand in the popular beliefs of his day, and on this canvas dashes down the vivid impressions and brooding ideas of which his soul is full; and that soul being a great one, the picture is great also.

In the case of Milton, on the other hand, we have an instance of a really great poet, who, "smit by the love of sacred song," derived his inspiration mainly from the Bible and from theology. And if theology acted thus powerfully on him, he in return reacted no less powerfully on it, for the conceptions of Adam and Eve, of paradise, of heaven and hell, and of the whole hierarchy of good and bad angels are derived mainly from his *Paradise Lost.* Specially that of Satan transformed from the grotesque, Pan-like devil of popular mythology into an heroic figure, not less than "archangel ruined," is purely Miltonic. The indomitable resolution with which he opposes his own personality and free will to the buffets of adverse fate, and the decrees of Omnipotence, elevates the horned and tailed "auld Clootie" of vulgar tradition into an heroic figure akin to the Prometheus of Greek tragedy. It may easily be seen from the example of Milton, how readily poetry may pass into mythology in uncritical ages. It was thought by some Greek philosophers that the gods of Olympus were a creation of Homer's. Had Milton's *Paradise Lost* been written before the invention of printing, and transmitted for centuries by the chants of itinerant bards, probably the same thing might have been said of many of the personifications of popular Christianity.

In contrasting the spirit of the Greek tragedians with that of modern poetry, it strikes me very forcibly how much more the element of morality enters into the former. The ground-note of Æschylus and Sophocles, and in a less degree of Euripides, is that of an inexorable and irresistible Fate, based mainly on a vindication of immutable moral laws. This all-powerful Fate grinds gods and mortals alike, regardless of individual lives, and of individual pains and sufferings, merits and demerits. The essence of tragedy lies in the heroic struggles of lofty souls to oppose this inexorable Fate, and either vindicate against it the more immediate laws of human justice and mercy, or, if defeated, to suffer and endure with unshaken resolution. Thus the Thyestian banquet entails a curse on the house of Atreus, which is visited from father to son to the third and fourth generation of those whose ancestor had violated one of the fundamental laws of human nature, and been guilty of cannibalism. The avenging Furies pursue Orestes to assert the eternal law against the unnatural crime of matricide, regardless of the extenuating circumstances which might have induced a modern jury to bring in a verdict of justifiable homicide. So also Œdipus undergoes the extreme of human suffering, regardless of the fact that the homicide of his father and marriage with his mother were committed in total ignorance, and without any taint of what may be called personal depravity. Antigone and Electra suffer, not only when they are free from guilt, but when their lives have been devoted to acts of natural piety. They suffer not for their own sins, but because circumstances have involved them in the train of events and family connections, for which the eternal moral laws require expiation. The spirit of modern poetry is very different. It is based less on Fate and more on nature; on nature as it is seen in the outward universe, conceived in the Pantheistic spirit of a living whole, and on nature as shown by the actual course of events and real characters and actions of actual men and women. Virtue is sometimes rewarded and vice punished, but not always; characters are partly good and partly bad, just as we see them in the real world; they do not stalk before us on the stage as heroes or demi-gods, in heroic mask and buskin, but tell their tale and act their parts as ordinary mortals, by the play of words, gesture, and of the human countenance. From Chaucer and Shakespeare downwards, the aim of all first-rate poets, dramatists, and novelists has been, not to preach sermons or illustrate views of "fate, free-will, foreknowledge absolute," but to hold up a mirror to nature and reflect it as it really is. Not partially, as in the modern French realistic school, which photographs only that which is ugly and obscene; nor as in society novels, which find nothing in the world but school-girl romance, and the rose-coloured trivialities of fashionable circles; but, as Shakespeare did in a supreme degree, the whole real world of nature, which lies within the domain of art, that is, which admits of being illuminated by genius into something which in its final impression is beautiful and not ugly, pleasing and not repulsive.

I have reserved for the last Tennyson, for he is the great poet of modern thought, who stands nearest to us, and who writes with the fullest knowledge of the discoveries of recent science, and of the problems which occupy the minds of the living generation. In writing of Tennyson I have to bear in mind that he has lived many days, and gone through many phases of thought, and might therefore probably object to be classed in any one category, or represented as consistently holding in his declining years the views which he expressed in his early youth or mature manhood. It is a long journey from the first *Locksley Hall*, where the poet of progress hails with exulting spirit the "wondrous mother age," and sees in his fellow-men

"Men my brothers, men the workers ever working something new,
What they have done but the earnest of the things that they shall do,"

to the *Locksley Hall, Sixty Years After*, of the mournful bard who, being old, "thinks gray thoughts," and walks from Dan to Beersheba, finding all things barren. It is not for us to complain that the sun is not always at its meridian splendour, but after having given us light and warmth for its appointed season, sinks, not in the softer glories of a glowing sunset, but behind the gray and clammy mists that obscure the horizon.

Let us rather take our great poet at his best and fullest, in the days when he poured out his inmost soul in *In Memoriam*, and gave the world his views on the deepest problems, in lines which dwell for ever in the minds of the foremost thinkers of his generation. No poet of any generation has struck a deeper or truer note than Tennyson in those noble stanzas in *In Memoriam*, in which he says—

"Are God and Nature then at strife,
That Nature lends such evil dreams?
So careful of the type she seems,
So careless of the single life;

"That I, considering everywhere
Her secret meaning in her deeds,
And finding that of fifty seeds
She often brings but one to bear;

"I falter where I firmly trod,
And falling with my weight of cares
Upon the great world's altar-stairs
That lead from darkness up to God;

"I stretch lame hands of faith, and grope,
And gather dust and chaff, and call
To Him I feel is Lord of all,
And faintly trust the larger hope.

"'So careful of the type?' but No!
From scarped cliff and quarried stone
She cries, 'A thousand types are gone:
I care for nothing, all shall go.

"'Thou makest thine appeal to me:
I bring to life, I bring to death:
The spirit doth but mean the breath:
I know no more.'—And He, shall He,

"Man, her last work, who looked so fair,
With splendid purpose in his eyes,
Who rolled the psalm to wintry skies,
And built him fanes of fruitless prayer;

"Who trusted God was love indeed
And Love Creation's final law—
Though Nature, red in tooth and claw
With ravine, shrieked against his creed;

"Who loved, who suffered countless ills,
And battled for the True and Just,
Be blown about the desert dust,
Or sealed within the iron hills?

"No more? a monster then, a dream,
A discord. Dragons of the prime,
Who tare each other in their slime,
Were mellow music matched with him.

"Oh life as futile, then, as frail!
Oh for thy voice to soothe and bless!
What hope of answer or redress?
Behind the Veil, behind the Veil!"

I never read those noble lines without almost a thrill of awe at the intense truthfulness with which they sum up the latest conclusions of the human intellect. Here at last is the *true* truth, based on the inexorable facts and laws of modern science, and on the ineradicable hopes, fears, and aspirations of human nature which underlie them in presence of the "unknowable" Tennyson has read his Darwin, and understands the facts of "Evolution" and the "struggle for existence." He has read his Lyell, and knows how the facts of geology show that what is true of individuals is true of types, and that all creation lives and dies, comes into existence, and is transformed, by immutable laws. He sees this as clearly as Herbert Spencer, but, like Spencer, he sees that this is not all, and that underlying these known or knowable facts and laws is a great unknowable, in presence of which we can only veil our faces and bow in reverent silence.

This much, at any rate, it teaches us—that the apprehensions are visionary which tell us that the progress of science and the light of reason will banish all poetry and all religion from the world, and reduce life to an arid and prosaic desert like that of a burnt-out planet. His science furnishes him with some of the most magnificently poetical similes ever penned by mortal poet. The struggle for existence, and apparent cruelty of Nature, is embodied as the wild eagle, dropping gore from beak and talon, and shrieking with ravine against the creed of love and mercy. The Ichthyosaurus and Plesiosaurus give him the

"Dragons of the prime,
Who tare each other in the slime."

The decay of the old simple paths, the slowly-dying creeds, translate themselves into a deep undertone of the "still, sad music of humanity." Men "falter where they firmly trod," doubt whether their churches and cathedrals are not "fanes of fruitless prayer," and their accepted creeds and solemn services but as the "cry of an infant in the night," and with "no language but a cry."

Tennyson's practical conclusion is very similar to that of Shakespeare and Goethe, viz. to place the centre of gravity of human life in the natural rather than in the supernatural. The advice of his Goddess of Wisdom is to cultivate "self-reverence, self-knowledge, self-control;" and without investigating too closely the origin of conscience, to accept it as a fact,

"And because right is right, to follow right."

In his *Two Voices*, after a deep philosophical disquisition on the Zoroastrian doctrine of polarity, or conflict of two principles, he finds the best solution of the problem in the spectacle of a man walking to the parish church between his wife and child.

This is apparently the last word of religions and philosophics. Work while it is day, for the night cometh when no man can work. Work well and wisely, and when your little day is over go to sleep calmly, accepting with an equal mind whatever fate, if fate there be, that may be in store for you

"BEHIND THE VEIL."

CHAPTER XII.

ARMED EUROPE.

Exhibition in Hyde Park—Predictions of Peace—Era of Great Wars—Increase of Armies—Difficulty of Disarmament—Diplomacy—Crimean War—Franco-Italian and Franco-German Wars—Results—Spirit of Nationality—France the Disturbing Element—England's Foreign Policy—Austria's Danger—Decay of Turkey—Its Inheritance—Possible Solutions—Constantinople—Balkan States—Russia's Policy.

WHAT an irony of fate the history of the latter half of the nineteenth century seems to one who can look back on the opening of the first Great Exhibition in Hyde Park. It seemed as if the beautiful glass fabric which the genius of Sir Joseph Paxton had raised amidst verdant turf and umbrageous elms, were a modern temple of Janus, in which the nations of the earth had met to celebrate the inauguration of an era of perpetual peace.

Nor were such anticipations altogether unreasonable. A quarter of a century had elapsed since the close of the Napoleonic wars in 1815, without a single war between first-rate powers. The revolutionary hurricane of 1830 had swept over Europe, prostrating for a time thrones and dynasties, but no great war had resulted from it. Even the thorny question of the separation of Belgium from Holland had been settled by diplomacy. Everything pointed to the conclusion that both nations and rulers had become wiser, and come to see that war was always a calamity and often a crime.

Where are those flattering visions now? "O cæca mens mortalium." How little is it given even to the most sagacious mortals to foresee the course of evolution, and how infinitely wise is the aphorism, "Never prophesy unless you know."

Instead of closing the temple of Janus, the Exhibition of 1851 seems to have been the signal for throwing its portals wide open, letting slip the dogs of war, and cheering them on with ever louder cries of havoc.

Since that date there have been eight first-class wars in which great powers have been engaged, large armies brought into the field, and battles fought on a scale equal to the greatest recorded in history.

WARS.	NATIONS ENGAGED.
The Crimean ...	Russia, France, England, and Turkey.
The 1st and 2nd Italian	Austria and Italy.
The Hungarian ...	Austria, Russia, and Hungary.
The 3rd Italian ...	France, Austria, and Italy.
The Austria-Prussian	Prussia, Austria, and the minor Germanic States.
The Turco-Russian ...	Russia and Turkey.
The Franco-German ...	France and Germany.
The American ...	Northern and Southern States of the United States.

And in addition a number of second-rate but still considerable wars, such as those of France in Mexico, Rome, and Tonquin; of Prussia and Austria against Denmark; of Russia in Poland and the Caucasus;

of Garibaldi in Italy; and of the United States in Mexico.

Of these minor wars England has had its full share. One indeed, the suppression of the Indian Mutiny, almost assumed the proportions of a great war; and in addition, we have had two Afghan wars, the Egyptian war, two Chinese wars, and at least four or five little wars in South Africa and New Zealand.

Confining our attention, however, to the great European wars, there are several remarkable facts to notice. They originated with the Crimean war, which first broke the long spell of peace, and introduced the element of uncertainty and distrust into the relations of the great military powers. They have gone on upon an increasing scale, the warfare of standing armies having developed into conflicts of armed nations. In talking of the armaments of nations, millions have come to mean what hundreds of thousands did fifty years ago, or even down to the date of Louis Napoleon's campaign in Italy. At Magenta and Solferino not above 100,000 men on each side actually confronted one another on the field of battle; while in the Austria-Prussian war, the two armies engaged in the campaign numbered together more than 500,000; and in the Franco-German war the effective force in the field of one power alone exceeded that number. And the process is still going on. The result of these great wars has not been to establish conditions of settled peace, but rather an armed truce, in which all the nations vie with one another in increasing armaments.

There are, or shortly will be, when the latest military organizations are carried out, not less than fifteen millions of soldiers drilled, disciplined, and to a certain extent taken from civil life in the five great military states alone, viz.—

Russia, in round figures	5,000,000
Germany	3,500,000
France	3,000,000
Austria	2,000,000
Italy...	1,500,000
	15,000,000

And the number still tends to increase, while vast sums are expended in new and improved forts, guns, and military railways. It is not surprising that all the countries whose resources are thus strained are accumulating debts and are increasing taxes, in some cases to an extent which threatens bankruptcy and general impoverishment.

And the worst of it is, that, as matters stand, there seems no issue from this *impasse* of progressive armaments and expenditure. Germany and Italy clearly cannot disarm unless France sets the example. Their recently acquired national unity and independence would be in serious danger, if France got so far ahead of them in military preparation as to be able, either alone or in alliance with Russia, to attack them with a superior force. France, again, cannot disarm without resigning herself to the loss of Alsace and Lorraine, and her chance of regaining her position as the leading state in Europe. Nor can Austria disarm unless Russia does so; and Russia cannot, without resigning all her national and traditional aspirations to be the head of the Greek Christian races whom she has emancipated from the Turkish yoke by a lavish expenditure of Russian blood and treasure, and seeing them and the inheritance of the fast dying Turkish empire, pass into alien and possibly hostile hands.

While this state of things continues, disarmament and permanent peace must remain a pious aspiration rather than a question of practical politics. The utmost that can be hoped is to prolong the precarious truce from year to year by the reluctance of any power to precipitate a conflict of such enormous dimensions and uncertain issue. In the meantime, the electricity is accumulating, and thunder-clouds rising ever blacker and higher above the horizon. Will the tension go on increasing, until some accident makes them explode in the thunder-peals and blood-rain deluges of the greatest war the world has ever seen? or may it be possible, by any diplomatic lightning-conductors, to draw the elements of danger noiselessly to the earth and avert the catastrophe?

This is a case in which it is peculiarly dangerous to prophesy, depending, as it does, on so many incidents and personalities on which no man can calculate. All that can be done is to appeal to past history, and from this "philosophy teaching by experience," endeavour to draw some deductions which may assist us in arriving at some conclusions as to the causes which have led to this enormous development of militarism among civilized nations, and the main conditions which tend to make any return to pacific relations so extremely difficult.

The first conclusion to be drawn is adverse to the chances of diplomacy being able to relieve the existing tension. For diplomacy was really the "fons et origo malorum." The Crimean war, which began the series of great wars, was essentially a diplomatic war. It was not a necessary war, or one arising from the conflict of great national interests, but distinctly a war made for diplomatic or personal objects by three men—the Emperor Nicholas, Louis Napoleon, and Lord Palmerston.

In the case of the Emperor Nicholas, a long reign of absolute power and uninterrupted success, acting on a strong and proud nature, had led to a feeling of arrogance, which made him incapable of yielding an inch in any pretensions which he had once put forward. He had posed too long as the divinely appointed champion of conservatism and protector of the Christian races and of Russian influence in the East, to let Lord Palmerston and Lord Stratford de Redcliffe score a point against him in the trumpery question of the holy places at Jerusalem, even when the manifest interest of Russia was to play a waiting rather than a forward game.

Louis Napoleon was actuated by purely personal motives. His empire, based on a *coup d'état,* and fusillades on the Boulevards, required the *éclat* of a successful war and the prestige of an English alliance to give it permanence and respectability.

Lord Palmerston, again, was actuated by purely diplomatic motives. He was the pupil of Canning, trained in the Foreign Office, and naturally high-spirited and liberal. For years he had been the champion of all liberal movements in the New and Old Worlds, and had everywhere found the Emperor Nicholas his foremost opponent. France under Louis Philippe had deserted him, and, as he thought, played him false, in the matter of the Spanish marriages. He was determined to have his revenge, and alone among English statesmen he hailed the accession of Louis Napoleon as a means of obtaining it. He saw his opportunity in an alliance between England, France, and Turkey to checkmate Nicholas in the East; and, like a true diplomatist, thought more of winning the next move, than of the real interests of the country and the permanent course of events. By his personal popularity, and the popular feeling against Nicholas as the champion of absolutism and destroyer of Polish and Hungarian liberty, he dragged the Court, the Cabinet, and the country with him, and involved us in the French alliance and the Crimean war. He won the game for the moment, but what were the permanent results? He seated a political adventurer in the saddle, who for the next fifteen years kept us and all Europe in hot water. He inaugurated the system of great wars and excessive armaments, and destroyed the feeling of security which had so long been the guarantee of peace. He raised the military prestige of France to the foremost place in Europe, and lowered that of England, for, notwithstanding the valour of our soldiers, their insufficient numbers and the miserable failure of our arrangements for recruiting and transport, made it palpable to the world that we were only playing second fiddle to France. He lowered it indeed to a degree that was to a great extent responsible for the Indian Mutiny, and for our ineffectual attempts to prevent the outbreak of subsequent wars. England, in fact, remained for many years almost a *quantité negligeable* in foreign politics; and Europe, as a witty Frenchman said, for a long time stood in the attitude of a poodle dog watching the eye of its master at the Tuileries.

On the other hand, the ostensible object of the war, the permanent renovation of Turkey as a substantial barrier against Russian encroachment, failed utterly, as it was bound to fail, against the irresistible current of events, which makes for decay of the Turkish Empire, and for the emancipation of the Christian races, who are so much more apt for progress and civilization. The old Foreign Office policy of bolstering up the Turkish rule over these races, and opposing Russia at every point in Europe and Asia, was not only a short-sighted, but what is worse, a cynical and immoral policy. It was a short-sighted policy, because it overlooked the disproportion between means and ends, and made us the catspaw to draw the chestnuts out of the fire for states like Austria, who had a far larger interest in the Eastern question than ourselves. It was a policy sure to fail in the long run, because the idea of regenerating Turkey was purely fallacious. It was a policy which directed our attention from real dangers nearer home from France, to remote, and to a great extent imaginary dangers from Russia in Central Asia. And it was a cynical and immoral policy, for even had it been possible, we had no right to say that Roumanians, Servians, Bulgarians, and Greeks should continue to groan for ever under the desolating rule of Turkish pachas, in order to give England some fancied better security against a remote danger of a Russian attack on India.

If we trace the action of diplomacy farther, we find it responsible not only for the first of the great modern wars, but for several of the succeeding ones. By diplomacy, meaning the personal action of the man or men who controlled foreign policy, as distinguished from great national interests or currents of national sympathy. Thus the Franco-Austrian war in Italy and the Franco-German war were distinctly due to the same cause as had been the principal cause of the

Crimean war—viz. the necessity felt by Louis Napoleon of giving France a sensational policy and military glory, in order to reconcile it to the loss of liberty. In the case of the Italian war, other motives may have conspired; such as the sympathy of Louis Napoleon with Italy from early recollections, and the fear of assassination by conspirators of the Orsini type. But the motives were purely personal. No one could say that, however desirable Italian independence might be in itself, France had any such interest as to justify spending French blood and treasure in promoting it. On the contrary, as the event has shown, the purely selfish interest of France was opposed to the creation of a strong power on her southern frontier, who might not improbably become a rival or an enemy.

But if there may have been some mixture of motives on the part of Louis Napoleon in commencing the Italian war, it remains certain that it was worked up to by diplomatic means, and that diplomacy failed signally in averting it, though every effort was used, and the war was never popular in France itself.

And there can be no question that the last and greatest of the great wars, the Franco-German war, was simply and solely a diplomatic war. The French Emperor had been for some time going down-hill. The startling Prussian victories in the campaign of Sadowa had dimmed the prestige of French military pre-eminence, and it had become apparent to himself and the whole world, that he had been overreached and overmastered by the superior genius of Bismarck. With this decline of his foreign prestige discontent at home had rapidly increased. Gambetta and a host of the best orators and writers of France were daily thundering philippics against his throne, and undermining it by sarcasms.

The Empress, who had acquired considerable ascendancy after the Emperor's surrender to her in order to avert the scandal of her flight to Edinburgh, saw clearly that victory alone could secure the dynasty, and place the crown on the head of her son. She was therefore keen for war, as indeed she had been for the Mexican war from religious motives, and the frivolous *entourage* of the Court followed her example. The carpet-generals, such as Lebouf, Frossard, and De Failly, were also all for war, and full of the Chauvinistic idea of the invincibility of the French army, and the marvels of the mitrailleuse and chassepôt. Louis Napoleon himself hesitated, for although he had grown lazy and lethargic with advancing years, he was still too much of a statesman not to realize the risks he ran in staking everything on the issue of a conflict with an army which had crushed Austria in a seven weeks' campaign. But he had lost his best adviser, the shrewd and cynical De Morny; Marshal Niel was also dead, and he had no military authority of sufficient weight to stem the tide. MacMahon was his best general, a gallant gentleman and good officer, but a man of no large views or force of character. Bazaine was a mere fighting bull-dog, of no more capacity than a common soldier.

Yet with all these unfavourable surroundings, war would hardly have been possible without the aid of the diplomatic machinery, which, in the hands of Grammont and Benedetti, envenomed trifling incidents, and led the Emperor step by step over the brink on the edge of which he was hesitating. If the communications between the courts of Paris and Berlin had been conducted through the Post Office by registered letter, instead of through ambassadors, it would have been impossible to inflame the Parisian populace by the invention of imaginary insults.

One reflection from a review of these great wars is, that although they originated in the purest personal motives of some three or four individuals, they led to far-reaching results, which their authors were as far as possible from contemplating. The Crimean war fixed Louis Napoleon on the throne. Louis Napoleon's position led him into further wars, the net result of which was to weld Germany and Italy into great nations. The principle of nationality was the great undercurrent of the age, on the surface of which Louis Napoleon, Palmerston, Cavour, and even Bismarck himself, were but as straws showing the direction of the movement they seemed to guide. Of Bismarck only can it be said that he foresaw the movement, and to a considerable extent by his personal character and action influenced the course of events. So true is it that there is a "Providence which shapes our ends, rough-hew them as we may." The modern spirit of nationality is, in fact, the ruling factor in European politics. The "Schoolmaster abroad," instead of inaugurating an era of peace, has, in fact, been the principal cause of the modern eras of great wars, and remains to the present day the chief element in the state of unstable equilibrium which necessitates excessive armaments. The press and education have taught all people who have a common race and language, to rush by an irresistible impulse, towards a common nationality. As in industrial enterprise railways tend to amalgamate, stores to supersede shops, and colossal companies to swallow up private undertakings, so in politics, populations who go to school and read books and newspapers, tend to rush together, according to affinities of race, language, literature, and past history, and either form great empires, or, at any rate, assert their independent nationalities. Even the smallest and most remote nationalities feel the impulse, and Greeks, Roumanians, Bulgarians, Servians, Magyars, Croats, and Czecks, agitate for greater independence or wider frontiers, introducing by their agitation an element of risk and instability in all the relations of the Austrian Empire and of Eastern Europe. Still more is the feeling of nationality paramount, where great civilized races, like the Germans and Italians, with a glorious common literature and great historical traditions, refuse to remain longer under foreign rule, or cut up into petty states, in order to give colossal neighbours the pleasure of bullying them with impunity, and insist on taking their natural place among the foremost nations of the world.

Another important fact results from an examination of recent wars. In three of the great wars—the Crimean, the Franco-Austrian, and the Franco-German—France has been the originator and principal party, while of the minor wars—those of Mexico and Tonquin—her aggression was the sole cause. If we follow the course of history farther back, we find this to be no isolated phenomenon, but that for more than two centuries France has been the principal disturber of the peace of Europe; and this in spite of repeated lessons, in opposition to the obvious interests of the French people, and in many cases even to the popular feeling of a majority of the nation, if it could have been fairly consulted. The whole series of wars of Louis XIV., Louis XV., and

Napoleon, were undertaken without any rational object, to gratify the vanity or ambition of rulers trading on the appetite of the French for military glory. In the recent wars of Louis Napoleon this was even more conspicuously the case, for it cannot be said of any one of them that it was necessary for any interest of the French nation, or otherwise than unpopular with the mass of the French people. The Crimean and Italian wars were never popular, though they resulted in victories. The Mexican war was so unpopular that it almost forced the Emperor into the last desperate risk of the war with Germany in order to retrieve his position. The latest war, that of Tonquin, was more than unpopular; it was so odious that it led to the return of a formidable minority of Royalists, and has estranged from power perhaps the ablest man of the Republican party, Jules Ferry, for the sole reason that he was responsible for it.

Such a series of historical events, extending over two centuries, cannot have occurred without great underlying causes in the character and institutions of the French nation, which have enabled individual rulers, and often mere court intriguers and courtesans, to lavish French blood and treasure in such senseless and, in the long run, disastrous enterprises. The causes are not far to seek. Since Cardinal Richelieu crushed the aristocracy and local liberties, France has been a country in which Central Administration was pushed to its extreme limits. The Revolution and the empire of Napoleon carried the levelling process still farther, and tightened the bands of centralization. Whoever gets hold of the War and Foreign Offices, and of the Telegraph, is, for the time being, master of France. Even this, however, would hardly suffice if there were not something in the character of the French nation on which ambitious rulers and aspiring adventurers could rely to give them, at any rate, a temporary support.

The French character remains essentially as it was described by Julius Cæsar—fickle, excitable, and vainglorious. Vanity, or the desire to shine, is the fundamental trait both of the personal and national character. Their emblem is still the Gallic cock.

> "Qui chante bien haut quand il est vainqueur,
> Plus haut encore quand il est vaincu."

I do not say this at all as a matter of reproach. Vanity is a quality which is at the bottom of a great deal that is good. To be amiable, polite, eager to shine and to excel, enthusiastic for ideas, open to novelties, may, within certain limits, be contrasted favourably with the opposite extreme into which we English, and other harder races like the Prussians, are apt to fall, of a surly, arrogant pride, which disdains to please, and looks down on all the world who are outside of their own limited set or nation as inferior mortals. The contrast may be summed up by saying, that France fights for ideas, England for interests.

But admitting that, as an abstract, ethical question, there may be much to be said in favour of the French, as contrasted with the Teutonic character, as a question of practical politics we must take things as we find them, and recognize that these traits of French character, which have been such a fruitful cause of wars in the past, remain so in the present and the future. In the case of other nations, we can, to a great extent, foresee and predict their course, if we understand rightly what are their interests, and their great currents of national aspirations and feelings. They are the planets of the European system revolving in more or less settled orbits by calculable forces; while France is a comet whose course may be retrograde, and which may blaze out suddenly at some unexpected moment. Who can tell whether, five years hence, France will be an Empire, a Monarchy, or a Republic, or whether she will be at peace or war with Germany, Italy, or England? This is a danger for all other States, but especially for England, for it must never be forgotten that France is the only enemy from whom we have anything serious to apprehend. Russia would in all probability let us alone in India if we let her alone in Europe; and if the worst came to the worst, a war with Russia would be, as Bismarck said, one between a whale and an elephant. Russia could not contest with us the empire of the seas, or threaten our coasts with invasion. All she could do would be to excite alarms on our Indian frontier, and put us to the expense of maintaining in India one or perhaps two army corps more than would otherwise be necessary.

But with France it would be a duel *à la mort*. In conceivable contingencies, under the unknown conditions of modern naval warfare, she might either command the Mediterranean and expel us from Egypt, or the Channel, at any rate for a time, and invade us with a superior force and capture London. In any case, she could inflict great injury on our maritime commerce, and transfer a large portion of it to neutral flags. She would certainly aim at one or all of these objects, and if possible at an invasion, setting off a victory on British soil, and the capitulation of London, as an offset against Waterloo and the occupations of Paris. In such a war we could not safely reckon on allies. In the absence of positive engagements, Germany would have no great interest in risking the bones of a Pomeranian Grenadier to defend England. On the contrary, a war between France and England would divert the attention of France from the recovery of her lost provinces. If adverse to France, the result would be to cripple her for a long period; if successful to her, it would lead to a scramble for naval and colonial supremacy, in which Germany might find her account, and in any event would throw England into the hostile camp, and ensure her seeking a German alliance on almost any terms that Bismarck might choose to dictate. The accession of England at once to the triple alliance would be a great security against these dangers, but it is a question of terms. Bismarck would undoubtedly act on his maxim, "Do ut des," and require positive engagements in exchange for those he gave. Lord Salisbury alone is in a position to know what those terms would be, but it is to be apprehended that they would be of such a nature that the British Parliament and public opinion would decline to ratify them. We should certainly be very reluctant to take engagements which obliged us to enter on a second Crimean war to bolster up Turkey, or to risk being drawn into a great war by the conflict of Austrian and Russian influences in the Balkan States. Moreover, while these dangers from France and Russia remain in the background, it is highly important for us to maintain friendly relations with those States as long as possible.

Our wisest course probably will be to avoid entangling alliances, and trust to our own strength; but in this case it is indispensable to put our naval and military defences—and especially our navy—on such a footing as to remove any temptation to make a sudden attack on us, in the hope that it might find us unprepared. But, in the mean time, it remains a primary factor in the European situation, that no general disarmament is possible, unless France sets the example.

This could only be accomplished in one of two ways—either by a great war, in which France was so utterly defeated as to be completely crippled, or by her being so isolated as to see that any attempt was hopeless, and so exhausted by increasing debt and taxation as to make some of the parties who, in the frequent vicissitudes of French politics, may come into power, see that peace was a safer card to stake upon than *la revanche* and military glory.

But this is hardly likely to come about as long as hopes remain of an alliance with Russia to redress the balance of force, and enable French armies to take the field with some reasonable chance of success. This, again, depends very much on the relations between Austria and Russia.

If the natural desire of France to regain her prestige and her lost provinces is one principal element in the European situation, the unstable equilibrium of the Austrian Empire is another. It has been said that "if Austria did not exist, it would be necessary to invent her." This is to a great extent true. Nothing but the tradition of loyalty to the Hapsburg dynasty, and the *esprit de corps* of a powerful army, keep together the heterogeneous elements of which Austria is composed. Half the population are of Slavonic and other alien races, who dislike the German and still more the Magyar elements, which are the dominant races in the dual empire. In the Cis-Leithian, or western half, where the Germans preponderate, it is a question of the nicest statesmanship to reconcile this German preponderance with the rival pretensions of the Czeks of Bohemia and Poles of Gallicia. Concessions to these make the Germans look towards Berlin, and concessions to the Germans make those look towards St. Petersburg. Still the situation is possible, for the colossal power of the German Empire stands behind, and makes it certain that a Slavonic Bohemia would not be tolerated in the heart of Germany. But in the eastern, or Hungarian, half of the empire, the situation is greatly aggravated. The Magyars are the ruling race, who, by their superior statesmanship, valour, and tenacity, have fairly won the foremost place; but they have one fatal defect—they are not sufficiently numerous. They are outnumbered by the Slavonic and Rouman races, alien to them by language, past history, and religion; and who, with the spread of education, and the rising feeling of nationality, resent more and more every day the attempts of the Magyars to consider them as mere appanages of the kingdom of St. Stephen. The great Croatian bishop, Strossmayer, is, as we have seen lately, a political force, who can treat almost on equal terms with popes and emperors. And well he may, for he represents the old Slavonic nation, who form a majority, and in many cases nearly the whole of the population in Croatia, Dalmatia, Carinthia, Southern Hungary, Bosnia, Herzegovina, Montenegro, Servia, and the western half of Macedonia. They are all of the same race, speak the same language, read—or are learning to read—the same books and newspapers, and are drawn together by so many affinities, that if all external pressure were withdrawn, they would almost certainly rush together, and re-form the great Servian kingdom which was shattered by the Turks at the battle of Kossova. And they are all animated by very much the same feeling, not to be Germanized, and above all not to be Magyarized. This is Austria's great difficulty, and, in case of a war, might readily become Russia's opportunity. While this state of things lasts Austria cannot disarm, and an armed Austria implies of necessity an armed Russia.

Is there any possible escape from this fatal circle, which compels all the great Powers not only to maintain, but to increase and improve those gigantic armies which have converted Europe into an armed camp, and passed 15,000,000 of men through the hands of the drill-sergeant? I can see only one possible alternative to that of a great war, which should definitely determine who was the strongest, and to a great extent remodel the map of Europe and the conditions of its equilibrium. It is this. If the "honest broker" at Berlin could negotiate such a compromise as should satisfy Russia without unduly weakening Austria, and by satisfying Russia should isolate France, and thus render a general disarmament possible. Such a compromise would have to be based on a partition of European Turkey.

A century ago the rivalries of Russia, Austria, and Prussia were settled by the partition of Poland. That was felt to be a political crime, for it extinguished the life of an historical nation, which, however turbulent and troublesome, had done signal service to Christendom under Sobieski at the siege of Vienna. But no such moral considerations would apply to the Turks, who have never been anything but a tribe of invading warriors, encamped on the soil of Europe, desolating its fairest provinces, and crushing out the civilization and progress of the conquered races. One has only to compare the present state of Roumania, Bulgaria, Servia, and Greece with what it was while they were governed by Turkish Pachas, to see what an immense boon to civilization it would be if Christian were established for Mahometan rule in the remaining provinces. And it would be the first step towards the establishment of a state of stable equilibrium in the east of Europe, for while the "sick man" is dying by inches, all sorts of interests are watching for his inheritance, each anxious either to secure the lion's share for themselves or to prevent others from appropriating it.

At the same time there are great practical difficulties in the way of a peaceable solution, to appreciate which it is necessary to understand the position of the principal parties interested. These are in the first place Russia, Austria, the new Balkan States, and Greece; and in a lesser degree Germany, England, and Italy. The interest of Germany is almost entirely Austrian. She cannot stand by and see a semi-German empire like that of Austria dismembered, and the formidable power of Russia, backed by Pan-Slavonic aspirations, preponderant over Eastern Europe almost up to the gates of Vienna. And although Constantinople is the back-door of Russia, it is also, to a considerable extent, the back-door of Austria and Southern Germany. The interest of England and Italy is almost exclusively confined to the question of Constantinople and the Dardanelles. It would be dangerous for us and Italy in the Mediterranean, if the Black Sea and Dardanelles were to become a sort of Russian *mare clausum*, inaccessible without her permission to commerce, and from which Russian fleets or privateers could issue as from an impregnable fortress, where they could not be attacked in return.

As regards the minor states—Roumania, Servia, Bulgaria, Greece, and Montenegro—though weak individually, they have all attained to a separate and growing nationality, and together cover too wide an extent of population and territory to be ignored. It is evidently a question of influence and protectorate rather than of annexation in the case both of these countries and of the remaining provinces of Turkey, which in the natural course of events must sooner or later fall to them. Thus Old Servia must gravitate towards Servia, Eastern Macedonia towards Bulgaria, the Macedonian sea-coast, Epirus, Crete, and the islands towards Greece, constituting in each case states large enough to be jealous of their independence, and averse to being annexed as provinces either of Russia or of Austria. But in the long run their leanings must be rather towards Russia than Austria, both from affinities of race and religion, and because the support of Russia is indispensable for them in order to obtain the natural extension of their frontiers and the liberation of their brethren who still remain under the chronic misgovernment of Turkey.

The solution of this problem must lie in the direction of a federation of these Eastern Christian States, and such a neutralization as prevents them from attacking one another, and from being used either as an outpost of Russia to attack Austria, or as an outpost of Austria against Russia to protract the agony of the Turkish Empire, and bar the way against any advance of Russia towards Constantinople. Under such conditions these new states might one and all disarm, and devote their energies to peaceful pursuits, instead of exhausting themselves by keeping up large armies and foreign military princes.

But after all Constantinople remains the chief difficulty. Unless some arrangement can be made respecting it, it must remain a constant source of antagonism between Russia and Austria, and a permanent element of unstable equilibrium in European politics. The prize is too valuable to be appropriated unconditionally by any one of the parties interested, except as the result of a great war which ended in the complete victory of one of the claimants.

To understand this fully we must endeavour to place ourselves impartially in the point of view of the principal parties. For Russia the question of Constantinople is absolutely vital. It is so both from material considerations, holding as it does the key of the back-door of her house, and in hostile hands barring the commerce of the southern half of her empire from its natural outlet, and enabling the enemy's fleets to enter the Black Sea while Russian ships of war are blockaded in it. And it is even more vital from the national and religious feelings of the entire Russian nation. Russia is the one remaining country in which religion still constitutes an important element in politics. The very phrase "Holy Russia" denotes the feeling of the immense majority of the 100,000,000 of its population. Devotion to the Christianity of the Greek Church, and to the Czar as its temporal representative, is the animating principle which makes the Moujik die in the trenches of Sebastopol, or storm the Balkan passes in the depths of winter. Add to this an hereditary hatred of Turks, bred by centuries of contests with them and Tartars.

To these simple, devoted Russians a war with Turkey for the emancipation of Christian races and places, is almost what a war with the infidel for Jerusalem was to the early Crusaders. And Constantinople is their Jerusalem, the cradle of their religion, the head-quarters of the orthodox faith, the afflicted elder sister of their own Moscow. To place the Cross above the Crescent on the dome of St. Sophia would be the dearest wish of every Russian, and the Czar who succeeded in realizing it would be for generations the object of almost divine veneration. The strength of this feeling was shown only the other day, when sympathy with Servians fighting against Turks attracted Russian volunteers of all classes, and finally developed into such an irresistible current of public opinion as swept away the Czar and his statesmen, and involved Russia in the last great war with Turkey.

The fact is that Russian politicians may avail themselves of this feeling for purposes of ambition, or restrain it for a time if the occasion does not seem opportune, but they cannot control it. Whether we like it or not, we must start with the fact that Russia will spend her last rouble and fight her last man rather than allow any other Power to seize Constantinople, or permanently bar the way towards it. Also, that although she may be content to remain passive and wait for a favourable opportunity, and for the approaching dissolution of the Turkish Empire, to strike a blow, she will not disarm, or allow any combination which might permanently debar her from her share of the inheritance, while the Eastern question remains in its present provisional state of unstable equilibrium.

This implies, that as long as the Eastern question

remains unsettled, Russia cannot allow France to be crushed, and thus leave herself without an ally, in presence of Austria backed by Germany. And Russia can afford to wait, for the course of events is tending steadily in her favour. Catholic Austria, with her conflicting nationalities, cannot in the long run compete for the protectorate or annexation of Eastern Christians of the Slav race and Greek Church with orthodox Russia, with her population of 100,000,000 of the same race and religion. Even a successful war would only add to the embarrassments of Austria by introducing a still larger Slavonic element into her empire, and making an equilibrium based on the preponderance of the Magyars still more difficult; while Russia, on the other hand, could keep whatever she got in the way of influence or territory without endangering the unity of her empire.

The result, therefore, is that in the present state of European politics disarmament is almost impossible, and the condition of precarious armed peace and ever-increasing armaments must go on, until some accident fires the match and it explodes in a great war. The only possible escape would be, as already suggested, by a settlement of the Eastern question at the expense of Turkey, in some way which would satisfy Russia without unduly crippling Austria. A federation of the Greek Christian States seems possible, as the first step towards a solution. Is any such solution possible as regards Constantinople? If nothing is done the course of events will probably, sooner or later, and after one or more wars, solve the problem by giving it to Russia.

A pacific settlement of the question of Constantinople would only be possible on the basis of making it, with the Dardanelles, a sort of neutralized and unarmed free city, open at all times to the commerce of the world, but precluded from taking any part in war, or allowing itself to be made a basis for hostile operations. This could be done either by neutralizing the whole of the Black Sea, or by allowing ships of war of all Powers to pass in or out, but not to remain within its limits, or to engage in hostilities within a limited distance of its ingress or egress; making the Dardanelles, in effect, a sort of Suez Canal.

Constantinople itself would have to be made a sort of Metropolitan city of the Greek Church, and its civil government vested in some council in which the interests of the guaranteeing Powers were fairly represented. The hereditary prince or president of such a council would have to be some one acceptable to Russia and professing the Greek religion.

Whether such a solution would be possible it is difficult to say, but the alternative seems to be a continuance of the present precarious state of things, involving constant alarms and the maintenance of excessive armaments, with the probable ultimate result of a still more complete protectorate or annexation by Russia. In fact the difficulties of any peaceful solution are so great that it seems probable that Europe cannot arrive at a state of stable equilibrium, making a general disarmament possible, without passing through the crisis of a great war, to ascertain by the rude test of the survival of the strongest, which conflicting interest has got might on its side, and which being the weaker must go to the wall. Some accident may precipitate such a crisis any day, but it would be rash to prophesy without knowing, and the outcome of the present state of tension must be relegated to the "Problems of the Future."

CHAPTER XIII.

TAXATION AND FINANCE.

New Departure in Finance—Increased Armaments—Foreign Policy and Finance—Russia and France—Policy of England—Home Defence—Army and Navy—Treasury responsible—How Budgets are framed—National Debt—Unpolitic to reduce Debt by under-insuring—Inefficient Administration—Want of Clear Responsibility—Incidence of Taxation—Proportions paid by Property and Labour—Unearned Increment—Income-tax—Succession Duty—Lines of Budget of the Future.

HAVING been practically conversant with financial subjects for the best part of half a century, I am naturally disposed to look at the questions of the day a good deal from the point of view of financial policy. It is clear to me that we are approaching a grave crisis as regards this policy. The necessity of placing the defences of the country in a state in which we can contemplate the enormous armaments of foreign nations and the menacing contingencies of European wars with tolerable security, has become so apparent, that a very large expenditure is inevitable in order to bring up the army and navy to a standard below which they never should have been allowed to fall. This of itself necessitates a departure from the principles on which Chancellors of the Exchequer have been accustomed to frame Budgets, viz., to pare down estimates, pay off National

Debt, and, if possible, reduce taxation: in a word, to make immediate popularity with the House of Commons and the country the primary condition in the art of Budget-making.

It is evident that this is incompatible with the necessity of making large and immediate expenditure on our armaments, and this of itself makes a new departure in finance inevitable.

To make a new departure we must also take into account the growing power of a vastly enlarged public opinion and electorate, which insists on applying rules of common-sense and natural equity to all institutions and all subjects of national policy, and will no longer be contented with authority and tradition. Finance, being a subject which comes home to every one in the unpleasant form of taxation, cannot escape from this influence; and if the country is called upon to incur larger expenditure, it will insist on two things: first, that it gets money's worth for its money; and, secondly, that the requisite taxation is levied fairly as between different classes.

Having thought much on these subjects, I have attempted, in the following article, to define some of the principal points which will have to be considered, and to indicate the lines upon which Budgets, suited to the altered circumstances of the times, will have to be framed. My conclusions may be right or wrong, but at any rate they are not those of a mere amateur, but of one who has in his time prepared two Indian, and assisted in preparing two English Budgets.

It has been said, "Give me a good foreign policy and I will give you good finance." There is much truth in this saying, for our foreign policy is responsible for a large portion of the national expenditure. Without going back to the great wars of the last century, or the struggle against the French Republic and Napoleon, respecting which opinions may differ, and confining ourselves to recent history, we may affirm with confidence that the Crimean, the Abyssinian, and the Afghan wars were diplomatic wars, and that our expenditure in Egypt, the Soudan, and South Africa is to a great extent attributable to a vacillating and unwise foreign and colonial policy. The surest test of the wisdom or unwisdom of a policy is to ask ourselves whether, if we had to do the thing over again, we should do it as it was done, or differently. Assuredly, in the cases above mentioned, we should not do it as it was done; and it is within the mark to say that at least £100,000,000 has been spent without necessity, without result, and with a loss rather than a gain of reputation.

At the same time there is a reverse to the medal, and it may be asserted with equal truth that bad finance often makes bad foreign policy. When I say bad finance I mean bad in the sense of neglecting the cardinal maxim that true economy is based on efficiency, and that a "penny-wise and pound-foolish" policy succeeds no better with a State than with an individual. Extravagance, rather than economy, is the certain result of living in a condition oscillating between periodical panic and periodical parsimony.

If we inquire what has been the cause of this state of things, the answer must be that we have felt ourselves to be unprepared, and being unprepared we have been nervous and afraid. Afraid of what? Practically there are only two Powers from whom any serious danger can be apprehended, Russia and France. The danger from Russia is remote, for she could neither invade our shores nor contest our naval supremacy. It resolves itself into the single apprehension that she might attack our Indian Empire. Now as to this, it is by no means certain that Russia would menace India if England abandoned the policy of bolstering up Turkey and thwarting Russia at every point in Eastern Europe. The Turkish rule in Europe is surely and speedily decaying, and the disposal of the inheritance is very much more the affair of Austria and Germany than of England. Any extension of the Russian Empire in this direction would tend to diminish rather than increase the chances of her undertaking a great war of aggression against India. But suppose the Russophobists are right, and that Russia really does entertain such a project, what is required to make our Indian frontier, humanly speaking, absolutely secure? Simply that we should be able to send there at a short notice 30,000 or 40,000 additional English troops fully equipped and ready for immediate service. With such a reinforcement added to the English and native armies already there, and the command of the frontier passes, no one but an amateur strategist planning campaigns on small-scale maps, can suppose that Russia would undertake such a tremendous enterprise as that of sending an army hundreds of miles from its base, across the rugged mountains and warlike tribes of Afghanistan, to attack us.

But the possibility of sending such a force in case of need to India is a question of English finance, for we cannot throw the cost exclusively on India without provoking widespread discontent, both by the sense of injustice and by the pressure of additional taxation.

The Indian question is, however, only one branch of the much larger question of the naval and military defence of the Empire. To feel secure, we ought to be in a position where we can command the seas and repel invasion from any probable enemy? If it is asked, From what possible or probable enemy? the reply must be —from France. France alone is in a position to menace our shores with invasion, or to contest our naval supremacy. It may be said that the interests of the two countries in preserving peace are so identical, and the consequences of war to both would be so disastrous, that a rupture between them is a remote contingency. So it is, no doubt, as far as England is concerned, but unfortunately the history of France leads to a different conclusion. The wars of Louis XIV. and of Napoleon were wars opposed to the true interests of France, and ended in disaster; but yet, in quite recent times, we have seen France engaged in four wars—the Crimean, the Italian, the Mexican, and the German, of each one of which it may be distinctly said that it was a dynastic war, undertaken for no substantial object affecting the well-being or safety of the French nation, but, on the contrary, involving a certain heavy sacrifice of treasure and blood for no sufficient reason, and with the net result of lowering the place of France in the scale of nations. They were wars undertaken in defiance of common-sense, for the sole purpose of consolidating the throne of a political adventurer.

What has happened once may happen again. Administration is so centralized in France that whoever gets hold of the War and Foreign Offices in Paris can plunge the nation into war almost without its knowing it and against its wish. The temptation to do so for a weak Government is always great, for although the majority of sober and sensible men and of rural electors might be opposed to war, there is always a turbulent and restless minority in Paris, the large towns, and the Press, whose influence is more immediately felt, with whom any measure appealing to the national Chauvinism and promising *la gloire* would for the moment be popular. The strong feeling of patriotism also, which is one of the honourable traits of the French character, would, for a time, induce all parties to lay aside their differences and support the Government of the day when once engaged in war.

There is always a danger, therefore, that under any form of government the man or men at the head of affairs might, if driven to extremities, follow the example of Louis Napoleon, and seek an escape from domestic difficulties by involving the country in war. Nor is there any security that if Germany and her allies seemed too strong to be attacked, England might not be selected as affording a less dangerous antagonist. The interests of France and England are in contact at so many points —in Egypt, Madagascar, Newfoundland, and the Pacific —that collisions frequently arise which are smoothed over with difficulty, as in the case of the New Hebrides, even when both Governments are sincerely desirous of peace, and which would easily furnish pretexts for war if either Government desired it.

The cardinal point, therefore, of English policy ought to be, while doing all that is possible to maintain friendly relations with France, to keep in view the possibility of a renewal of the old historical wars between England and its restless and rival neighbour. To avert such a calamity the same measures are needed as to protect ourselves from serious dangers in case we are attacked. Our naval supremacy should be so assured that there is no temptation to attack us, and our home defences such, that the risk of invasion, in case some of the untried contingencies of modern warfare gave the enemy a temporary command of the Channel, is reduced to a minimum.

As regards the home defences the question resolves itself into a better organization of the reserve forces, fortifying our principal ports and arsenals, and an increase of the regular army. Above all, we want such an organization as would insure us against surprise, and enable every man and gun which appear on paper to find their place at once, and take the field in a state of efficiency in case of any sudden emergency. As regards the regular army, the best military authorities seem to agree that the two army corps, of which we have often heard, in a state of immediate readiness, either for home or foreign service, with proper transport, artillery, and other appliances, are about what would be sufficient to give reasonable security. Of these one is a question not of additional expense, but of Irish policy. Without discussing the merits or demerits of this policy, it is an obvious fact that as long as we maintain a policy hostile to a great majority of the Irish race at home and abroad, we must support it by a force of not less than 30,000 soldiers and 15,000 military police, who, in case of war or apprehension of war, could not be withdrawn, and are for all practical purposes non-existent for the defence of the Empire.

In addition to the two army corps there is no doubt that we require more artillery and better organization for the Reserve, Militia, and Volunteer forces, and stronger fortifications to protect our more important arsenals and seaport towns against sudden attacks. All this costs money, but after all the main question is to insure our naval supremacy. It is evident that this is not the case at present. We may be a little stronger than France if the whole naval force of the two countries could be arrayed against each other in a single engagement; but it is a question whether we could command, at the same time, both the Channel and the Mediterranean. Probably the command of the latter would depend on the side which Italy took in the war, and our safety ought not to depend on foreign alliances, which we shall be likely to gain if we are strong and lose if we are weak. But in any case it is pretty clear that with our present force we could not hope to maintain a permanently efficient blockade of four or five ports at once, and prevent portions of the French fleet and cruisers and privateers from escaping and inflicting immense damage on our commerce, and possibly on our coast towns and colonies. It is the most reckless extravagance to be remitting taxes and paying off National Debt while this state of things continues.

Who is responsible for it? The answer may seem to be paradoxical, but it is nevertheless true: the fault lies mainly with the Treasury.

The Chancellor of the Exchequer is always a powerful, and often the most powerful, member of the Cabinet, and his interests and prepossessions all lie in the direction of cutting down estimates and bringing in popular Budgets. He is surrounded by officials whose business it is to criticize all expenditure that admits of being cut down or postponed. It is a useful and necessary function of Government, and ably discharged by men of great intelligence and experience at the Treasury, whose lives have been devoted to it. It requires a strong man as Chancellor of the Exchequer to emancipate himself from this influence and take a large and statesmanlike view of necessary expenditure. And it takes a still stronger man to escape the temptation to earn for himself the character of a sound financier, and for his Government and party a certain immediate popularity, and to brave the attacks sure to be made upon him by ultra-economists and political opponents, for the sake of the ultimate and probably remote results of a really national statesmanship. It is not a question of party; the same influences affect Conservative as Liberal Governments; and it has been reserved for the party which is nothing if not Imperialist, to furnish some of the most recent and extreme instances of this sacrifice of efficiency to economy, as in the reduction of the Horse Artillery.

There is a mischievous superstition at the Treasury, that the test of a sound financier is to pay off the National Debt. This question of a National Debt affords a good illustration of the axiom for which I often contend, that complicated social problems do not admit of hard-and-fast solutions. Even the primary proposition that a National Debt is an evil, obvious as it seems, is by no means necessarily true. The few remaining countries of the world which have no debts, such as Persia and Morocco, are scarcely countries with which we should wish to exchange conditions. The example of the United States shows that a surplus may be almost a greater embarrassment than a deficit, and more calculated to produce alternations of artificial stringency and plethora in the money market. The fact is that a National Debt has become almost one of the necessities of a progressive and civilized country. As in the case of a railway company, if traffic expands, money must be spent on increased plant and appliances, and if the capital account is rigidly closed, this can only come out of revenue, and increasing prosperity may mean diminishing dividends. The question is, what is the amount of debt compared with the resources of the nation; and how the money is spent, whether unprofitably in useless wars, or wisely in prudent precautions against inevitable risks, and on objects such as education and sanitation, which promote the welfare and ultimately the wealth of the community. For it must be always remembered that the amount of a National Debt is a relative quantity, depending not on absolute figures, but on the ratio which the annual charge bears to the annual income of the country. Thus a debt of £700,000,000 at 3 per cent., of which the capital cannot be called in, is practically a smaller debt than one of £400,000,000 at 6 per cent. The rate of interest payable on a debt is, however, a very important factor in deciding whether it is or is not wise to increase taxation for the purpose of paying it off. Thus in the case of the United States, which affords the principal instance of large repayment of debt by excessive taxation, the repayment has not been effected without great sacrifices. From being the cheapest the United States have become one of the dearest countries in the world, the mercantile marine has been almost annihilated, and protected industries have grown up which threaten serious difficulties. Experience shows that Protection may succeed as well as Free Trade in its earlier stages, while the demand of the home market is more than sufficient to meet the production. But the time comes when the home market is glutted, and manufacturers must look to foreign markets for the sale of part of their commodities. In such markets they cannot compete with the cheaper products of Free Trade countries, and the United States have already approached this stage.

Notwithstanding these drawbacks the policy pursued by the United States was probably a wise one, for this decisive consideration predominated, that at the end of the war their enormous debt carried interest at 6 per cent., while now they can borrow any amount at 3 per cent. Every £1 therefore redeemed by taxation practically paid off £2 of debt.

In the case of England this consideration does not apply. The rate of interest now paid, especially since the recent Conversion, is so low that there is little to hope from further reductions, and the question of repaying debt may be treated on its own merits, and as one of raising £1 by taxes to pay off £1 of debt. There are two ways of reducing debt—one by actual repayment, the other by outgrowing it. Thus, if we take Mr. Giffen's estimate that the national income, which in 1843 was £515,000,000 a year, is now £1,200,000,000, while the annual charge for the National Debt has remained stationary, or rather diminished, we have practically paid off more than half our debt. The total charge may be taken at about £25,000,000 a year for interest, and £5,000,000 for sinking funds in the form of terminable annuities or otherwise. That is to say, taking the nominal capital of the debt at £750,000,000, we were in 1843 in the position of a man who with an income of £500 a year owes £750, or a year and a half's income; and are now in the position of one who, with £1200 a year owes the same £750, or less than three-quarters of a year's income. If the same comparison were carried back to the close of the war in 1815, it would show that the burden of the National Debt is practically four or five times less now than it was then.

In making this comparison it must be remembered also that even if the ratio of debt to income remains the same, a large debt with a correspondingly large income is a much lighter burden than in the converse case of a small debt and small resources. Thus, to take an illustration from private life, a debt of £200 is a very serious affair for a clerk living, perhaps with a wife and family to support, on a salary of £200 a year; while a debt of £20,000 is a mere trifle to a man of £20,000 a year. The latter can pay it off with ease out of revenue, and renew it or repay it by a fresh loan, without the slightest difficulty and at a very moderate rate of interest; while to the former it may mean ruin, or a bill of sale of his effects and usurious interest.

It is clearly, therefore, better for a country to remain with a fixed debt and outgrow it, than to attempt to pay it off by taxes which fetter trade and retard the development of industry and wealth. This was substantially the policy of the great Sir Robert Peel when he imposed the Income-tax, not for the purpose of paying off debt, but to repeal oppressive taxes and inaugurate the system of Free Trade under which the Empire has made such marvellous strides in prosperity that, as Mr. Giffen shows, its aggregate annual income has increased in forty-five years from £515,000,000 to £1,200,000,000 a year. No one can say that the country would have been as well off if Sir Robert Peel had adopted the opposite policy, which a good many amateur financiers and half-informed journalists now call sound finance, and applied the proceeds of his Income-tax as a sinking fund. Even Mr. Gladstone, rigid economist as he is, has practically adopted the same policy as Sir Robert Peel, and his splendid financial reforms have been carried out by applying surpluses to reduce and simplify taxation, instead of appropriating them to large repayments of debt.

In fact, it is sufficient for a Chancellor of the Exchequer to aim at avoiding any permanent increase of debt in times of peace. To insure this, as experience shows that with our extended empire, and the growing wants of an increasing population, the necessity of occasional drafts on capital account cannot be avoided, it is wise to frame estimates on the safe side, and make a moderate provision in the way of sinking funds, so as to have surpluses in ordinary years to apply in counteracting this tendency towards increase. But this is a very different thing from opposing an inflexible *non possumus* to all demands for increased expenditure on capital account, however indispensable they may be for national safety and welfare.

If, for instance, it should be clearly established that an outlay of, say, £50,000,000 in addition to the ordinary estimates is absolutely necessary in order to bring our army and navy up to the standard necessary to give us reasonable security, there should be no hesitation in raising it by a loan. The charge for it would not exceed £1,500,000 a year, or less than one penny in the pound of Income-tax, and the existing sinking funds are ample to secure us against its being a permanent addition to the debt. Surely this is better than remaining with our eyes open, only half insured, risking being involved in great wars menacing our very existence, and in all probability having to do expensively in a panic what might have been done efficiently and economically by prudent and timely preparation.

In view of the necessity for larger expenditure to provide for the security of the Empire, it is important to consider whether the system of taxation by which the revenue is raised is such as commends itself to the intelligence and good sense of the community, and taxes the different classes fairly in proportion to their several interests. The main argument of demagogues is to represent the army and navy as institutions by which poor men are taxed to provide outdoor relief for scions of the aristocracy. This is a gross exaggeration, and on the whole there is no civilized country in which taxation is less unfair and less oppressive than in our own. A country in which the total effective taxation for Imperial purposes does not exceed 5 or 6 per cent. of the national income, and in which the money wages of labour have doubled and their spending power increased in the last forty years, cannot justly be described as groaning under excessive taxation. Still there is a certain substratum of truth in the assertion that the enormous unearned wealth of the country does not pay as much as it ought towards the defence of the Empire and the maintenance of law and order, on which its very existence depends. In order to form any just opinion on this subject it is indispensable to keep clearly in view the fundamental distinction, which has been too much overlooked, between earned and unearned income. The former is a creation of natural, the latter of artificial law. The former commands a market all over the world wherever muscles and brains are in request. The latter depends to a great extent on rights and privileges, secured by laws which differ in different ages and countries, and are in this country exceptionally favourable to the extreme rights of property.

The real difficulty in carrying out such a loan as has been suggested is not so much in the amount required, as in the impression which prevails that there is no security for the money being properly spent, and the feeling that our system of taxation is inequitably assessed. As regards the first point, it is unfortunately only too true that under our present system of administration we cannot depend on getting money's worth for our money. How can it be otherwise when we consider what that system has been and to a great extent still is? A long experience of administration, both in the affairs of the State and of private companies, has taught me this lesson—the great secret both of efficiency and economy is to have a clear chain of responsibility, so that, if anything goes wrong, you can at once put your finger on the man who is accountable for it. Having this, and a clear system of accounts, so as to be able to see at a glance what the results really are, give your officials a free hand and let them feel that they are sure of your support as long as the results come out right. And above all avoid frequent changes, and let there be a reasonable degree of permanence in your policy, so that the heads of departments may know what work they have to do, and how much they will have to spend, with some tolerable assurance of certainty.

Our existing system violates all these rules. Governments change on the average every three or four years, and with every change of Ministry new men come into power at the Admiralty and War Office, who are selected by Parliamentary considerations, and are as a rule totally inexperienced in the work of the departments over which they preside. With new men at the head and changes in many of the principal officials, new views and a new policy are introduced, and a programme is hardly laid down before it is either expanded to meet some momentary panic, or more probably cut down to enable the new Chancellor of the Exchequer to introduce a Budget contrasting favourably with that of his predecessor. The object seems to be not to define responsibility, but to conceal it. The Admiralty, for instance, seems to be constituted with curious ingenuity for making it impossible to fix responsibility on any definite individual. Does a new ironclad refuse to answer her helm or rolls so that she cannot fire her guns in a seaway, who is to blame? Is it the naval constructor?—but perhaps he was overruled by the Sea Lords, or the First Sea Lord by the Board, or the Board by the First Lord, or the First Lord by the Treasury. Very probably the design was sanctioned and the construction begun in Lord Northbrook's time, and the ship was finished and her defects discovered under Lord George Hamilton. And what reasonable man could hold either one First Lord or the other responsible for not being a heaven-born naval architect, and for adopting plans laid before them by presumably competent officials?

So, again, if guns burst, or ships and forts lie idle for want of guns, whose fault is it? Scarcely that of the Admiralty, who do not even manufacture or buy their own guns, but most probably that of the Chancellor of the Exchequer and Cabinet, who refuse to sanction the necessary expenditure. Or, again, in the case of dockyards, who knows exactly what the work costs, and how that cost compares with that of other countries and of private establishments; and who is responsible for detecting and preventing waste and extravagance?

I often think what the result would be if the railway companies managed their affairs on the same principles as the nation applies to its naval and military expenditure. Suppose the Brighton Board were turned out every three years, and a new Board came in with new views, a new policy, and new men at the head of the locomotive, traffic, and other great spending departments, how long would it be before expenses went up and dividends down?

One great advantage of the system which I advocate would be that such a loan would almost of necessity introduce a better system of administration. It would not be sanctioned without a definite and well-considered programme of the purposes to which it was to be applied. So many ironclads; so many cruisers and torpedo-vessels, of such tonnage and speed, and at such estimated cost per annum, until the required number was completed; and so forth for forts, batteries, and other requisites for an efficient army. And this definite expenditure would have to be carried out by individuals, or by small special commissions, which would be selected for their fitness and practical experience in their respective departments, whose tenure of office was independent of Parliamentary changes, and who knew beforehand for five or six years what work they were expected to do and what money they would have to do it with. Of course the general supervision and control would remain of the Cabinet Ministers at the head of the departments, and the ultimate control of Parliament would not be affected. But there would be a practical assurance that so long as the programme was being properly carried out it would not be interfered with; and with a clear system of accounts showing the results year by year, the control of Parliament would really be greater than when matters are so muddled up that it is almost impossible to say what the actual results are, and, if they are unsatisfactory, who is responsible.

The next question is, whether the burden of taxation is equitably assessed on the different classes and interests.

Taking Mr. Giffen's estimate of the national income and its sources, in 1884 the total was £1,200,000,000 a year, of which £400,000,000 was unearned income from capital, and £800,000,000 working income, £180,000,000 of the latter being derived from professional and trading incomes above £150 a year included in the Income-tax, and £620,000,000 from working incomes of lower amount, principally consisting of wages. Measured by income, therefore, the unearned is one-third, and the earned two-thirds of the total amount. But it must be remembered that the unearned third is derived from realized property, and is worth on the average perhaps twenty years' purchase, while the unearned two-thirds is precarious, depending on life, health, employment, and a hundred other contingencies. Without attempting any detailed estimate, it is evident that the value of the unearned property, which requires a higher insurance against risks, far exceeds that of the property which is earned by work, and that it ought to pay its fairly corresponding share of the premium which is required to cover those risks adequately.

Let us see now how the national revenue to provide

for national expenditure is actually raised. Taking the average expenditure of the last three or four years in round figures, it is about £90,000,000 a year, of which

£30,000,000	is for National Debt interest and sinking fund,
30,000,000	for naval and military defence,
20,000,000	for civil administration,
10,000,000	for expenses of collection of revenue.
£90,000,000	

This is met by

Post Office, telegraphs, &c., which are mainly payments for services rendered	£10,000,000
Crown lands and interest on advances, &c., which are not taxes	1,500,000
Miscellaneous, which are mainly matters of account, and fees for services rendered	3,500,000
Revenue which is not taxation	£15,000,000

Leaving in round figures £75,000,000, which is raised by taxes as follows, viz.—

Customs	£21,000,000
Excise	27,000,000
Stamps and taxes, including probate and succession duties	15,000,000
Income-tax	12,000,000
	£75,000,000

Continuing the analysis more closely we find—

TAXES MAINLY PAID BY THE NON-PROPERTIED CLASSES.

Alcohol—	Home spirits	£14,000,000	
	Foreign spirits	4,500,000	
	Beer	8,500,000	
	Licences	3,500,000	
			£30,500,000
Tobacco			9,000,000
Tea and coffee			5,000,000
Total			£44,000,000

TAXES PAID EXCLUSIVELY OR PRINCIPALLY BY THE PROPERTIED CLASSES.

Income-tax	£12,000,000	(but of this nearly one-half according to Giffen's estimate, is paid by trading, professional, and other working incomes).
Probate and succession duties	£8,000,000	
Deeds	2,000,000	
Assessed taxes... ...	3,000,000	
Wine	2,000,000	
	£27,000,000	

Leaving about £3,500,000, which is raised mainly by taxes affecting trade, such as bills of exchange, receipt stamps, railways, marine insurances, &c.

As far as can be ascertained by the aid of Mr. Giffen's figures, the amount paid specially by unearned income does not exceed £15,000,000 to £20,000,000 a year out of a total Imperial taxation of £75,000,000.

The mere statement of the figures is sufficient to show that this is not a sufficient proportion. Without proposing any Radical or Socialistic change in our fiscal system, it is evident that such a tax as that on tea ought not to be maintained to enable unearned income to escape from paying a larger share of taxation. The tea duty combines almost every conceivable disadvantage. It discourages temperance, restricts the development of an important industry in our colonies, and presses with special severity on the unrepresented and weaker female half of the population, whose interests we are bound to consider. The first step towards a really national Budget of the future ought to be to repeal this tax, and make up the deficiency by equalizing and increasing the duties on all property alike, real or personal, which passes by gift or succession, and is therefore clearly unearned. The additional cost of providing for an efficient navy and army, including the interest and sinking fund of any loan raised for the purpose, ought also to fall mainly on this class, though a portion of it might properly be provided by a temporary reduction of the large amount of sinking fund applied to the redemption of debt.

As regards the manner in which taxation should reach this class of unearned incomes there are two ways possible: one to reform the Income-tax on the broad, simple principle of observing a distinction between earned and unearned income, and making the latter pay at a higher rate; the other, that of making a large addition to the succession duties, especially on all property which did not go to make a moderate provision for widows and children. Or perhaps both plans might be adopted, though I incline to think that the greater part of any increased taxation on unearned property should take the form of a heavier duty when it passes and repasses for the first time into the hands of those who have done nothing to earn it. A higher rate of Income-tax on unearned than on precarious income would be fairer in principle, and would remove much of the discontent with the tax which makes Chancellors of the Exchequer court popularity by reducing it, and it would be very desirable to introduce it.

On the other hand, a heavy succession duty is paid once for all in a lifetime, and those who come into land or money by the fortunate accident of having been born, have no reason to complain if their windfall turns out to be somewhat less than it would have been if they could have kept the whole and transferred the burden to their less fortunate brethren who have nothing but what they have worked for.

It is, however, in regard to local taxation that the distinction between earned and unearned income is of most importance. Let me give a practical instance of what is meant by the "unearned increment."

There is a mountain valley in Wales the value of which for agricultural purposes might be at the outside £800 a year. But coal and iron were discovered in it; a set of capitalists took a lease, sunk pits, and erected works, and a town sprang up. The first and second set of capitalists lost their money; and about £1,000,000 was sunk in the concern, which ultimately passed into the hands of a third set for about £200,000, and with this reduced capital is now a fairly flourishing company. But all the time wages were paid, and the population increased until it numbered over 8000.

As regards the landlord the result was this: that his £800 was converted into £8000 a year, which has been punctually paid through good times and bad, and represents a capitalized value of probably £160,000. This is as purely a stroke of luck as if he had won the amount at Monte Carlo or by backing a Derby winner; indeed, more so, for in that case he must have stood to lose as well as to win, while in this actual instance he risked nothing. Again, he would not have received this windfall if the law of England had been like that of many other countries, in which minerals below the soil belong to the State or the Commune. Surely in such a case as this the unearned increment ought to contribute largely towards the local rates for providing sewers, water supply, schools, and other requisites of civilized existence in the town to which the owner of the soil was indebted for this enormous increase of his wealth.

The same thing applies with equal force to the immense unearned increment which has accrued to the fortunate owners of the soil from the growth of industry and population in large towns. It ought to contribute largely towards local rates, and be held under conditions not fixed solely by the landlord's right to make the most he can of his own, but by a due regard for the welfare of the community by which the additional value of the property has been created.

To sum up: if, to use a bold figure of speech, I were Chancellor of the Exchequer, I should look forward to framing a "Budget of the future" on something like the following lines:

1. To equalize the succession duties on real and personal property, and raise the amount to a sufficient figure to enable me to repeal the tax on tea.
2. To reform the Income-tax on the principle of charging a higher rate on unearned than on earned income.
3. To assign the "unearned increment" in towns and from mines and royalties to Local Boards, as a subject for local taxation within equitable limits, in aid of rates for local purposes.
4. To raise by loan a sufficient sum (say £50,000,000) to be spent over five or six years in placing the army and navy, but especially the navy, on a footing which, according to a programme prepared by practical authorities, would be sufficient to place the defences of the Empire on a reasonably secure footing.
5. To intrust the carrying out of this programme, under the supervision of Government and of

Parliament, to permanent Commissions of the best practical men in each department, with large powers and clearly-defined responsibilities.

6. To provide for the interest and sinking fund of this loan by appropriating to it the saving from the recent Conversion of National Debt and a slight reduction of the sinking fund now appropriated towards paying off its capital.

CHAPTER XIV.

POPULATION AND FOOD.

Malthusian Theory—Seems Self-evident—But is Contradicted by Experience—England—United States—Canada—Reserves of Wheat-growing Land—Increase of Urban and Industrial Population—Emigration—Working of Malthus' Law—Prospect of Increasing Supply of Food in Old Countries—Checks on Population—Wars—Pestilence—Famine—Example of Ireland—England Safe for the Present—Free Trade and Competition—Cannot go on Indefinitely—Prospects for Future Generations—It is a "Problem of the Future."

THE Malthusian theory that population tends to increase faster than food is one which, at first sight, seems to commend itself by the mere statement. The particular ratio of increase may not be exactly that of geometrical to arithmetical progression, but the general fact appears to be incontestible that a single pair, whether of the human or of any other animal race, would in a comparatively short time increase and multiply beyond any conceivable increase in the supply of food available for their support from a limited area. It is, in fact, only a particular instance of that struggle for life, which Darwin has shown to be going on throughout all branches of creation, and which ends in the weaker going to the wall, and the survival of the fittest. It is illustrated clearly in the animal world, as by the swarms of rabbits which, in a few years, have overrun the pastures of Australia and New Zealand, from the progeny of single pairs.

And yet when we come to test the theory by facts, nothing can be more evident than that in the recent history of the civilized nations of Europe and America, the direct contrary has taken place, and food has increased faster than population. Take the instance of England. The population of Great Britain has increased in less than a century from fifteen to over thirty millions, and yet it is clearly demonstrated by statistics that each one of the thirty millions gets a far larger average share of food and other commodities than fell to the lot of the smaller number.

Bread, the staff of life, has fallen with the price of wheat to a far lower level than it stood at when the population was half the present amount, and what is even more important, instead of fluctuating widely from year to year, the price remains nearly uniform at this low level. The quantity of wheat and flour imported from foreign countries has risen in less than 50 years from 42 lbs. per head of the smaller, to 220 lbs. per head of the larger population; that of bacon and hams from almost nothing to 14 lbs. per head; of cheese, from 1 to 6 lbs.; of eggs, from 4,000,000 to 22,000,000; and of other articles of consumption, such as tea, sugar, butter, and rice, in proportion. Butcher's meat alone has slightly risen in price, and this is being reduced by the importation of frozen carcases from the United States, Canada, the Argentine Republic, Australia and New Zealand.

At the same time, while prices have greatly fallen, the purchasing power of the community has been largely augmented. The average money wages of the labouring classes have nearly doubled, deposits in savings banks have increased from £16,000,000 to over £80,000,000, and each 1*d.* in the pound of income-tax produces £2,000,000 instead of £1,000,000.

In America, the refutation of the Malthusian theory has been even more signal. The population of the United States has increased in little more than a century from six to sixty millions, fully realizing the rate of increase by geometrical progression assumed by Malthus. And yet the production of food has increased so much more rapidly, that not only are the sixty millions better fed than any other nation in the world, but a surplus remains for exportation, which feeds probably not less than fifteen or twenty millions of mouths in foreign countries.

How is a fact to be explained, which stands in such flat contradiction to what seems at first sight an almost self-evident theory? The answer is obvious. The increased command over the powers of nature given by the practical applications of modern science, has not only increased the productiveness of limited areas, but what is more important, has by means of railways, steamers, and telegraphs, enormously extended the area from which supplies can be drawn. Wheat grown and cattle reared one thousand miles west of Chicago, reach Liverpool and London as cheaply as they used to do from an English or Scotch county.

India, Australia, New Zealand, California, and the Argentine States pour their surplus food products into the markets of Europe. The compound marine engine cheapens freights, and lower freights bring with them lower rents, agricultural depression, and a serious aggravation of the Irish question. At the same time the same agencies triple and quadruple the power of producing commodities wherewith to buy food, by the consuming nations which no longer grow enough on their own soil to feed their population.

As long as this goes on, progress continues; a larger number of human souls live in the world, and the vast majority of them live better. We can afford to dismiss Malthus and his theory to a remote future, and look on it as a bogey no more affecting practical action than the prospect of the world coming to an end by the dissipation in space of solar heat.

But is it really so? The Irish famine is there to teach us a sharp lesson, that under given circumstances three millions out of eight of a population may disappear from the effects of famine and pestilence brought about by overcrowding. True the circumstances were exceptional, and traceable to a considerable extent to bad laws and bad government, but when we come to look closer into the matter we shall find that this inexorable law of Malthus is not reversed or repealed, but merely suspended, and hangs like the sword of Damocles by a thread over the head of future generations.

Behind the steam-ploughs and reaping-machines, behind the railways and steamers, lies the fundamental fact that there must be a reserve of unoccupied land on which to employ them.

Suppose all North America west of the Great Lakes and the Mississippi had been an arid desert like the Sahara, where would have been the food-products on which so many millions in the Old and New Worlds depend for their daily bread? No competition of railways, no improvement of steamers, could have brought wheat, flour, beef, and pork from regions where they were not produced. Nor could they be exported in continually increasing quantities from countries where surplus land was getting scarce, and the native population was already beginning to press closely on the means of subsistence. In a comparatively few years cultivation will have spread up to the base of the Rocky Mountains, and there will be an urban population of ten or fifteen millions to feed in Chicago, St. Louis, and other cities of the West; while New York, Boston, Philadelphia, and the great manufacturing and mining eastern and middle States will constantly require larger supplies.

It is stated in a recent article in the *Century Magazine*, that the total arable and pasture land in the United States is estimated at 960,000,000 acres, of which 700,000,000 has been already taken up, leaving only 260,000,000 acres, which will certainly be all appropriated in a few years, while the population, at the present rate of increase, will be 120,000,000 by the year 1920. The United States, therefore, will in a very few years be brought face to face with the difficult problem, "a rapidly increasing population, and all the arable land in the hands of private owners."

When we come to survey the extent of the remaining reserve of land on which the fabric of progressive civilized society so mainly depends, it is startling to find how little of it is left. By far the greater portion of the earth's surface is excluded, either by climate or by prior occupation. In the Old World scarcely anything is left. Tropical regions are, for obvious reasons, unavailable, either as fields for emigration or for producing a supply of the staple foods required for the support of the principal white races. The highlands of Central Africa might possibly support a white population, but they are already occupied by native races. So also is South

Africa, except to a limited extent at its southern extremity. Central and Eastern Asia are either desert and mountain, or occupied by the already swarming millions of India and China. The vast territory of the Russian empire is wanted for the rapidly increasing population of Russians, which, in Russia in Europe alone, has risen in less than a century from thirty-five to eighty-eight millions. The climate of Siberia is too rigorous, the distance by land too great, and the Arctic Ocean too inaccessible for it to become a great grain-exporting country. Western Asia, formerly the seat of a dense population, great cities, and active commerce, remains a comparative desert under Turkish rule. But even here the difficulty of prior occupation exists, and although the Governments might be got rid of, it would not be so easy to dispose of the twenty or thirty millions of Turks, Arabs, and Persians who already occupy, however sparsely, the regions which, down to the fall of the Roman Empire, supported such a vast population.

In Europe it is obvious that all the principal States are already overcrowded, in the sense of having no reserves of land, and a larger population than the soil can supply with food. It is only on the Lower Danube, and in some parts of European Turkey, that some reserves still remain, and these are to an extent quite inappreciable as compared with the wants of Western Europe, and not more than will be filled up in a generation or two by the Bulgarian and other native Christian races.

America and Australia remain; but here it must be observed, that for providing the surplus population of Europe with food, only those districts are available which produce what may be called the staff of life. Practically this means wheat-growing districts. Thus

Brazil may produce coffee and sugar, Florida oranges, and Southern California grapes and peaches; but valuable as these are as luxuries, and as articles of commerce, people cannot live on them; and to go on as we are going, we want cheap and ever-increasing supplies of bread and meat. Even the great ranges of pasture land which support vast herds and flocks in America and Australia supply a very small per centage of food per acre, compared with the arable lands which grow the cereals and fatten pigs and cattle.

These may be defined generally as the wheat-growing belt. When this is exhausted, no increase of tropical products, and no extension of commerce and manufactures can arrest the inevitable result of an increasing scarcity of food to meet the wants of an increasing population.

Now of this the area is limited and is rapidly being filled up. The largest supply has hitherto come from the states east of the Mississippi, Ohio, Illinois, Indiana, and Michigan. But these, which used to be the western, have now become Central States, and the mass of food-products, of which Chicago is the centre, comes from new Western States, such as Iowa, Minnesota, Wisconsin, and Kansas.

The agricultural portion of the United States is every year advancing further west, through new states and territories, towards the base of the Rocky Mountains, and has overleapt these, and brought California, Oregon, and Washington territory into the position occupied by the older states not twenty years ago of food-exporting districts.

The centre of gravity, as it has been called, of the population of the United States, which a century ago was almost on the Atlantic, is now west of Cincinnati, and is moving uniformly westwards at the average rate of about five miles per annum; while the advanced guard of cultivation is moving still more rapidly towards the Rocky Mountains on a broad frontage from Texas to Dakota; while the Pacific states, California and Oregon, are filling up with even greater rapidity. As we have already seen, the United States will in a very few years be brought face to face with Malthus's theory of a population growing by geometrical progression to an amount which no longer leaves any unoccupied land available for the production of surplus food.

Fortunately a very large reserve of land remains in the north-western districts of Canada, for experience has shown that, owing to the bending of the isothermal lines to the south, an immense extent of territory, reaching almost to the Polar Sea, which was recently thought to be as barren as the tundras of Siberia, is in reality capable of producing fine crops of wheat. The report of the Canadian Senate Committee of 1888 estimates the area adapted for the cultivation of wheat in this territory at 202,240,000 acres, and that adapted for pasture 512,000,000 acres, making a total reserve equal to that of the whole original territory of the United States, and promising a long respite before the inexorable pinch of Malthus's law is fully felt.

But the growth of an urban and manufacturing population is increasing with such rapidity in the New World, that the home market will soon absorb the greater part of the home produce. Chicago does not add 100,000 to its population every ten years without consuming more of the bread and meat which would be otherwise exported; and the same may be said of St.

Louis, Pittsburg, Cincinnati, Cleveland, and the numerous large cities and industrial centres which are everywhere springing up in the States, which have been reclaimed from the Indian and the buffalo. And in Canada itself the same process is going on, though not so rapidly. Say that the United States will, in the next fifty years, have increased its population from 60,000,000 to 120,000,000, how much surplus food will remain over for exportation to Europe?

The tendency of population to accumulate in towns, and the increasing proportion of industrial to agricultural pursuits which are such marked features in England, are already producing similar effects in America. A century ago less than four per cent. of the total population of the United States lived in towns, the rest living in the country, and being mainly agricultural. To-day about twenty-five per cent. of the population of the United States is urban, and of the remainder a large and increasing number live by industrial pursuits other than agriculture. In all the older States, such as Pennsylvania, New York, and New England, the number of food-consumers far exceeds that of food-producers, and a large proportion of the native population migrates westwards every year in search of land on which to settle. Even Central States, like Ohio, are becoming too densely settled for an agricultural population, and sending out contingents to swell the flood of westward emigration.

Europe also continues to pour in an enormous flood of emigration. During the last fifty years upwards of 10,000,000 of European emigrants have landed in the United States, of whom about 3,500,000 have come from Germany, and an equal number from Ireland. Many of these have settled on land, or become agricultural labourers, while others have taken the places of native-born Americans who have become food-producers. Thus the existence of this vast field for emigration has benefited the old countries, both by affording an outlet for their surplus population, and by increasing the production of the world's surplus food. But this, again, depends on the existence of surplus land, and the operation of such powerful causes tends every day to use it up.

Already the approaching scarcity of land is showing itself by a great rise in the market value of real estate throughout the United States. It is not too much to say, that the price per acre of cultivated land, or land fit for cultivation from soil, climate, and proximity to any one of the four or five great railways which now span the continent, has risen on the average thirty or forty per cent. in the last three or four years, and in California the rise has been even greater. Railways are to a great extent responsible for this result; but while they tend, in the first instance, to increase largely the area of emigration and production, they accelerate the process by which reserves are used up, and the progress of population overtakes that of surplus food. Assuming with Malthus that the ratio between the two is that of geometrical to arithmetical progression, it is certain that, although with a large common difference, the latter may at first outstrip the former, it will soon be left far behind. Thus if we take the series

Population . . .	2,	4,	8,	16,	32,	64,	128
Food . . .	2,	12,	22,	32,	42,	52,	62

it is evident that while for the first five terms of the series food keeps ahead, and the condition of the population improves, after the fifth term the proportion between them is reversed, and very soon becomes one in which existence would be impossible without some very severe and far-reaching checks on the natural rate of increase of births over deaths. It is probable that we are not very far removed now from the third or fourth stage of this progression, and the next generation or the one after will have to face very seriously the question of what checks nature has provided, and what measures it will be necessary to take to prevent, or mitigate as far as possible, the inevitable results of the struggle for existence.

In the first place, however, it is necessary to consider what prospect there may be of increasing the supply of food produced in the older countries. I am afraid it is very little. England might conceivably support a larger agricultural population if it were cut up into small holdings of five or ten acres each. But manifestly this could only be done by lowering the general average scale of living, and descending from wheat to potatoes. To support a family by farming in decency and comfort, and have a surplus produce to sell, it is essential, under our conditions of soil and climate, that farms should be large enough to admit of cultivation by the plough and a rotation of crops. This means that there must be at least five or six fields of five or six acres each—two in grain, one in green crops, one in sown grass or clover for hay, and two in permanent or second year grass or fallow. Thirty or forty acres is therefore the minimum size of farms on which an agricultural population can live up to the standard of well-paid labourers and artisans, unless in a few exceptional cases of market gardens and holdings near large towns; and any further subdivision on an extensive scale would only land us in the state of Ireland.

Cottage allotments are often excellent things as a supplement to labour, but as the sole support of a large population they can only lead to one result, that of semi-starvation on half rations of potatoes. Moreover, the question is not one of food only, but of surplus food. If four or five millions more could live on the soil of England if cut up into small holdings by consuming their own produce, what would become of the remaining millions who are not agriculturists, and half of whom are now fed by the surplus produce of British agriculture? Large farms may not produce so much in the aggregate as the same area would do in small holdings, though this is doubtful, but it is beyond doubt that they produce more surplus for sale, after feeding those who are actually employed. And there is no doubt also that, as in Ireland, a population living poorly on small holdings tends to increase more rapidly than the normal rate under more favourable conditions.

It is a formidable question also, how long we can depend on the outlet for a surplus population which is afforded by emigration. Already the countries which have given a hospitable reception to so many millions of the poorer class of emigrants are beginning to show an unwillingness to receive an unlimited amount of cheap labour. The United States prohibit the importation of Chinese, and are becoming more particular every day as to the admission of destitute European emigrants. The Australian colonies are ceasing to tax themselves in order to assist emigration, and Canada and New Zealand are almost the only colonies left where a farther influx of emigrants seems to be desired. Even here there is no opening for the pauperized classes whom, in our own interest, we should be most anxious to get rid of.

Emigration will doubtless go on, for, as we see in the case of Ireland, with many millions of Irish already settled in new countries, and the passage reduced to a question of ten days in time and £5 in money, there is an irresistible tendency impelling the Irish of old Ireland to follow in the footsteps of their friends and relations. Labour, like water, seeks to find its level, and nothing but invincible barriers of ignorance and repressive legislation can prevent men going from a country where wages are a shilling to one where they are a dollar a day. But there is danger that by this process the old countries may be gradually drained of the most able-bodied, intelligent, and energetic portion of their population, and left with more and more of an unmanageable residuum. We must recollect also that the rapid rate of increase which has tripled the population of England during the present century has gone on concurrently with this tide of emigration, and unless it were to flow with increased rapidity, each succeeding generation would find us with an ever-increasing surplus of mouths to feed, unless either the death-rate or the birth-rate were materially altered. And the same thing applies not to England only, but to every European country except France. Russia is rapidly filling up her immense empire; Germany, Italy, and Belgium are full to overflowing, and send out swarms of emigrants; Spain sends a surplus to Buenos Ayres; Portugal to Brazil.

We must look, therefore, to external checks to maintain the balance between food and population in the not far distant time when the world's reserves of arable land are approaching towards exhaustion.

Of such checks the general remark may be made,

that in modern history they all tend to operate with diminishing force, so that the natural increase of population is continually accelerated. What were the checks which, in retracing the history of the human race, we find to have been principally operative? Infanticide, war, pestilence, and famine. Infanticide has long since died out, except among a few savage tribes, though it can be traced as once an important operating cause in the traditions of polyandry and descent through the female line, which point to a deficiency in the female population only to be accounted for by female infanticide. But we can no more look to it as a possible check in the future than we can to a reversion to the stone implements of our palæolithic ancestors. War has been in all ages a principal, and is still an important check. But apart from the outcome of the growing feeling that war is for the most part a mistake, the conditions of modern warfare have so greatly changed that even great wars no longer play the important part they once did in checking population. In the first place they are much shorter. A thirty years' war devastating all Central Europe, and throwing its civilization back for a couple of generations, is no longer possible. Invasions of barbarians like those of Goths, Huns, and Turks, which reduced populous provinces to deserts, are no longer to be feared. Contrast the invasion of Attila which rolled westward to Chalons, over the plains of Champagne, with the advance of the German army only the other day over the same line of march. Burning towns and villages, slaughtered heaps of their inhabitants, droves of captive women and children, marked the line of the Hunnish advance; while in the Franco-German war we read of the peasant girls of Champagne standing unalarmed at their cottage doors to gaze on the Crown Prince and his brilliant staff.

Even the gigantic wars of the first Napoleon produced no very perceptible or permanent effect on the population of Europe. A certain number of able-bodied men were swept away, but their removal left room for others, the rising generation married a little earlier, and the population grew up almost as rapidly as the grass over the blood-stained fields of Borodino and Waterloo. It is a remarkable fact that the rate of increase of the population of England was highest in the first twenty years of the present century, during fifteen years of which we were engaged in a gigantic war with Napoleon. The vast standing armies of recent years are fed mainly by recruits taken at an early age from the population, and restored to it, invigorated in mind and body, after a short service of three or four years. Germany, with its three millions of soldiers, may show signs of financial pressure, but none hitherto of declining population.

Pestilence in times past has played a great part in keeping down excess of population. The black death in the reign of Edward III. is reported to have swept away nearly a third of the population of England, and to this day large parish churches, often standing within a stone's throw of one another, in Norfolk, testify to the existence of a dense population where now there are only a few large farmers and agricultural labourers. The sweating sickness, plague, and small-pox also counted their victims by millions, and defective sanitary arrangements kept the death-rate high almost down to the present day. But science and sanitation, aided by better food, clothing, and lodging, have of late years rapidly brought down and are still bringing down the death-rate, and even since the tables of the principal Life Assurance Offices were framed, the average duration of life has been lengthened by from five to ten per cent.

Famine remains, and in some of the Eastern countries of old civilization and dense population it is still the main check by which Nature asserts the inexorable law of the struggle for existence. But in European countries generally, the establishment of settled order, the accumulation of wealth, and above all the improvement of communications, have for a long time past prevented scarcity from degenerating into famine. England especially, as long as present conditions continue, and there is surplus food left anywhere in the world, is not likely to see famine an effective operating cause in checking the advance of population.

In one instance, however, at our own doors and in our own days, we have seen that Nature, "red in tooth and claw," asserts its inevitable laws even by this extreme and cruel remedy. The redundant population of Ireland has been reduced from eight to five millions by famine, and its results, fever and forced emigration. There has been no such destruction of life and arrest of the progress of population since the black death. The causes no doubt were exceptional, and we can place our fingers on them. Long years of oppression, bad legislation, and a vicious land system led to the multiplication of a pauper population reduced to the very lowest subsistence on the precarious potato, with the life-blood of the country, which should have accumulated capital in the form of the multiplied little savings of individual cultivators, drained from them by alien and absentee landlords; and all hope, providence, and energy crushed out of them by the knowledge that they were liable to be rack-rented on their own improvements.

This important lesson may be learned from the experience of Ireland, that when a population is brought down, by unfavourable circumstances, to a very low standard of food and comfort, the immediate tendency is not to retard but to accelerate the rate of increase. Where there is nothing to look forward to from providence, prudential restraints on early marriages cease to operate. In fact, as far as they operate at all they operate the other way, for the only chance for a man plunged in hopeless poverty is to have grown-up sons and daughters able to help him when he is getting old and past work. Half the population of Ireland who are tenants of small holdings, have for years past only been able to live and pay rents so as to keep a roof over their heads, by the aid of remittances from members of the family who had emigrated to America and Australia. Thus, as the condition of a people deteriorates, and food and employment become scarce, the pent-up fires accumulate more rapidly until nature relieves itself by some great explosion.

This brings us to the practical consideration of what is likely to happen in the future. As long as present conditions continue, and the reserve of food-producing land remains unexhausted, it is probable that England will progress and prosper. The condition of Ireland will improve by better legislation, and England and Scotland have such resources in their mineral wealth, their facilities of communication, their accumulated capital, and in the character and industrial aptitudes of their people, that as long as there is any surplus food in the world they will get the lion's share of it. The rate of progress is not even likely to slacken until we approach more nearly to the exhaustion of the world's reserves. More poverty there may be, for if five per cent. be a fair average of failures in the struggle for existence, owing to weakness of mind or body, unfavourable surroundings, and ill luck, five per cent. on forty millions is a larger figure than five per cent. was on twenty millions. But I see no reason to doubt that for many years to come the mass of the population will eat as good or better food, be paid as high or higher wages, work as short or shorter hours, deposit as much or more money in Savings Banks and Provident Societies, as they do at present. And although it is never safe to prophesy unless you know, it is not a very hazardous prediction that each 1*d.* in the pound of income-tax will give future Chancellors of the Exchequer a larger rather than a smaller contribution to the national revenue.

Foreign competition does not much alarm me, for the inevitable tendency of manufacturing and mining labour in France, Germany, and Belgium, must be to level up towards our standard, or else to explode in strikes and socialist revolutions, to which they are all much nearer than we are in this country. If we are behind any other nation, as for instance Germany, in technical education, we can and will apply a remedy, and with equal brains and more money we are not likely to be long outstripped in anything which intelligence and capital can cure.

The only really formidable competition I can imagine in the near future would be from the United States of America, if they ever came to adopt President Cleveland's policy of taxing no free citizen for more than his fair share of necessary national expenditure, and thus incidentally were brought to abandon Protection. We should then have to compete in foreign markets with a people fully equal to our own in all essential qualities, and with the advantage of being more adaptable, more inventive, more eager to get on, and less under the influence of routine and prejudice; while in certain respects nature gives them an advantage, as in growing their own cotton, and having larger reserves of land and larger deposits of coal and iron. Even here, however, it is probable that competition would lead rather to the diversion of certain branches of our foreign trade into other channels and the substitution of others, than to a diminution of its aggregate amount, and there would be a large compensation to us from throwing open a market of sixty millions of people which is now greatly restricted by prohibitory tariffs.

It is not therefore in the near future that I anticipate any of the dangers and difficulties of a redundant population, but the present rate of progress cannot last for ever, and our posterity, if not in one, then in a few generations, will eventually have to face them. The latest statistics, those of Professor Levasseur, show that since 1800 the population has increased

	Millions.
In the United Kingdom from . . .	16½ to 37.
Russia in Europe	35 to 88.
German Empire	27 to 47.

while he estimates that between 1810 and 1874 the entire population of the world increased from 682,000,000 to 1,391,000,000, or about doubled. If anything like this rate of increase were maintained for another century, nature will obviously have to provide remedies.

Can we foresee what these remedies will be when

reserves of land are approaching exhaustion and supplies of food begin to fail? Scarcely, for in these cases evolution works by its own laws rather than by any logical deduction of philosophers or politicians, and all we know is that there will be a "struggle for existence," and that "the fittest will survive." Still we may gather dimly from present and past experience, that there are two directions from which the inevitable checks may be expected to come. One from a diminution of the birth-rate owing to fewer and later marriages, as the result of education and improved conditions. We have seen in the case of Ireland that poverty tends to accelerate the birth-rate, and commonly the well-to-do and upper classes scarcely keep up their numbers unless recruited from below. As the mass of the population rise to a higher standard of respectability and comfort they will be less ready to risk falling below that standard by contracting early and imprudent marriages. The possession of property also, especially of property in land, is, as we may see in France, a powerful factor in checking the progress of population. In that country, while the population of England, Germany, and Russia has more than doubled, the increase has only been during the same time from 33,000,000 to 38,000,000. Should these checks prove insufficient, I confess I can see no other outcome than an increase of the death-rate on a large scale, such as might come from the combination of war, pestilence, and famine, which would result from a general upheaval of the dangerous and discontented classes of the community, owing to distress and demagogic excitement. Society is safe enough against any irruption of outer barbarians, but it is not so safe against its own barbarians, who are accumulating in the slums of its great cities. Or rather, it is safe as long as it has only these barbarians to deal with, but not so safe if these are reinforced by multitudes of honest and well-intentional men, who are driven desperate by the difficulty of getting "a fair day's wages for a fair day's work."

The history of the Commune in Paris may be a lesson to us of the amount of death and destruction which might be occasioned by such an uprising. A month of such a Commune in London would bring about such a destruction of capital and credit as would throw millions out of employment, and reduce them to the dire necessity of cutting one another's throats or starving. Fortunately such a result is far distant, and at any rate we have the consolation of knowing, that if the States of civilized Europe are to be swallowed up by such a Polyphemus, our lot, like that of the man of many resourses, the wise and much-enduring Ulysses, will probably come last. The tide of empire and civilization has hitherto followed the sun and flowed westward. It has reached the shores of the Atlantic and crossed over to the New World. When that New World is fully occupied, and the human tide reaches the Pacific, what will happen? Will it, like the army of the lemmings in Lapland, march ever westward until it topples over into the ocean? We can only answer, it is a "Problem of the Future."

THE END.

INDEX.

Richard Clay & Sons, Limited, London & Bungay.

HUMAN ORIGINS:

EVIDENCE FROM HISTORY AND SCIENCE.

By SAMUEL LAING.

With Illustrations. Demy 8vo, 3s. 6d. Tenth Thousand.

SOME OPINIONS OF THE PRESS.

The Times says:—"Mr. Laing's present purpose is to use recent researches into the history of the ancient nations of the world as a rough 'measuring-rod' for gauging the duration of the pre-historic periods of human existence. . . . This is Mr. Laing's plan, and its execution is divided into two parts; in the first Mr. Laing recapitulates the results, carried up to the latest date, of the labours of Egyptologists and Assyriologists, with a glance at civilization which, like the Chinese, cannot be traced back to such a remote era. In the second he reviews once more the evidence for the Neolithic and Palæolithic ages, and the Quaternary and Tertiary periods. . . . Mr. Laing performs an operation of great utility to the general reader."

The Globe says:—"Mr. Laing's theme is the antiquity of man, and his treatment of it is pleasant and profitable. . . . The various views concerning the origin of man receive unbiassed attention in the readable volume."

The Daily Graphic says:—"Mr. Samuel Laing's books have a peculiar interest for all those who have neither time nor opportunity to be in the vanguard of scientific progress, but yet like to follow within measurable distance prompted by some reliable coach. But Mr. Laing is more than a coach, he is never anything but interesting. He never seeks to obscure his own point of view; but few people have the requisite moderation for writing on controversial matters in his uncontroversial way."

The Review of Reviews says:—"In the present volume Mr. Laing describes, with great lucidity of expression and clearness of exposition, the theories which now obtain as to the origin and evolution of the human race. . . . Stimulating and informative to the last degree, and sufficiently popular to be understood by the veriest tyro in science. He presents the latest results arrived at by modern science in as clear and lucid a manner as possible, and every one who has mastered his volumes will be as much abreast of scientific thought as the general reader can ever hope to be."

Black and White says:—"Mr. Laing has been inspired to put his great knowledge to a laudable use, and compose for us clearly, in one volume, the teaching of modern science upon the origin and antiquity of man. . . . The book is wonderfully interesting and surprisingly cheap."

The National Observer says:—"It is clearly written, the facts it records are of stupendous interest."

Science (New York) says:—"This is an exceedingly well-written and interesting summary of all the theories, facts, and mysterious questions connected with the origin of mankind on earth, by a somewhat remarkable man whose previous works met with a wide circulation in England. . . . His various publications present the results of wide and discriminating reading and research, in a logical, concise, yet comprehensible style, for the benefit of those who have not the time to look into such matters for themselves."

The Glasgow Herald says:—"The author of this book is a well-known writer upon various philosophical, scientific, and historical subjects. Not only is he possessed of a large fund of information, but his style is generally clear and intelligent, while the arrangement of his matter is almost invariably satisfactory. The present work is devoted mainly to support and illustration of the theory of evolution. Although here and there he seems to take too much for granted, the volume is fresh, attractive, and pleasing. Especially should it be found valuable by the young and those who, from the pressure of the ordinary avocations of life, have neither the time nor the education to fit them for becoming specialists. The advances made in modern days in the investigation of the various problems of science and history have been so great that Mr. Laing is entitled to all praise for publishing so many of their results in a semi-popular form which is intelligible without being slipshod or slovenly. It should be mentioned that *Human Origins* is profusely and successfully illustrated. It is a book to awaken thought and to stimulate criticism."

The Daily Chronicle says:—"In the present work Mr. Laing attacks some theories of the highest importance, and on the whole with a fullness of knowledge and an ability which leaves little to be desired. . . . The reader will find many questions canvassed in Mr. Laing's pages with a clearness, a literary power, and an amount of scientific knowledge which is seldom insufficient and never wearisome."

The Birmingham Gazette says:—"Mr. S. Laing, who is well known to a large circle of readers as an original thinker on problems relating to the history, unwritten and recorded, of this globe, has just published a volume on *Human Origins*, which is likely to be the cause of considerable controversy. His subject is a vast one, and only his wide reading and erudition justify him in choosing it. Mr. Laing is a wonderfully lucid writer, and he marshals his immense array of facts with the utmost precision. No one can rise from the reading of one of his chapters without feeling that a great amount of useful and valuable information has been gained, and that light has been plentifully cast upon obscure subjects."

Truth says:—"More admirable than its admirable predecessors."

The Guardian says:—"The immediate object of this book is to demonstrate the immense antiquity of man. This is done by following up in detail the two distinct lines of evidence from history and science. . . . All this is set forth in broad outline with lucidity; and these chapters will be found very useful to any one who desires to know in a general way what are the conclusions which modern investigation has reached on these topics, and on what sort of arguments these conclusions rest."

The Melbourne Argus says:—"Mr. Laing writes with much force and clearness, so plainly as to be easily understanded of the people."

The Newcastle Chronicle says:—"Mr. Laing condenses a great mass of various information; his matter is admirably arranged; and his style is as admirably adapted to its purpose by its clearness and simplicity. *Human Origins* is an absorbing and stimulating book, and it will well repay the perusal even of those who may dissent upon this point or that from the author's conclusions and suggestions."

From the Illustrated London News.

"A veteran public man, who has during forty years past, after gaining high mathematical honours at Cambridge, been constantly employed in Board of Trade official administration, or in the duties of a member of Parliament, or as Finance Minister of India, or as Chairman of the Brighton Railway Co., is not likely to be a dreamy idealist, or a scholastic bigot. He has a vast acquired fund of practical common-sense, a trained faculty of induction and experimental reasoning, the habit of looking all round the different sides of a question, and then hitting the central point, and that of reducing to just proportion and co-ordination.

"Accordingly, Mr. Laing's summaries of the general propositions concerning the physical constitution of the visible universe, the origin and development of the forms of organic life, and the laws determining the progress of mankind to secular welfare, added to our stock of knowledge within the last thirty-four years, are very useful reading. . . . This instructive volume also contains several essays of a political character, on the huge military armaments of Europe, on taxation and finance, on the increase of population and the prospects of an adequate food-supply, concerning which the opinions of Mr. Laing should have considerable weight."

From Vanity Fair.

"Mr. Laing resembles Mr. Gladstone in one thing—that in spite of having been immersed all his life in business, politics, and economics, he seems always to have time to write essays *de omnibus rebus et quibusdam aliis;* but there is this difference, that while Mr. Gladstone claims to be original and critical and fails to establish such a position—Mr. Laing merely claims to expound in forcible and lucid terms to semi-scientific readers, the results of modern science and thinking, and in such his purpose succeeds admirably. . . . *Armed Europe, Taxation and Finance,* and *Population and Food,* are written by one who has thought before he writes, and writes well what he thinks, and this remark applies broadly to the whole work."

From the Literary Guide.

"Always interesting, Mr. Laing attracts our attention; always relevant, he retains it; always accurate, he gains our confidence; always cheerful, he quickens our hopes. Mr. Laing is no closet philosopher, and he refuses to hold whispered conclaves with metaphysicians and esoteric teachers. Every subject he exposes to the free air and the sun, and he discusses it in terse, forcible, comprehensible, and yet elegant language."

From the Birmingham Daily Gazette.

"The recent attainments of science, and the vistas of future knowledge opened by those attainments, are so numerous and far-reaching, that any one who undertakes successfully the task of clearly defining the present position of the world of science is deserving of thanks. Mr. Laing not only achieves the task, but provides a most engrossing volume. . . . The volume throughout is engrossing reading, and the chapter in which the possible future sympathy between Christians and Agnostics is dealt with, contains much that is reassuring."

From the Literary World.

"Under the title of *Problems of the Future*, we have to welcome another book by the author of *Modern Science and Modern Thought*, and *A Modern Zoroastrian*, admirable alike in the amount of information it conveys, the style in which it is written, and the earnest love of truth by which it is throughout pervaded."

From the Sheffield Telegraph.

"This is almost too comprehensive a book to review. Its range is wide, and its mode of treatment so suggestive of food for reflection and themes for discussion, that at most but a glimpse can be afforded to the reader of its immense capacity for affording information and instruction. The volume is made up of essays from the pen of Mr. Laing, whose previous works in a kindred vein have had so powerful an effect on the public mind, and have been so widely read. Mr. Laing's method has the charm of leading the reader to a point where he can carry the line of thought on for himself. It is impossible for the reader to avoid thinking out the subjects in which Mr. Laing's immense store of knowledge has been placed before him. The communication of knowledge is effected with such consummate skill that the reader is hardly aware of his indebtedness to the author."

From the Nottingham Daily Guardian.

"The problems and essays fall roughly into three divisions—the purely scientific, the theological, and the political. All are most ably written, but we think Mr. Laing's readers will consider that the author's wonderful power of clear exposition is most evident in the essays in which he deals with the conflicts of Science and Theology. Whether we agree with him or no, we must at least grant that he states his case with marvellous lucidity, and that his conclusions are reached by close reasoning, and not by the hop-skip-and-jump style of argument too common with writers who are more anxious to bolster up a particular solution than to fairly face the difficulties of a problem with the sole desire to arrive at truth. . . . *Problems of the Future* is emphatically a book to set the reader thinking, and what is more, thinking to some purpose. The essays, as we have said, range over a wide variety of subjects, but whatever the subject, the reader may always calculate with certainty on finding it handled in a thorough and vigorous fashion, and with an obvious intention to ascertain and set forth the honest truth, if it be discoverable."

From Secular Thought, Toronto.

"The present handsomely printed volume will, we venture to think, be found by most readers the most engrossing of Mr. Laing's works. The grasp it exhibits of the true bearings of the most difficult problems in nearly all departments of knowledge, is only equalled by the clearness and interest with which they are placed before the reader.

"Of the four hundred pages forming the volume, there seems to be scarcely one which does not exhibit in a favourable light the author's power of research and ability to place the results of that research before the reader in a plain but most instructive form. We are sure every purchaser of the volume will consider himself amply repaid by the new light which Mr. Laing's pen throws upon so many of the debatable questions of the intellectual world of to-day, and which comprise scientific, theological, poetical, biblical, and political subjects."

MODERN SCIENCE AND MODERN THOUGHT.

By SAMUEL LAING.

Demy 8vo, 3s. 6d. Seventeenth Thousand.

SOME OPINIONS OF THE PRESS.

From the Times.

"Mr. Laing is a man of active mind, and he has had a busy and rather multifarious life. He is on good terms with his work, his fellow-workers, and his fellow-thinkers. He treads a beaten path. If he does not pretend to originality it is because science is not original except to those who devote themselves to some one field of investigation. He reports what is known, or believed, or not believed, by the majority of scientific men. The character of the work is foreshadowed in its divisions and titles. Two hundred pages are given to 'Science,' followed by about one hundred professedly given to 'Thought.' The thought, however, is scientific, and it is science that dominates from the first page to the last. In the first part Mr. Laing exhibits with much power and effect the immense discoveries of science, and its numerous victories over old opinions whenever they have had the rashness to challenge conclusions with it. These discoveries are not so familiar to the world at large but that any ordinary reader may learn much from a writer combining matter and style, and conveying solid information in simple yet striking language. In a comparatively small compass are here displayed the results of recent inquiries into the composition and constitution of the earth and of the universe, into the nature and laws of matter, into the development of organized and animated existence, into the history of man, into the myths of all races and the faiths of all people; into force, motion, electricity, light, and heat. As one turns over the glowing pages one is tempted to lament that a man so qualified to instruct and to illustrate should have been almost exclusively occupied in absorbing official and practical duties."

From the Pall Mall Gazette.

"Apart from the uselessness and undue multiplication of all such books, Mr. Laing's brief statement of an agnostic creed is good enough and sensible enough in its own way. It is the expression of a sensible, well-read, compromise-loving Briton's final conclusions upon religious matters. The first part is a rapid and clearly written *résumé* of all that modern science and modern criticism have done to sap the foundation of current theologies and the current dogmas. This *résumé* is admirably done. Mr. Laing manages to condense into a few short chapters an amount of salient information on matters astronomical, geological, archæological, and historical; and withal he condenses it cleverly. . . . The evidence of geology against the Mosaic cosmogony; the evidence of biology, and especially of evolutionism, against the story of creation; the evidence of palæolithic flints, and the reindeer age cave-men, against the naïve history of Adam and Eve; the evidence of human development against the entire Biblical conception of man's importance in the scheme of nature—all marshalled with considerable skill, and enforced by excellent and typical examples. The anxious but unlearned inquirer who really wishes to know how much recent researches have effected towards undermining the ground-work of the existing creeds, cannot do better than turn to Mr. Laing's pleasantly written pages."

From the Morning Post.

"Mr. Laing's book is of a useful and reliable character. . . . There is much in the book which will be useful to those for whom it is intended."

From the Westminster Review.

"From the first page to the last the book is charmingly written, with temperance and wisdom that will win a hearing for the author from many who may not share his views."

From the Scotsman.

"His task Mr. Laing has discharged in a business-like manner. In his first part he has presented the chief results of modern scientific investigations with singular terseness and lucidity, and what is more, he has contrived to indicate by simple and impressive illustrations the methods by which those results have been attained. And this he has done in a style so simple and elementary, yet so sufficient for his purpose, that any fairly intelligent reader who has never been able to give attention to scientific subjects may fully grasp the sum of the knowledge he seeks to impart. . . . These chapters are of genuine interest, as showing the conclusions of a practical man who belongs neither to the philosophers nor the theologians, but has intelligently studied the researches and reasonings of both, and has formed his judgment between them on the principles of common sense, and by means of the ordinary rules by which men weigh evidence in ordinary life."

A MODERN ZOROASTRIAN.

By SAMUEL LAING.

Demy 8vo, 3s. 6d. Seventh Thousand.

SOME OPINIONS OF THE PRESS.

From the Morning Post.

"The present volume is rather the complement of certain parts of its predecessor than a sequel. In the former work on *Modern Science and Modern Thought* Mr. Laing dealt with the more popular, because more easily comprehensible, divisions of science, while now he applies his faculty of lucid condensation to such abstruse subjects as the atomic theory, the laws of energy and heredity, the constitution of the mind, and other matters of which most of us know little indeed. The discoveries and illustrations of recognized authorities are set out clearly and briefly."

From the Westminster Review.

"In the strictly scientific part of the work the exposition is admirable, such as any great teacher might be glad to have written, marked by breadth of grasp and clearness. . . . From its clearly written, able, and sympathetic discussion of so many of the great problems of existence, the book cannot fail to exercise a great influence on a large number of readers."

From Progress.

"This volume may be taken as a continuation of Mr. Laing's former work, *Modern Science and Modern Thought*. It exhibits the same power of lucid statement. Even the most abstruse questions are elucidated by apt illustrations, and if we occasionally feel that a difficult point is too easily summed up, we are compelled to admit that Mr. Laing has mastered the questions, and presented the most essential features."

From the Scotsman.

"Whether it be admitted or denied that Mr. Laing has explained all he attempts to explain in *A Modern Zoroastrian*, it may be granted that he has, as he claims to have, a faculty of lucid condensation."

From the St. Stephen's Review.

"We undertake to predict that this work will be acknowledged as the most remarkable scientific and philosophical work of 1887. It is abstruse in abstract thought, but it is simple enough in intelligent language to be understood by the people. Those who have read Mr. Laing need scarcely be told how deftly he weaves the crushing proof round and round his arguments, or how skilfully he advances new dogma, which on the face of it is unanswerable. There are points in the book which are of supreme interest to the non-scientific reader, and, indeed, it teems with topical matter."

From the Literary World.

"The reception given to Mr. Laing's former work, *Modern Science and Modern Thought*, has induced him to publish this most interesting and fascinating book, which will be read and re-read by all students of philosophy and science with avidity and delight."

From the Secular Review.

"If our readers would like to form an acquaintance with a book replete with the grandest generalizations and most daring speculations that do not spring from imaginative ideation, but rest on the foundation of physical phenomena, we recommend them to a thoughtful perusal of Mr. Laing's *Modern Zoroastrian*. They will find the pith and epitome of much that is most salient in the writings of Darwin, Huxley, and Haeckel put with a terseness and perspicuity for which Mr. Laing stands unrivalled."

From the Ayrshire Post.

"We wonder if any of our readers have ever read that most interesting and fascinating book entitled *A Modern Zoroastrian*. If not, we would recommend them to make an early acquaintance with it."

From the Literary Guide.

"The work is executed with a strong, supple, and well-trained hand. Mr. Laing's philosophy is peculiarly practical. While metaphysicians and theologians are straying into bogs here and pathless wilds there, he keeps resolutely to the beaten path of common sense."

From the Popular Science Monthly (New York).

"This book, like many others of late date, is evidence of two facts: first, that the traditional religion has lost its hold on most scientifically educated minds, and secondly, that such minds are not content without some religion. The author's views are well expressed, and readers having a taste for this class of subjects will find the book an interesting one."

11, Henrietta Street, Covent Garden, W.C.
August, 1893.

A

Catalogue of Books

PUBLISHED BY

CHAPMAN & HALL,

LIMITED.

A separate Illustrated Catalogue is issued, containing

Drawing Examples, Diagrams, Models, Instruments, etc.,

ISSUED UNDER THE AUTHORITY OF

THE SCIENCE AND ART DEPARTMENT,
SOUTH KENSINGTON,

FOR THE USE OF SCHOOLS AND ART AND SCIENCE CLASSES.

NEW AND FORTHCOMING BOOKS.

COLONEL A. B. ELLIS.

HISTORY OF THE GOLD COAST OF WEST AFRICA. By A. B. Ellis, Col. 1st West Indian Regiment. Demy 8vo. [*In the Press.*

ROBERT KEMPT.

CONVIVIAL CALEDONIA: Inns and Taverns of Scotland, and some Famous People who have frequented them. By Robert Kempt. Crown 8vo. [*In the Press.*

CHARLES DIXON.

JOTTINGS ABOUT BIRDS. By Charles Dixon, Author of "The Migration of Birds." Crown 8vo. [*In the Press.*

MAJOR ARTHUR GRIFFITHS.

SECRETS OF THE PRISON HOUSE. By Major Arthur Griffiths. 2 vols. Demy 8vo. [*In the Press.*

A. F. MACDONALD.

OUR OCEAN RAILWAYS; or, The Rise, Progress, and Development of Ocean Steam Navigation, etc. etc. By A. F. Macdonald. With Maps and Illustrations. Demy 8vo. [*In the Press.*

F. BOYLE.

ABOUT ORCHIDS. By F. Boyle. Crown 8vo. [*In the Press.*

J. BRINTON.

A TOUR IN PALESTINE AND SYRIA. By J. Brinton. With Illustrations and Map. Large crown 8vo. [*In the Press.*

W. CLARK RUSSELL, W. E. NORRIS, J. M. BARRIE, AND OTHERS.

STORIES FROM "BLACK AND WHITE." By W. Clark Russell, W. E. Norris, Julian Hawthorne, Mrs. L. B. Walford, F. C. Philips, J. M. Barrie, Mrs. Alexander, and William Westall. With numerous Illustrations. Crown 8vo. (*Second Series.*) [*In the Press.*

LADY ISABEL BURTON.

THE LIFE OF SIR RICHARD FRANCIS BURTON, K.C.M.G., etc. By his Wife, Isabel Burton. With numerous Portraits, Coloured and other Illustrations, and Maps. 2 vols. Demy 8vo, 42s.

THE REV. H. N. HUTCHINSON.

EXTINCT MONSTERS. A Popular Account of some of the larger forms of Animal Life. By the Rev. H. N. Hutchinson. With numerous Illustrations by J. Smit and others, and a Preface by Dr. Henry Woodward, F.R.S. Third Thousand, Revised and Enlarged. Demy 8vo, 12s.

EDWARD DELILLE.

SOME FRENCH WRITERS. By Edward Delille. Crown 8vo, 5s. Contents: Bourget, Pierre Loti, Baudelaire, Guy de Maupassant, Verlaine, Maurice Barres, &c.

PHYSIOGNOMICAL SKETCHES.

CHARACTER IN THE FACE: Our Looks and what they mean. Crown 8vo, 5s.

ASHMORE RUSSAN AND FREDERICK BOYLE.

THE ORCHID SEEKERS: A Story of Adventure in Borneo. By Ashmore Russan and Frederick Boyle. Illustrated by Alfred Hartley. Crown 8vo, 7s. 6d.

BOOKS

PUBLISHED BY

CHAPMAN & HALL, LIMITED.

ABOUT (EDMOND)—

HANDBOOK OF SOCIAL ECONOMY; OR, THE WORKER'S A B C. From the French. With a Biographical and Critical Introduction by W. FRASER RAE. Second Edition, revised. Crown 8vo, 4s.

AFLALO (F. G.) and PASKE (SURGEON-GENERAL C. T.)—

THE SEA AND THE ROD. With Illustrations. Crown 8vo, 4s. 6d.

AFRICAN FARM, STORY OF AN. By OLIVE SCHREINER (Ralph Iron). New Edition. Crown 8vo, 1s.; in cloth, 1s. 6d.

AGRICULTURAL SCIENCE (LECTURES ON), AND OTHER PROCEEDINGS OF THE INSTITUTE OF AGRICULTURE, SOUTH KENSINGTON, 1883-4. Crown 8vo, sewed, 2s.

ANDERSON (ANDREW A.)—

A ROMANCE OF N'SHABÉ: Being a Record of Startling Adventures in South Central Africa. With Illustrations. Crown 8vo, 5s.

TWENTY-FIVE YEARS IN A WAGGON IN THE GOLD REGIONS OF AFRICA. With Illustrations. Second Edition. Demy 8vo, 12s.

ANDERSON (CAPTAIN LINDSAY)—

THE STORY OF ALLAN GORDON. With Illustrations. Crown 8vo, 5s.

AMONG TYPHOONS AND PIRATE CRAFT. With Illustrations by STANLEY WOOD. Crown 8vo, 5s.

A CRUISE IN AN OPIUM CLIPPER. With Illustrations. Crown 8vo, 6s.

AVELING (EDWARD), D.Sc., Fellow of University College, London—

MECHANICS AND EXPERIMENTAL SCIENCE. As required for the Matriculation Examination of the University of London.

MECHANICS. With numerous Woodcuts. Crown 8vo, 6s.
Key to Problems in ditto, crown 8vo, 3s 6d.

CHEMISTRY. With numerous Woodcuts. Crown 8vo, 6s.
Key to Problems in ditto, crown 8vo, 2s. 6d

MAGNETISM AND ELECTRICITY. With Numerous Woodcuts. Crown 8vo. 6s

LIGHT AND HEAT. With Numerous Woodcuts. Crown 8vo, 6s.

Keys to the last two volumes in one vol. Crown 8vo, 5s.

BAILEY (JOHN BURN)—

FROM SINNER TO SAINT; OR, CHARACTER TRANSFORMATIONS. Crown 8vo, 6s.

MODERN METHUSELAHS; or, Short Biographical Sketches of a few advanced Nonagenarians or actual Centenarians. Demy 8vo, 10s. 6d.

BAKER (W. L.), A.M.I.C.E.—

THE BEAM; OR, TECHNICAL ELEMENTS OF GIRDER CONSTRUCTION. Crown 8vo, 4s.

BEATTY-KINGSTON (W.)—

A JOURNALIST'S JOTTINGS. 2 vols. Demy 8vo, 24s.

BEATTY-KINGSTON (W.)—continued—

A WANDERER'S NOTES. 2 vols. Demy 8vo, 24s.

MONARCHS I HAVE MET. 2 vols. Demy 8vo, 24s.

MUSIC AND MANNERS: Personal Reminiscences and Sketches of Character. 2 vols. Demy 8vo, 30s.

BELL (JAMES, Ph.D., &c.), Principal of the Somerset House Laboratory—

THE CHEMISTRY OF FOODS. With Microscopic Illustrations.

PART I. TEA, COFFEE, COCOA, SUGAR, ETC. Large crown 8vo, 2s. 6d.
PART II. MILK, BUTTER, CHEESE, CEREALS, PREPARED STARCHES, ETC. Large crown 8vo, 3s.

BENTLEY (H. CUMBERLAND)—

SONGS AND VERSES. Illustrated by FINCH MASON, and dedicated to J. G. WHYTE MELVILLE. Crown 8vo, 4s.

BIRDWOOD (SIR GEORGE C. M.), C.S.I.—

THE INDUSTRIAL ARTS OF INDIA. With Map and 174 Illustrations. New Edition. Demy 8vo, 14s.

BLACKIE (JOHN STUART), F.R.S.E.—

THE SCOTTISH HIGHLANDERS AND THE LAND LAWS. Demy 8vo, 9s.

ALTAVONA: FACT AND FICTION FROM MY LIFE IN THE HIGHLANDS. Third Edition. Crown 8vo, 6s.

BLATHERWICK (C.)—

IN THE SHADE OF SCHIEHALLION. With 8 Illustrations. 1 vol. Crown 8vo, 1s.

CYNTHIA. With 4 Illustrations. Crown 8vo, 1s.

BLEUNARD (A.)—

BABYLON ELECTRIFIED: The History of an Expedition undertaken to restore Ancient Babylon by the Power of Electricity, and how it Resulted. Translated from the French. Illustrated. Royal 8vo, 12s.

BLOOMFIELD'S (BENJAMIN LORD), MEMOIR OF—
MISSION TO THE COURT OF BERNADOTTE. With Portraits. 2 vols. Demy 8vo, 28s.

BONVALOT (GABRIEL)—

THROUGH THE HEART OF ASIA OVER THE PAMIR TO INDIA. Translated from the French by C. B. PITMAN. With 250 Illustrations by ALBERT PÉPIN. Royal 8vo, 32s.

BOWERS (G.)—

HUNTING IN HARD TIMES. With 61 coloured Illustrations. Oblong 4to, 12s.

BOYLE (FREDERICK)—

ABOUT ORCHIDS. Crown 8vo. [*In the Press.*

BOYLE (FREDERICK) and ASHMORE (RUSSAN)—

THE ORCHID SEEKERS: A Story of Adventure in Borneo. Illustrated by ALFRED HARTLEY. Crown 8vo, 7s. 6d.

BRACKENBURY (COL. C. B.)—

FREDERICK THE GREAT. With Maps and Portrait. Large crown 8vo, 4s.

BRADLEY (THOMAS), Royal Military Academy, Woolwich—

ELEMENTS OF GEOMETRICAL DRAWING. In Two Parts, with Sixty Plates. Oblong folio, half bound, each Part 16s.

BRIDGMAN (F. A.)—

WINTERS IN ALGERIA. With 62 Illustrations. Royal 8vo, 10s 6d.

BRINTON (J.)—

A TOUR IN PALESTINE. With Illustrations and Map. Crown 8vo.

BRITISH ARMY, THE. By the Author of "Greater Britain." Demy 8vo, 12s.

BROCK (DR. J. H. E.), Assistant Examiner in Hygiene, Science and Art Department—

ELEMENTS OF HUMAN PHYSIOLOGY FOR THE HYGIENE EXAMINATIONS OF THE SCIENCE AND ART DEPARTMENT. Crown 8vo, 1s. 6d.

BROMLEY-DAVENPORT (the late W.), M.P.—

SPORT: Fox Hunting, Salmon Fishing, Covert Shooting, Deer Stalking. With numerous Illustrations by General CREALOCK, C.B. New Cheap Edition. Post 8vo, 3s. 6d.

BROWN (J. MORAY)—

POWDER, SPEAR, AND SPUR: A Sporting Medley. With Illustrations. Crown 8vo, 10s. 6d.

BUCKLAND (FRANK)—

LOG-BOOK OF A FISHERMAN AND ZOOLOGIST. With Illustrations. Sixth Thousand. Crown 8vo, 3s. 6d.

BUFFEN (F. F.)—

MUSICAL CELEBRITIES: Portraits and Biographies. Second Series. Crown 4to, 21s.

BURCHETT (R.)—

DEFINITIONS OF GEOMETRY. New Edition. 24mo, cloth, 5d.

LINEAR PERSPECTIVE, for the Use of Schools of Art. New Edition. With Illustrations. Post 8vo, cloth, 7s.

PRACTICAL GEOMETRY: The Course of Construction of Plane Geometrical Figures. With 137 Diagrams. Eighteenth Edition. Post 8vo, cloth, 5s.

BURGESS (EDWARD)—

ENGLISH AND AMERICAN YACHTS. Illustrated with 50 Beautiful Photogravure Engravings. Oblong folio, 42s.

BURTON (LADY ISABEL)—

THE LIFE OF SIR RICHARD FRANCIS BURTON. With Portraits, numerous Coloured and other Illustrations, and Maps. 2 vols. Demy 8vo, 42s.

BUTLER (A. J.)—

COURT LIFE IN EGYPT. Second Edition. Illustrated. Large crown 8vo, 12s.

CARLYLE (THOMAS), WORKS BY—*See pages* 29, 30.

THE CARLYLE BIRTHDAY BOOK. Second Edition. Small fcap. 8vo, 3s.

CARSTENSEN (A. RIIS)—

TWO SUMMERS IN GREENLAND: An Artist's Adventures among Ice and Islands in Fjords and Mountains. With numerous Illustrations by the Author. Demy 8vo, 14s.

CHAPMAN & HALL'S SHILLING SERIES.

HOW TO BUY A HORSE. With Hints on Shoeing and Stable Management. By PELAGIUS. Crown 8vo.

IN THE SHADE OF SCHIEHALLION. By CHARLES BLATHERWICK, Author of "Peter Stonnor." With 4 Illustrations. Crown 8vo.

CYNTHIA. By CHARLES BLATHERWICK. With 4 Illustrations. Crown 8vo.

IN AND OUT OF THE PIGSKIN. By G. F. UNDERHILL. With Illustrations. Crown 8vo.

THE CHUMPLEBUNNYS AND SOME OTHER ODDITIES. Sketched from Life. By W. BEATTY-KINGSTON. Illustrated. Crown 8vo.

A SUBURB OF YEDO. By the late THEOBALD A. PURCELL. Illustrated.

ARCADIAN LIFE. By S. S. BUCKMAN, F.G.S. With Illustrations. Crown 8vo.

JOHN DARKE'S SOJOURN IN THE COTTESWOLDS AND ELSEWHERE. By S. S. BUCKMAN. Illustrated.

SINGER'S STORY, A. Related by the Author of "Flitters, Tatters, and the Counsellor.'

A POPPY'S TEARS. By MANNINGTON CAFFYN.

NOTCHES ON THE ROUGH EDGE OF LIFE. By LYNN CYRIL D'OYLE.

WE TWO AT MONTE CARLO. By ALBERT D. VANDAM.

WHO IS THE MAN? A Tale of the Scottish Border. By J. S. TAIT.

THE CHILD OF STAFFERTON. By CANON KNOX LITTLE.

THE BROKEN VOW. By CANON KNOX LITTLE.

THE STORY OF AN AFRICAN FARM. By OLIVE SCHREINER.

PADDY AT HOME. By BARON E. DE MANDAT-GRANCEY.

CHARACTER IN THE FACE. Physiognomical Sketches. Our Looks and what they mean. Crown 8vo, 5s.

CHARLOTTE ELIZABETH, LIFE AND LETTERS OF, Princess Palatine and Mother of Philippe d'Orléans, Regent of France, 1652–1722. With Portraits. Demy 8vo, 10s. 6d.

CHARNAY (DÉSIRÉ)—

THE ANCIENT CITIES OF THE NEW WORLD. Being Travels and Explorations in Mexico and Central America, 1857—1882. With upwards of 200 Illustrations. Super Royal 8vo, 31s. 6d.

CHURCH (PROFESSOR A. H.), M.A. Oxon.—

FOOD GRAINS OF INDIA. With numerous Woodcuts. Small 4to, 6s.

ENGLISH PORCELAIN. A Handbook to the China made in England during the Eighteenth Century, as illustrated by Specimens chiefly in the National Collection. With numerous Woodcuts. Large crown 8vo, 3s.

ENGLISH EARTHENWARE. A Handbook to the Wares made in England during the 17th and 18th Centuries, as illustrated by Specimens in the National Collections. With numerous Woodcuts. Large crown 8vo, 3s.

PLAIN WORDS ABOUT WATER. Illustrated. Crown 8vo, sewed, 6d.

FOOD: Some Account of its Sources, Constituents, and Uses. A New and Revised Edition. Large crown 8vo, cloth, 3s.

PRECIOUS STONES: considered in their Scientific and Artistic Relations. With a Coloured Plate and Woodcuts. Second Edition. Large crown 8vo, 2s. 6d.

COBDEN, RICHARD, LIFE OF. By the RIGHT HON. JOHN MORLEY, M.P. With Portrait. New Edition. Crown 8vo, 7s. 6d.

Popular Edition, with Portrait, 4to, sewed, 1s.; cloth, 2s.

COLLIER (THE HON. MARGARET), Madame Galetti di Cadilliac—

RACHEL AND MAURICE, and OTHER TALES. Crown 8vo, 3s. 6d.

COLLINS (WILKIE) and DICKENS (CHARLES)—

THE LAZY TOUR OF TWO IDLE APPRENTICES; NO THOROUGHFARE; THE PERILS OF CERTAIN ENGLISH PRISONERS. With 8 Illustrations. Crown 8vo, 5s.

COOKERY—

DINNERS IN MINIATURE. By MRS. EARL. Crown 8vo, 2s. 6d.

HILDA'S "WHERE IS IT?" OF RECIPES. Containing many old CAPE, INDIAN, and MALAY DISHES and PRESERVES; also Directions for Polishing Furniture, Cleaning Silk, etc.; and a Collection of Home Remedies in Case of Sickness. By HILDAGONDA J. DUCKITT. Fourth Thousand. Crown 8vo, 4s. 6d.

THE PYTCHLEY BOOK OF REFINED COOKERY AND BILLS OF FARE. By MAJOR L——. Fifth Edition. Large crown 8vo, 8s. [*In the Press.*

BREAKFASTS, LUNCHEONS, AND BALL SUPPERS. By MAJOR L——. Crown 8vo, 4s.

OFFICIAL HANDBOOK OF THE NATIONAL TRAINING SCHOOL FOR COOKERY. Containing Lessons on Cookery, forming the Course of Instruction in the School. Compiled by "R. O. C." Twenty-first Thousand. Large crown 8vo, 6s.

BREAKFAST AND SAVOURY DISHES. By "R. O. C." Ninth Thousand. Crown 8vo, 1s.

THE ROYAL CONFECTIONER: English and Foreign. By C. E. FRANCATELLI. With Illustrations. Sixth Thousand. Crown 8vo, 5s.

COOPER-KING (LT.-COL.)—

GEORGE WASHINGTON. Large crown 8vo. With Portrait and Maps. [*In the Press.*

COUPERUS (LOUIS)—

ELINE VERE. Translated from the Dutch by J. T. GREIN. Crown 8vo, 5s.

COURTNEY (W. L.), M.A., LL.D., of New College, Oxford—

STUDIES AT LEISURE. Crown 8vo, 6s.

STUDIES NEW AND OLD. Crown 8vo, 6s.

CONSTRUCTIVE ETHICS: A Review of Modern Philosophy and its Three Stages of Interpretation, Criticism, and Reconstruction. Demy 8vo, 12s.

CRAIK (GEORGE LILLIE)—

ENGLISH OF SHAKESPEARE. Illustrated in a Philological Commentary on "Julius Cæsar." Eighth Edition. Post 8vo, cloth, 5s.

OUTLINES OF THE HISTORY OF THE ENGLISH LANGUAGE. Eleventh Edition. Post 8vo, cloth, 2s. 6d.

CRAWFURD (OSWALD)—

ROUND THE CALENDAR IN PORTUGAL. With numerous Illustrations. Royal 8vo, 18s.

BEYOND THE SEAS; being the surprising Adventures and ingenious Opinions of Ralph, Lord St. Keyne. Second Edition. Crown 8vo, 3s. 6d.

CRIPPS (WILFRED JOSEPH), M.A., F.S.A.—

COLLEGE AND CORPORATION PLATE. A Handbook for the Reproduction of Silver Plate. With numerous Illustrations. Large crown 8vo, cloth, 2s. 6d.

CURZON (LOUIS HENRY)—

A MIRROR OF THE TURF; or, The Machinery of Horse-racing Revealed, showing the Sport of Kings as it is to-day. Crown 8vo, 8s.

DAIRY FARMING—

DAIRY FARMING. To which is added a Description of the Chief Continental Systems. With numerous Illustrations. By JAMES LONG. Crown 8vo, 9s.

DAIRY FARMING, MANAGEMENT OF COWS, etc. By ARTHUR ROLAND. Edited by WILLIAM ABLETT. Crown 8vo, 5s.

DALY (J. B.), LL.D.—

IRELAND IN THE DAYS OF DEAN SWIFT. Crown 8vo, 5s.

DAS (DEVENDRA N.)—

SKETCHES OF HINDOO LIFE. Crown 8vo, 5s.

DAUBOURG (E.)—

INTERIOR ARCHITECTURE. Doors, Vestibules, Staircases, Anterooms, Drawing, Dining, and Bed Rooms, Libraries, Bank and Newspaper Offices, Shop Fronts and Interiors. Half-imperial, cloth, £2 12s. 6d.

DAVIDSON (ELLIS A.)—

PRETTY ARTS FOR THE EMPLOYMENT OF LEISURE HOURS. A Book for Ladies. With Illustrations. Demy 8vo, 6s.

DAY (WILLIAM)—

THE RACEHORSE IN TRAINING, with Hints on Racing and Racing Reform. Seventh Edition. Demy 8vo, 9s.

DE BOVET (MADAME)—

THREE MONTHS' TOUR IN IRELAND. Translated and Condensed by MRS. ARTHUR WALTER. With Illustrations. Crown 8vo., 6s.

DE CHAMPEAUX (ALFRED)—

TAPESTRY. With numerous Woodcuts. Cloth, 2s. 6d.

DE FALLOUX (THE COUNT)—

MEMOIRS OF A ROYALIST. Edited by C. B. PITMAN. 2 vols. With Portraits. Demy 8vo, 32s.

DE KONINCK (L. L.) and DIETZ (E.)—

PRACTICAL MANUAL OF CHEMICAL ASSAYING, as applied to the Manufacture of Iron. Edited, with notes, by ROBERT MALLET. Post 8vo, cloth, 6s.

DE LESSEPS (FERDINAND)—

RECOLLECTIONS OF FORTY YEARS. 2 vols. Demy 8vo, 24s.

DELILLE (EDWARD)—

SOME FRENCH WRITERS. Crown 8vo, 5s.
Contents: Bourget, Pierre Loti, Baudelaire, Guy de Maupassant, Verlaine, Maurice Barres, &c.

DE LISLE (MEMOIR OF LIEUTENANT RUDOLPH), R.N., of the Naval Brigade. By the Rev. H. N. OXENHAM, M.A. Third Edition. Crown 8vo, 7s. 6d.

DE MANDAT-GRANCEY (BARON E.)—

PADDY AT HOME; OR, IRELAND AND THE IRISH AT THE PRESENT TIME, AS SEEN BY A FRENCHMAN. Fifth Edition. Crown 8vo, 1s.; in cloth, 1s. 6d.

DE STAËL (MADAME)—

MADAME DE STAËL: Her Friends, and Her Influence in Politics and Literature. By LADY BLENNERHASSETT. Translated from the German. With a Portrait. 3 vols. Demy 8vo, 36s.

DE WINDT (H.)—

SIBERIA AS IT IS. With numerous Illustrations. Demy 8vo, 18s.

DE WINDT (H.)—continued—

FROM PEKIN TO CALAIS BY LAND. With numerous Illustrations by C. E. FRIPP from Sketches by the Author. New and Cheaper Edition. 7s. 6d.

A RIDE TO INDIA ACROSS PERSIA AND BELUCHISTAN. With numerous Illustrations. Demy 8vo, 16s.

DICKENS (CHARLES), WORKS BY—See pages 31—37.

THE CHARLES DICKENS BIRTHDAY BOOK. With Five Illustrations. In a handsome fcap. 4to volume, 12s.

THE HUMOUR AND PATHOS OF CHARLES DICKENS. By CHARLES KENT. With Portrait. Crown 8vo, 6s.

THE DICKENS DICTIONARY. A Key to the Characters and Principal Incidents in the Tales of Charles Dickens. New Edition, uniform with the "Crown" Edition of Dickens's Works. Large crown 8vo, 5s.

DICKENS (MARY A.)—

CROSS CURRENTS: a Novel. A New Edition in 1 vol. Crown 8vo, 3s. 6d.

DILKE (LADY)—

ART IN THE MODERN STATE. With Facsimile. Demy 8vo, 9s

DINARTE (SYLVIO)—

INNOCENCIA: A Story of the Prairie Regions of Brazil. Translated from the Portuguese and Illustrated by JAMES W. WELLS, F.R.G.S. Crown 8vo, 6s.

DIXON (CHARLES)—

JOTTINGS ABOUT BIRDS. Crown 8vo. [*In the Press.*

THE NESTS AND EGGS OF BRITISH BIRDS: When and Where to Find Them. Being a Handbook to the Oology of the British Islands. Crown 8vo, 6s.

THE GAME BIRDS AND WILD FOWL OF THE BRITISH ISLANDS Illustrated by A. T. ELWES. Demy 8vo, 18s.

THE MIGRATION OF BIRDS: An Attempt to Reduce the Avian Season-flight to Law. Crown 8vo, 6s

THE BIRDS OF OUR RAMBLES: A Companion for the Country. With Illustrations by A. T. ELWES. Large Crown 8vo, 7s. 6d.

IDLE HOURS WITH NATURE. With Frontispiece. Crown 8vo, 6s.

ANNALS OF BIRD LIFE: A Year-Book of British Ornithology. With Illustrations. Crown 8vo, 7s. 6d.

DOUGLAS (JOHN)—

SKETCH OF THE FIRST PRINCIPLES OF PHYSIOGRAPHY. With Maps and numerous Illustrations. Crown 8vo, 6s.

DRAYSON (MAJOR-GENERAL A. W.)—

THIRTY THOUSAND YEARS OF THE EARTH'S PAST HISTORY. Large Crown 8vo, 5s.

EXPERIENCES OF A WOOLWICH PROFESSOR during Fifteen Years at the Royal Military Academy. Demy 8vo, 8s.

PRACTICAL MILITARY SURVEYING AND SKETCHING. Fifth Edition. Post 8vo, cloth, 4s. 6d.

DUCKITT (HILDAGONDA J.)—

HILDA'S "WHERE IS IT?" OF RECIPES. Containing many old CAPE, INDIAN, and MALAY DISHES and PRESERVES; also Directions for Polishing Furniture, Cleaning Silk, etc. Third Edition. Crown 8vo, 4s. 6d.

DUCOUDRAY (GUSTAVE)—

THE HISTORY OF ANCIENT CIVILISATION. A Handbook based upon M. Gustave Ducoudray's "Histoire Sommaire de la Civilisation." Edited by REV. J. VERSCHOYLE, M.A. With Illustrations. Large crown 8vo, 6s.

THE HISTORY OF MODERN CIVILISATION. With Illustrations. Large crown 8vo, 9s.

DYCE (WILLIAM), R.A.—

DRAWING-BOOK OF THE GOVERNMENT SCHOOL OF DESIGN. Fifty selected Plates. Folio, sewed, 5s.; mounted, 18s.

ELEMENTARY OUTLINES OF ORNAMENT. Plates I. to XXII., containing 97 Examples, adapted for Practice of Standards I. to IV. Small folio, sewed, 2s. 6d.

SELECTION FROM DYCE'S DRAWING BOOK. 15 Plates, sewed, 1s. 6d.; mounted on cardboard, 6s. 6d.

TEXT TO ABOVE. Crown 8vo, sewed, 6d.

EARL (MRS.)—

DINNERS IN MINIATURE. Crown 8vo, 2s. 6d.

EDWARDS (MRS. SUTHERLAND)—

THE SECRET OF THE PRINCESS. A Tale of Country, Camp, Court, Convict, and Cloister Life in Russia. Crown 8vo, 3s. 6d.

ELLIOT (FRANCES)—

OLD COURT LIFE IN SPAIN. Demy 8vo.

ELLIS (A. B., Major 1st West India Regiment)—

HISTORY OF THE GOLD COAST OF WEST AFRICA. Demy 8vo. [*In the Press.*

THE EWE-SPEAKING PEOPLE OF THE SLAVE COAST OF WEST AFRICA. With Map. Demy 8vo, 10s. 6d.

THE TSHI-SPEAKING PEOPLES OF THE GOLD COAST OF WEST AFRICA: their Religion, Manners, Customs, Laws, Language, &c. With Map. Demy 8vo, 10s. 6d.

SOUTH AFRICAN SKETCHES. Crown 8vo, 6s.

THE HISTORY OF THE WEST INDIA REGIMENT. With Maps and Coloured Frontispiece and Title-page. Demy 8vo, 18s.

THE LAND OF FETISH. Demy 8vo, 12s.

ENGEL (CARL)—

MUSICAL INSTRUMENTS. With numerous Woodcuts. Large crown 8vo, cloth, 2s. 6d.

ENGLISHMAN IN PARIS: NOTES AND RECOLLECTIONS DURING THE REIGN OF LOUIS PHILIPPE AND THE EMPIRE. New Edition in one volume. Crown 8vo, 7s. 6d.

ESCOTT (T. H. S.)—

POLITICS AND LETTERS. Demy 8vo, 9s.

ENGLAND: ITS PEOPLE, POLITY, AND PURSUITS. New and Revised Edition. Demy 8vo, 3s. 6d.

EUROPEAN POLITICS, THE PRESENT POSITION OF. By the Author of "Greater Britain." Demy 8vo, 12s.

FANE (VIOLET)—

AUTUMN SONGS. Crown 8vo, 6s.

THE STORY OF HELEN DAVENANT. Crown 8vo, 3s. 6d.

QUEEN OF THE FAIRIES (A Village Story), and other Poems. Crown 8vo, 6s.

ANTHONY BABINGTON: a Drama. Crown 8vo, 6s.

FIELD (HENRY M.)—

GIBRALTAR. With numerous Illustrations Demy 8vo, 7s. 6d.

FITZGERALD (PERCY), F.S.A.—

HENRY IRVING: A Record of Twenty Years at the Lyceum. With Portrait. Demy 8vo, 14s.

CHRONICLES OF BOW STREET POLICE OFFICE. New and Cheaper Edition. [*In the Press.*

THE HISTORY OF PICKWICK. An Account of its Characters, Localities, Allusions, and Illustrations. With a Bibliography. Demy 8vo, 8s.

FLEMING (GEORGE), F.R.C.S.—

ANIMAL PLAGUES: THEIR HISTORY, NATURE, AND PREVENTION. 8vo, cloth, 15s.

PRACTICAL HORSE-SHOEING. With 37 Illustrations. Fifth Edition, enlarged. 8vo, sewed, 2s.

RABIES AND HYDROPHOBIA: THEIR HISTORY, NATURE, CAUSES, SYMPTOMS, AND PREVENTION. With 8 Illustrations. 8vo, cloth, 15s.

FORSTER (JOHN)—

THE LIFE OF CHARLES DICKENS. Original Edition. Vol. I., 8vo, cloth, 12s. Vol. II., 8vo, cloth, 14s. Vol. III, 8vo, cloth, 16s.

Uniform with the Illustrated Library Edition of Dickens's Works. 2 vols Demy 8vo, 20s.

Uniform with the Library and Popular Library Editions. Post 8vo, 10s. 6d. each.

Uniform with the "C. D." Edition. With Numerous Illustrations. 2 vols. 7s.

Uniform with the Crown Edition. Crown 8vo, 5s.

Uniform with the Household Edition. With Illustrations by F. Barnard. Crown 4to, cloth, 5s.

Uniform with the Pictorial Edition. With 40 Illustrations by F. Barnard and others. Royal 8vo, 3s. 6d.

FORSTER, THE LIFE OF THE RIGHT HON. W. E. By T. Wemyss Reid. With Portraits. Fourth Edition. 2 vols. Demy 8vo, 32s. FIFTH EDITION, in one volume, with new Portrait. Demy 8vo, 10s. 6d.

FORSYTH (CAPTAIN)—

THE HIGHLANDS OF CENTRAL INDIA: Notes on their Forests and Wild Tribes, Natural History and Sports. With Map and Coloured Illustrations. A New Edition. Demy 8vo, 12s.

FORTNIGHTLY REVIEW (*see page* 40)—

FORTNIGHTLY REVIEW.—First Series, May, 1865, to Dec. 1866. 6 vols. Cloth, 13s. each.

New Series, 1867 to 1872. In Half-yearly Volumes. Cloth, 13s. each.

From January, 1873, to the present time, in Half-yearly Volumes. Cloth, 16s. each.

CONTENTS OF FORTNIGHTLY REVIEW. From the commencement to end of 1878. Sewed, 2s.

FORTNUM (C. D. E.), F.S.A.—

MAIOLICA. With numerous Woodcuts. Large crown 8vo, cloth, 2s. 6d.

BRONZES. With numerous Woodcuts. Large crown 8vo, cloth, 2s. 6d.

FOUQUÉ (DE LA MOTTE)—

UNDINE: a Romance translated from the German. With an Introduction by JULIA CARTWRIGHT. Illustrated by HEYWOOD SUMNER. Crown 4to, 5s.

FRANCATELLI (C. E.)—

THE ROYAL CONFECTIONER: English and Foreign. A Practical Treatise. With Illustrations. Sixth Thousand. Crown 8vo, 5s.

FRANKS (A. W.)—

JAPANESE POTTERY. Being a Native Report, with an Introduction and Catalogue. With numerous Illustrations and Marks. Large crown 8vo, cloth, 2s. 6d.

FROBEL, FRIEDRICH; a Short Sketch of his Life, including Frobel's Letters from Dresden and Leipzig to his Wife, now first Translated into English. By EMILY SHIRREFF. Crown 8vo, 2s.

GALLENGA (ANTONIO)—

ITALY: PRESENT AND FUTURE. 2 vols. Dmy. 8vo, 21s.

EPISODES OF MY SECOND LIFE. 2 vols. Dmy. 8vo, 28s.

IBERIAN REMINISCENCES. Fifteen Years' Travelling Impressions of Spain and Portugal. With a Map. 2 vols. Demy 8vo, 30s.

GARDNER (J. STARKIE)—

IRONWORK. From the Earliest Time to the End of the Mediæval Period. With 57 Illustrations. Large crown 8vo, 3s.

GASNAULT (PAUL) and GARNIER (ED.)—

FRENCH POTTERY. With Illustrations and Marks. Large crown 8vo, 3s.

GILLMORE (PARKER)—

THE HUNTER'S ARCADIA. With numerous Illustrations. Demy 8vo, 10s. 6d.

GIRL'S LIFE EIGHTY YEARS AGO (A). Selections from the Letters of Eliza Southgate Bowne, with an Introduction by Clarence Cook. Illustrated with Portraits and Views. Crown 4to, 12s.

GLEICHEN (COUNT), Grenadier Guards—

WITH THE CAMEL CORPS UP THE NILE. With numerous Sketches by the Author. Third Edition. Large crown 8vo, 9s.

GORDON (GENERAL)—

LETTERS FROM THE CRIMEA, THE DANUBE, AND ARMENIA. Edited by DEMETRIUS C. BOULGER. Second Edition. Crown 8vo, 5s.

GORST (SIR J. E.), Q.C., M.P.—

AN ELECTION MANUAL. Containing the Parliamentary Elections (Corrupt and Illegal Practices) Act, 1883, with Notes. Third Edition. Crown 8vo, 1s. 6d.

GOWER (A. R.), Royal School of Mines—

PRACTICAL METALLURGY. With Illustrations. Crown 8vo, 3s.

GRESWELL (WILLIAM), M.A., F.R.C.I.—

OUR SOUTH AFRICAN EMPIRE. With Map. 2 vols. Crown 8vo, 21s.

GRIFFIN (SIR LEPEL HENRY), K.C.S.I.—

THE GREAT REPUBLIC. Second Edition. Crown 8vo, 4s. 6d.

GRIFFITHS (MAJOR ARTHUR), H.M. Inspector of Prisons—

SECRETS OF THE PRISON HOUSE. 2 vols. Demy 8vo. With numerous Illustrations. [*In the Press.*

FRENCH REVOLUTIONARY GENERALS. Large crown 8vo, 6s.

GRIFFITHS (MAJOR ARTHUR), H.M. Inspector of Prisons—continued—

CHRONICLES OF NEWGATE. Illustrated. New Edition. Demy 8vo.

MEMORIALS OF MILLBANK: or, Chapters in Prison History. With Illustrations. New Edition. Demy 8vo, 12s.

GRIMBLE (A.)—

SHOOTING AND SALMON FISHING: HINTS AND RECOLLECTIONS. Second Edition. With Illustrations. Demy 8vo. 16s.

HALL (SIDNEY)—

A TRAVELLING ATLAS OF THE ENGLISH COUNTIES. Fifty Maps, coloured. New Edition, including the Railways, corrected up to the present date. Demy 8vo, in roan tuck, 10s. 6d.

HARRIS (FRANK)—

ELDER CONKLIN, and Other Stories. Crown 8vo. [*In the Press.*

HAWKINS (FREDERICK)—

THE FRENCH STAGE IN THE EIGHTEENTH CENTURY. With Portraits. 2 vols. Demy 8vo, 30s.

ANNALS OF THE FRENCH STAGE: FROM ITS ORIGIN TO THE DEATH OF RACINE. 4 Portraits. 2 vols. Demy 8vo, 28s.

HILDEBRAND (HANS), Royal Antiquary of Sweden—

INDUSTRIAL ARTS OF SCANDINAVIA IN THE PAGAN TIME. With numerous Woodcuts. Large crown 8vo, 2s. 6d.

HILL (MISS G.)—

THE PLEASURES AND PROFITS OF OUR LITTLE POULTRY FARM. Small 8vo, 3s.

HOLBEIN—

TWELVE HEADS AFTER HOLBEIN. Selected from Drawings in Her Majesty's Collection at Windsor. Reproduced in Autotype, in portfolio. £1 16s.

HOLMES (GEORGE C. V.), Secretary of the Institution of Naval Architects—

MARINE ENGINES AND BOILERS. With Sixty-nine Woodcuts. Large crown 8vo, 3s.

HOPE (ANDRÉE)—

THE VYVYANS; or, The Murder in the Rue Bellechasse. Crown 8vo, 3s. 6d.

CHRONICLES OF AN OLD INN; or, a Few Words about Gray's Inn. Crown 8vo, 5s.

HOUSSAYE (ARSÈNE)—

BEHIND THE SCENES OF THE COMÉDIE FRANCAISE, AND OTHER RECOLLECTIONS. Translated from the French. Demy 8vo, 14s.

HOVELACQUE (ABEL)—

THE SCIENCE OF LANGUAGE: LINGUISTICS, PHILOLOGY, AND ETYMOLOGY. With Maps. Large crown 8vo, cloth, 3s. 6d.

HOW TO BUY A HORSE. With Hints on Shoeing and Stable Management. By PELAGIUS. Crown 8vo, 1s.

HOZIER (H. M.)—

TURENNE. With Portrait and Two Maps. Large crown 8vo, 4s.

HUDSON (W. H.), C.M.Z.S. Joint Author of "Argentine Ornithology"—

BIRDS IN A VILLAGE. Square crown 8vo, buckram, 7s. 6d.

IDLE DAYS IN PATAGONIA. With numerous Illustrations by J. SMIT and A. HARTLEY. Demy 8vo, 14s.

THE NATURALIST IN LA PLATA. With numerous Illustrations by J. SMIT. Second Edition. Demy 8vo, 16s.

HUEFFER (F.)—

HALF A CENTURY OF MUSIC IN ENGLAND. 1837—1887. Demy 8vo, 8s.

HUGHES (W. R.), F.L.S.—

A WEEK'S TRAMP IN DICKENS-LAND. With upwards of 100 Illustrations by F. G. KITTON, HERBERT RAILTON, and others. Second Edition. Demy 8vo, 7s. 6d.

HUNTLY (MARQUIS OF)—

TRAVELS, SPORTS, AND POLITICS IN THE EAST OF EUROPE. With Illustrations. Large Crown 8vo, 12s.

HUTCHINSON (REV. H. N.)—

EXTINCT MONSTERS. A Popular Account of some of the larger forms of Ancient Animal Life. With numerous Illustrations by J. SMIT and others. Third Thousand, Revised and Enlarged. Demy 8vo, 12s.

INDUSTRIAL ARTS: Historical Sketches. With numerous Illustrations. Large crown 8vo, 3s.

IRVING, HENRY. A Record of Twenty Years at the Lyceum. By PERCY FITZGERALD, M.A., F.S.A. With Portrait. Demy 8vo, 14s.

JACKSON (FRANK G.), Master in the Birmingham Municipal School of Art—

DECORATIVE DESIGN. An Elementary Text Book of Principles and Practice. With numerous Illustrations. Second Edition. Crown 8vo, 7s. 6d.

JAMES (HENRY A.), M.A.—

HANDBOOK TO PERSPECTIVE. Crown 8vo, 2s. 6d.

PERSPECTIVE CHARTS, for use in Class Teaching. 2s.

JARRY (GENERAL)—

OUTPOST DUTY. With TREATISES ON MILITARY RECONNAISSANCE AND ON ROAD-MAKING. By Major-Gen. W. C. E. NAPIER. Third Edition. Crown 8vo, 5s.

JEAFFRESON (J. CORDY), B.A.—

MIDDLESEX COUNTY RECORDS. Indictments, Recognizances, Coroners' Inquisitions Post Mortem, Orders, Memoranda, and Certificates of Convictors of Conventiclers, *temp.* 19 Charles II. to 4 James II. 4 vols. With Portraits, Illustrations, and Facsimiles. Demy 8vo, 25s. each.

JEANS (W. T.)—

CREATORS OF THE AGE OF STEEL. Memoirs of Sir W. Siemens, Sir H. Bessemer, Sir J. Whitworth, Sir J. Brown, and other Inventors. Second Edition. Crown 8vo, 7s. 6d.

JOKAI (MAURICE)—

PRETTY MICHAL. Translated by R. NISBET BAIN. Crown 8vo, 5s.

JONES (CAPTAIN DOUGLAS), R.A.—

NOTES ON MILITARY LAW. Crown 8vo, 4s.

JOPLING (LOUISE)—

HINTS TO AMATEURS. A Handbook on Art. With Diagrams. Crown 8vo, 1s. 6d.

JUNKER (DR. WM.)—

TRAVELS IN AFRICA. Translated from the German by Professor KEANE.

Vol. I. DURING the YEARS 1875 to 1878. Containing 38 Full-page Plates and 125 Illustrations in the Text and Map. Demy 8vo, 21s.

Vol. II. DURING the YEARS 1879 to 1883. Containing numerous Full-page Plates, and Illustrations in the Text and Map Demy 8vo, 21s.

Vol. III. DURING the YEARS 1882 to 1886. Containing numerous Full-page Plates and Illustrations in the Text and Maps. Demy 8vo, 21s.

KELLY (JAMES FITZMAURICE)—

THE LIFE OF MIGUEL DE CERVANTES SAAVEDRA: A Biographical, Literary, and Historical Study, with a Tentative Bibliography from 1585 to 1892, and an Annotated Appendix on the "Canto de Caliope." Demy 8vo, 16s.

KEMPT (ROBERT)—

CONVIVIAL CALEDONIA: Inns and Taverns of Scotland, and some Famous People who have frequented them. By ROBERT KEMPT. Crown 8vo. [*In the Press.*

KENNARD (EDWARD)—

NORWEGIAN SKETCHES: FISHING IN STRANGE WATERS. Illustrated with 30 beautiful Sketches. Second Edition. 14s.

KENNARD (H. MARTYN)—

PHILISTINES AND ISRAELITES: A New Light on the World's History. Demy 4to, 6s.

KING (LIEUT.-COL. COOPER)—

GEORGE WASHINGTON. Large crown 8vo. [*In the Press.*

KLACZKO (M. JULIAN)—

TWO CHANCELLORS: PRINCE GORTCHAKOF AND PRINCE BISMARCK. Translated by MRS. TAIT. New and cheaper Edition, 6s.

LACORDAIRE (PÈRE)—

JESUS CHRIST; GOD; AND GOD AND MAN. Conferences delivered at Notre Dame in Paris. Seventh Thousand. Crown 8vo, 3s. 6d.

LAINÉ (J. M.), R.A.—

ENGLISH COMPOSITION EXERCISES. Crown 8vo, 2s. 6d.

LAING (S.)—

HUMAN ORIGINS: EVIDENCE FROM HISTORY AND SCIENCE. With Illustrations. Tenth Thousand. Demy 8vo, 3s. 6d.

PROBLEMS OF THE FUTURE AND ESSAYS. Tenth Thousand. Demy 8vo, 3s. 6d.

MODERN SCIENCE AND MODERN THOUGHT. Fifteenth Thousand. Demy 8vo, 3s. 6d.

A MODERN ZOROASTRIAN. Seventh Thousand. Demy 8vo, 3s. 6d.

LAMENNAIS (F.)—

WORDS OF A BELIEVER, and THE PAST AND FUTURE OF THE PEOPLE Translated from the French by L. E. MARTINEAU. With a Memoir of Lamennais. Crown 8vo, 4s.

LANDOR (W. S.)—

LIFE AND WORKS. 8 vols.

VOL. 1. Out of print.
VOL. 2. Out of print.
VOL. 3. CONVERSATIONS OF SOVEREIGNS AND STATESMEN, AND FIVE DIALOGUES OF BOCCACCIO AND PETRARCA. Demy 8vo, 14s.
VOL. 4. DIALOGUES OF LITERARY MEN. Demy 8vo, 14s.
VOL. 5. DIALOGUES OF LITERARY MEN (*continued*). FAMOUS WOMEN. LETTERS OF PERICLES AND ASPASIA. And Minor Prose Pieces. Demy 8vo, 14s.
VOL. 6. MISCELLANEOUS CONVERSATIONS. Demy 8vo, 14s.
VOL. 7. GEBIR, ACTS AND SCENES AND HELLENICS. Poems. Demy 8vo, 14s.
VOL. 8. MISCELLANEOUS POEMS AND CRITICISMS ON THEOCRITUS, CATULLUS, AND PETRARCH. Demy 8vo, 14s.

LANIN (E. B.)—

RUSSIAN CHARACTERISTICS. Reprinted, with Revisions, from *The Fortnightly Review*. Demy 8vo, 14s.

LAVELEYE (EMILE DE)—

THE ELEMENTS OF POLITICAL ECONOMY. Translated by W. POLLARD, B.A., St. John's College, Oxford. Crown 8vo, 6s.

LE CONTE (JOSEPH), Professor of Geology and Natural History in the University of California—

EVOLUTION: ITS NATURE, ITS EVIDENCES, AND ITS RELATIONS TO RELIGIOUS THOUGHT. A New and Revised Edition. Crown 8vo, 6s.

LEFÈVRE (ANDRÉ)—

PHILOSOPHY, Historical and Critical. Translated, with an Introduction, by A. W. KEANE, B.A. Large crown 8vo, 3s. 6d.

LE ROUX (H.)—

ACROBATS AND MOUNTEBANKS. With over 200 Illustrations by J. GARNIER. Royal 8vo, 16s.

LEROY-BEAULIEU (ANATOLE)—

PAPACY, SOCIALISM, AND DEMOCRACY. By ANATOLE LEROY-BEAULIEU, Member of the Institute of France. Translated by Professor B. L. O'DONNELL. Crown 8vo, 7s. 6d.

LESLIE (R. C.)

THE SEA BOAT: HOW TO BUILD, RIG, AND SAIL HER. With numerous Illustrations by the Author. Crown 8vo, 4s. 6d.

OLD SEA WINGS, WAYS, AND WORDS, IN THE DAYS OF OAK AND HEMP. With 135 Illustrations by the Author. Demy 8vo, 14s.

LIFE ABOARD A BRITISH PRIVATEER IN THE TIME OF QUEEN ANNE. Being the Journals of Captain Woodes Rogers, Master Mariner. With Notes and Illustrations by ROBERT C. LESLIE. A New and Cheaper Edition. Large crown 8vo. [*In the Press.*

A SEA PAINTER'S LOG. With 12 Full-page Illustrations by the Author. Large crown 8vo, 12s.

LETOURNEAU (DR. CHARLES)—

SOCIOLOGY. Based upon Ethnology. Large crown 8vo, 3s. 6d.

BIOLOGY. With 83 Illustrations. A New Edition. Demy 8vo, 3s. 6d.

LILLY (W. S.)—

ON SHIBBOLETHS. Demy 8vo, 12s.

ON RIGHT AND WRONG. Second Edition. Demy 8vo, 12s.

A CENTURY OF REVOLUTION. Second Edition. Demy 8vo, 12s.

CHAPTERS ON EUROPEAN HISTORY. With an Introductory Dialogue on the Philosophy of History. 2 vols. Demy 8vo, 21s.

ANCIENT RELIGION AND MODERN THOUGHT. Second Edition. Demy 8vo, 12s.

LINEHAM (WILFRID J.), late Professor of Engineering at the School of Science and Art and Technical College, Newcastle-on-Tyne; Member of the Institute of Mechanical Engineers, &c., &c.—

A TEXT-BOOK OF MECHANICAL ENGINEERING. [*In the Press.*

LITTLE (THE REV. CANON KNOX)—

THE CHILD OF STAFFERTON: A Chapter from a Family Chronicle. New Edition. Crown 8vo, boards, 1s.; cloth, 1s. 6d.

THE BROKEN VOW. A Story of Here and Hereafter. New Edition. Crown 8vo, boards, 1s. cloth, 1s. 6d.

LLOYD (COLONEL E.M.), R.E.—
VAUBAN, MONTALEMBERT, CARNOT: ENGINEER STUDIES. With Portraits. Crown 8vo, 5s.

LLOYD (W. W.), late 24th Regiment—
ON ACTIVE SERVICE. Printed in Colours. Oblong 4to, 5s.

SKETCHES OF INDIAN LIFE. Printed in Colours. 4to, 6s.

LONG (JAMES)—
DAIRY FARMING. To which is added a Description of the Chief Continental Systems. With numerous Illustrations. Crown 8vo, 9s.

LOVELL (ARTHUR)—
THE IDEAL OF MAN. Crown 8vo, 3s. 6d.

LOW (WILLIAM)—
TABLE DECORATION. With 19 Full Illustrations. Demy 8vo, 6s.

MACDONALD (A. F.)—
OUR OCEAN RAILWAYS; or, The Rise, Progress, and Development of Ocean Steam Navigation, etc., etc. With Maps and Illustrations. Demy 8vo. [*In the Press.*

McCOAN (J. C.)—
EGYPT UNDER ISMAIL: a Romance of History. With Portrait and Appendix of Official Documents. Crown 8vo, 7s. 6d.

MALLESON (COL. G. B.), C.S.I.—
PRINCE EUGENE OF SAVOY. With Portrait and Maps. Large crown 8vo, 6s.

LOUDON. A Sketch of the Military Life of Gideon Ernest, Freicherr von Loudon. With Portrait and Maps. Large crown 8vo, 4s.

MALLET (ROBERT)—
PRACTICAL MANUAL OF CHEMICAL ASSAYING, as applied to the Manufacture of Iron. By L. L. DE KONINCK and E. DIETZ. Edited, with notes, by ROBERT MALLET. Post 8vo, cloth, 6s.

MALLOCK (W. H.)—
A HUMAN DOCUMENT. Crown 8vo, 3s. 6d.

MARCEAU (SERGENT)—
REMINISCENCES OF A REGICIDE. Edited from the Original MSS. of SERGENT MARCEAU, Member of the Convention, and Administrator of Police in the French Revolution of 1789. By M. C. M. SIMPSON. Demy 8vo, with Illustrations and Portraits, 14s.

MASKELL (ALFRED)—
RUSSIAN ART AND ART OBJECTS IN RUSSIA. A Handbook to the Reproduction of Goldsmiths' Work and other Art Treasures. With Illustrations. Large crown 8vo, 4s. 6d.

MASKELL (WILLIAM)—
IVORIES: ANCIENT AND MEDIÆVAL. With numerous Woodcuts. Large crown 8vo, cloth, 2s. 6d.

HANDBOOK TO THE DYCE AND FORSTER COLLECTIONS. With Illustrations. Large crown 8vo, cloth, 2s. 6d.

MASPÉRO (G.), late Director of Archæology in Egypt, and Member of the Institute of France—
LIFE IN ANCIENT EGYPT AND ASSYRIA. Translated by A. P. MORTON. With 188 Illustrations. Third Thousand. Crown 8vo, 5s.

MORLEY (THE RIGHT HON. JOHN), M.P.—
RICHARD COBDEN'S LIFE AND CORRESPONDENCE. Crown 8vo, with Portrait, 7s. 6d.
Popular Edition. With Portrait. 4to, sewed, 1s. Cloth, 2s.

MURRAY (ANDREW), F.L.S.—
ECONOMIC ENTOMOLOGY. APTERA. With numerous Illustrations. Large crown 8vo, 3s. 6d.

NECKER (MADAME)—
THE SALON OF MADAME NECKER. By VICOMTE D'HAUSSONVILLE. 2 vols. Crown 8vo, 18s.

NESBITT (ALEXANDER)—
GLASS. With numerous Woodcuts. Large crown 8vo cloth, 2s. 6d.

NICOL (DAVID)—
THE POLITICAL LIFE OF OUR TIME. Two vols. Demy 8vo, 24s.

NORMAN (C. B.)—
TONKIN; OR, FRANCE IN THE FAR EAST. With Maps. Demy 8vo, 14s.

O'BYRNE (ROBERT), F.R.G.S.—
THE VICTORIES OF THE BRITISH ARMY IN THE PENINSULA AND THE SOUTH OF FRANCE from 1808 to 1814. An Epitome of Napier's History of the Peninsular War, and Gurwood's Collection of the Duke of Wellington's Despatches. Crown 8vo, 5s.

O'GRADY (STANDISH)—
TORYISM AND THE TORY DEMOCRACY. Crown 8vo, 5s.

OLIVER (PROFESSOR D.), F.R.S., &c.—
ILLUSTRATIONS OF THE PRINCIPAL NATURAL ORDERS OF THE VEGETABLE KINGDOM, PREPARED FOR THE SCIENCE AND ART DEPARTMENT, SOUTH KENSINGTON. With 109 Plates. Oblong 8vo, plain, 16s.; coloured, £1 6s.

OLIVER (E. E.), Under-Secretary to the Public Works Department, Punjaub—
ACROSS THE BORDER; or, PATHAN AND BILOCH. With numerous Illustrations by J. L. KIPLING, C.I.E. Demy 8vo, 14s.

PAPUS—
THE TAROT OF THE BOHEMIANS. The most ancient book in the world. For the exclusive use of the Initiates. An Absolute Key to Occult Science. Translated by A. P. MORTON. With numerous Illustrations. Large crown 8vo, 7s. 6d.

PATERSON (ARTHUR)—
A PARTNER FROM THE WEST. Crown 8vo, 5s.

PAYTON (E. W.)—
ROUND ABOUT NEW ZEALAND. Being Notes from a Journal of Three Years' Wandering in the Antipodes. With Twenty original Illustrations by the Author. Large crown 8vo, 12s.

PELAGIUS—
HOW TO BUY A HORSE. With Hints on Shoeing and Stable Management. Third Thousand. Crown 8vo, 1s.

PERROT (GEORGES) and CHIPIEZ (CHARLES)—

A HISTORY OF ANCIENT ART IN PERSIA. With 254 Illustrations, and 12 Steel and Coloured Plates. Imperial 8vo, 21s.

A HISTORY OF ANCIENT ART IN PHRYGIA—LYDIA, AND CARIA—LYCIA. With 280 Illustrations. Imperial 8vo, 15s.

A HISTORY OF ANCIENT ART IN SARDINIA, JUDÆA, SYRIA, AND ASIA MINOR. With 395 Illustrations. 2 vols. Imperial 8vo, 36s.

A HISTORY OF ANCIENT ART IN PHŒNICIA AND ITS DEPENDENCIES. With 654 Illustrations. 2 vols. Imperial 8vo, 42s.

A HISTORY OF ART IN CHALDÆA AND ASSYRIA. With 452 Illustrations. 2 vols. Imperial 8vo, 42s.

A HISTORY OF ART IN ANCIENT EGYPT. With 600 Illustrations. 2 vols. Imperial 8vo, 42s.

PETERBOROUGH (THE EARL OF)—

THE EARL OF PETERBOROUGH AND MONMOUTH (Charles Mordaunt): A Memoir. By Colonel FRANK RUSSELL, Royal Dragoons. With Illustrations. 2 vols. demy 8vo. 32s.

PILLING (WILLIAM)—

THE EMANCIPATION OF SOUTH AMERICA. Being a Condensed Translation of "The History of San Martin." By GENERAL DON BARTOLOME MITRE, First Constitutional President of the Argentine Republic. Demy 8vo, with Maps, 12s.

LAND TENURE BY REGISTRATION. Second Edition of "Order from Chaos," Revised and Enlarged. Crown 8vo, 5s.

PITT TAYLOR (FRANK)—

THE CANTERBURY TALES. Selections from the Tales of GEOFFREY CHAUCER rendered into Modern English. Crown 8vo, 6s.

POLLEN (J. H.)—

GOLD AND SILVER SMITH'S WORK. With numerous Woodcuts. Large crown 8vo, cloth, 2s. 6d.

ANCIENT AND MODERN FURNITURE AND WOODWORK. With numerous Woodcuts. Large crown 8vo, cloth, 2s. 6d.

POOLE (STANLEY LANE), B.A., M.R.A.S.—

THE ART OF THE SARACENS IN EGYPT. Published for the Committee of Council on Education. With 108 Woodcuts. Large crown 8vo, 4s.

POYNTER (E. J.), R.A.—

TEN LECTURES ON ART. Third Edition. Large crown 8vo, 9s.

PRATT (ROBERT), Headmaster School of Science and Art, Barrow-in-Furness—

SCIOGRAPHY, OR PARALLEL AND RADIAL PROJECTION OF SHADOWS. Being a Course of Exercises for the use of Students in Architectural and Engineering Drawing, and for Candidates preparing for the Examinations in this subject and in Third Grade Perspective conducted by the Science and Art Department. Oblong quarto, 7s. 6d.

PURCELL (the late THEOBALD A.), Principal Medical Officer to the Japanese Government—

A SUBURB OF YEDO. With numerous Illustrations, Crown 8vo, 2s. 6d.

PUSHKIN (A. S.)—

THE QUEEN OF SPADES AND OTHER STORIES. With a Biography. Translated from the Russian by Mrs. SUTHERLAND EDWARDS. Illustrated. Crown 8vo, 3s. 6d.

RADICAL PROGRAMME, THE. From the *Fortnightly Review*, with additions. With a Preface by the RIGHT HON. J. CHAMBERLAIN, M.P. Thirteenth Thousand. Crown 8vo, 2s. 6d.

RAE (W. FRASER)—

AUSTRIAN HEALTH RESORTS THROUGHOUT THE YEAR. A New and Enlarged Edition. Crown 8vo, 5s.

RAMSDEN (LADY GWENDOLEN)—

A BIRTHDAY BOOK. Containing 46 Illustrations from *Original Drawings.* Royal 8vo, 21s.

RANKIN (THOMAS T.), C.E.—

SOLUTIONS TO THE QUESTIONS IN PURE MATHEMATICS (STAGES 1 AND 2) SET AT THE SCIENCE AND ART EXAMINATIONS FROM 1881 TO 1886. Crown 8vo, 2s.

RAPHAEL: his Life, Works, and Times. By EUGENE MUNTZ. Illustrated with about 200 Engravings. A New Edition, revised from the Second French Edition. By W. ARMSTRONG, B.A. Imperial 8vo, 25s.

READE (MRS. R. H.)—

THE GOLDSMITH'S WARD; A Tale of London City in the Fifteenth Century. With 27 Illustrations by W. BOWCHER. Crown 8vo, 6s.

REDGRAVE (GILBERT)—

OUTLINES OF HISTORIC ORNAMENT. Translated from the German. Edited by GILBERT REDGRAVE. With numerous Illustrations. Crown 8vo, 4s.

REDGRAVE (RICHARD), R.A.—

MANUAL OF DESIGN. With Woodcuts. Large crown 8vo, cloth, 2s. 6d.

ELEMENTARY MANUAL OF COLOUR, with a Catechism on Colour. 24mo, cloth, 9d.

REDGRAVE (SAMUEL)—

A DESCRIPTIVE CATALOGUE OF THE HISTORICAL COLLECTION OF WATER-COLOUR PAINTINGS IN THE SOUTH KENSINGTON MUSEUM. With numerous Chromo-lithographs and other Illustrations. Royal 8vo, £1 1s.

RENAN (ERNEST)—

THE FUTURE OF SCIENCE: Ideas of 1848. Demy 8vo, 18s.

HISTORY OF THE PEOPLE OF ISRAEL.

FIRST DIVISION. Till the time of King David. Demy 8vo, 14s.
SECOND DIVISION. From the Reign of David up to the Capture of Samaria. Demy 8vo, 14s.
THIRD DIVISION. From the time of Hezekiah till the Return from Babylon. Demy 8vo, 14s.

RECOLLECTIONS OF MY YOUTH. Translated from the French, and revised by MADAME RENAN. Second Edition. Crown 8vo, 3s. 6d.

RIANO (JUAN F.)—

THE INDUSTRIAL ARTS IN SPAIN. With numerous Woodcuts. Large crown 8vo, cloth, 4s.

RIBTON-TURNER (C. J.)—

A HISTORY OF VAGRANTS AND VAGRANCY AND BEGGARS AND BEGGING. With Illustrations. Demy 8vo, 21s.

ROBERTS (MORLEY)—

IN LOW RELIEF: A Bohemian Transcript. Crown 8vo, 3s 6d.

ROBINSON (JAMES F.)—

BRITISH BEE FARMING. Its Profits and Pleasures. Large crown 8vo, 5s.

ROBINSON (J. C.)—

ITALIAN SCULPTURE OF THE MIDDLE AGES AND PERIOD OF THE REVIVAL OF ART. With 20 Engravings. Royal 8vo, cloth, 7s. 6d.

ROBSON (GEORGE)—

ELEMENTARY BUILDING CONSTRUCTION. Illustrated by a Design for an Entrance Lodge and Gate. 15 Plates. Oblong folio, sewed, 8s.

ROCK (THE VERY REV. CANON), D.D.—

TEXTILE FABRICS. With numerous Woodcuts. Large crown 8vo, cloth, 2s. 6d.

ROLAND (ARTHUR)—

FARMING FOR PLEASURE AND PROFIT. Edited by WILLIAM ABLETT. 8 vols. Crown 8vo, 5s. each.

DAIRY-FARMING, MANAGEMENT OF COWS, etc.
POULTRY-KEEPING.
TREE-PLANTING, FOR ORNAMENTATION OR PROFIT.
STOCK-KEEPING AND CATTLE-REARING.
DRAINAGE OF LAND, IRRIGATION, MANURES, etc.
ROOT-GROWING, HOPS, etc.
MANAGEMENT OF GRASS LANDS, LAYING DOWN GRASS, ARTIFICIAL GRASSES, etc.
MARKET GARDENING, HUSBANDRY FOR FARMERS AND GENERAL CULTIVATORS.

ROOSE (ROBSON), M.D., F.C.S.—

THE WEAR AND TEAR OF LONDON LIFE. Second Edition. Crown 8vo, sewed, 1s.

INFECTION AND DISINFECTION. Crown 8vo, sewed, 6d.

ROOSEVELT (BLANCHE)—

ELISABETH OF ROUMANIA: A Study. With Two Tales from the German of Carmen Sylva, Her Majesty Queen of Roumania. With Two Portraits and Illustration. Demy 8vo, 12s.

ROSS (MRS. JANET)—

EARLY DAYS RECALLED. With Illustrations and Portrait. Crown 8vo, 5s.

RUSSAN (ASHMORE) and BOYLE (FREDERICK)—

THE ORCHID SEEKERS: a Story of Adventure in Borneo. Illustrated by ALFRED HARTLEY. Crown 8vo, 7s. 6d.

SCHAUERMANN (F. L.)—

WOOD-CARVING IN PRACTICE AND THEORY, AS APPLIED TO HOME ARTS. With Notes on Designs having special application to Carved Wood in different styles. Containing 124 Illustrations. Second Edition. Large crown 8vo, 5s.

SCOTT (JOHN)—

THE REPUBLIC AS A FORM OF GOVERNMENT; or, The Evolution of Democracy in America. Crown 8vo, 7s. 6d.

SCOTT-STEVENSON (MRS.)—

ON SUMMER SEAS. Including the Mediterranean, the Ægean, the Ionian, and the Euxine, and a voyage down the Danube. With a Map. Demy 8vo, 16s.

OUR HOME IN CYPRUS. With a Map and Illustrations. Third Edition. Demy 8vo, 14s.

OUR RIDE THROUGH ASIA MINOR. With Map. Demy 8vo, 18s.

SEEMAN (O.)—

THE MYTHOLOGY OF GREECE AND ROME, with Special Reference to its Use in Art. From the German. Edited by G. H. BIANCHI. 64 Illustrations. New Edition. Crown 8vo, 5s.

SETON-KARR (H. W.), F.R.G.S., etc.—

BEAR HUNTING IN THE WHITE MOUNTAINS; or, Alaska and British Columbia Revisited. Illustrated. Large Crown, 4s. 6d.

TEN YEARS' TRAVEL AND SPORT IN FOREIGN Lands; or, Travels in the Eighties. Second Edition, with additions and Portrait of Author. Large crown 8vo, 5s.

SHEPHERD (MAJOR), R.E.—

PRAIRIE EXPERIENCES IN HANDLING CATTLE AND SHEEP. With Illustrations and Map. Demy 8vo, 10s. 6d.

SHIRREFF (EMILY)—

A SHORT SKETCH OF THE LIFE OF FRIEDRICH FROBEL; a New Edition, including Frobel's Letters from Dresden and Leipzig to his Wife, now first Translated into English. Crown 8vo, 2s.

HOME EDUCATION IN RELATION TO THE KINDERGARTEN. Two Lectures. Crown 8vo, 1s. 6d.

SHORE (ARABELLA)—

DANTE FOR BEGINNERS: a Sketch of the "Divina Commedia." With Translations, Biographical and Critical Notices, and Illustrations. With Portrait. Crown 8vo, 6s.

SIMKIN (R.)—

LIFE IN THE ARMY: Every-day Incidents in Camp, Field, and Quarters. Printed in Colours. Oblong 4to, 5s.

SIMMONDS (T. L.)—

ANIMAL PRODUCTS: their Preparation, Commercial Uses, and Value. With numerous Illustrations. Large crown 8vo, 3s. 6d.

SINNETT (A. P.)—

ESOTERIC BUDDHISM. Annotated and enlarged by the Author. Seventh Edition. Crown 8vo, 3s. 6d.

KARMA. A Novel. New Edition. Crown 8vo, 3s.

SMITH (MAJOR R. MURDOCK), R.E.—

PERSIAN ART. With Map and Woodcuts. Second Edition. Large crown 8vo, 2s.

STANLEY (H. M.): HIS LIFE, WORKS, AND EXPLORATIONS. By the Rev. H. W. LITTLE. Demy 8vo, 10s. 6d.

STATHAM (H. H.—)

FORM AND DESIGN IN MUSIC: A Brief Outline of the Æsthetic Conditions of the Art, addressed to General Readers. With Musical Examples. Demy 8vo, 2s. 6d.

MY THOUGHTS ON MUSIC AND MUSICIANS. Illustrated with Frontispiece of the Entrance-front of Handel's Opera House and Musical Examples. Demy 8vo, 18s.

STODDARD (C. A.)—

SPANISH CITIES: with Glimpses of Gibraltar and Tangiers. With 18 Illustrations. Large crown 8vo, 7s. 6d.

ACROSS RUSSIA FROM THE BALTIC TO THE DANUBE. With Numerous Illustrations. Large crown 8vo, 7s. 6d.

STOKES (MARGARET)—

EARLY CHRISTIAN ART IN IRELAND. With 106 Woodcuts. Demy 8vo, 7s. 6d.

Cheaper Edition, Crown 8vo, 4s.

STORIES FROM "BLACK AND WHITE." By GRANT ALLEN, MRS. LYNN LINTON, J. M. BARRIE, MRS. OLIPHANT, W. CLARK RUSSELL, THOMAS HARDY, W. E. NORRIS, and JAMES PAYN. With numerous Illustrations. Crown 8vo, 5s.

STORIES FROM "BLACK AND WHITE." Second Vol. By W. CLARK RUSSELL, W. E. NORRIS, JULIAN HAWTHORNE, MRS. L. B. WALFORD, F. C. PHILIPS, J. M. BARRIE, MRS. ALEXANDER, and WILLIAM WESTALL. With numerous Illustrations. Crown 8vo. [*In the Press.*

STORY (W. W.)—

CASTLE ST. ANGELO. With Illustrations. Crown 8vo, 10s. 6d.

SUTCLIFFE (JOHN)—

THE SCULPTOR AND ART STUDENT'S GUIDE to the Proportions of the Human Form, with Measurements in feet and inches of Full-Grown Figures of Both Sexes and of Various Ages. By Dr. G. SCHADOW Plates reproduced by J. SUTCLIFFE. Oblong folio, 31s. 6d.

SUVÓROFF, LIFE OF. By LIEUT.-COL. SPALDING. Crown 8vo, 6s.

SWIFT: THE MYSTERY OF HIS LIFE AND LOVE. By the Rev. JAMES HAY. Crown 8vo, 6s.

SYMONDS (JOHN ADDINGTON)—

ESSAYS, SPECULATIVE AND SUGGESTIVE.

TANNER (PROFESSOR), F.C.S.—

HOLT CASTLE; or, Threefold Interest in Land. Crown 8vo, 4s. 6d.

JACK'S EDUCATION; OR, HOW HE LEARNT FARMING. Second Edition. Crown 8vo, 3s. 6d.

TAYLOR (EDWARD R.), Head Master of the Birmingham Municipal School of Art—

ELEMENTARY ART TEACHING: An Educational and Technical Guide for Teachers and Learners, including Infant School-work; The Work of the Standards; Freehand; Geometry; Model Drawing; Nature Drawing; Colours; Light and Shade; Modelling and Design. With over 600 Diagrams and Illustrations. Imperial 8vo, 10s. 6d.

TEMPLE (SIR RICHARD), BART., M.P., G.C.S.I.—

COSMOPOLITAN ESSAYS. With Maps. Demy 8vo, 16s.

THOMSON (D. C.)—

THE BARBIZON SCHOOL OF PAINTERS: Corot, Rousseau, Diaz, Millet, and Daubigny. With 130 Illustrations, including 36 Full-Page Plates, of which 18 are Etchings. 4to, cloth, 42s.

THRUPP (GEORGE A.) and FARR (WILLIAM)—

COACH TRIMMING. With 60 Illustrations. Crown 8vo, 2s. 6d.

THRUPP (THE REV. H. W.), M.A.—

AN AID TO THE VISITATION OF THOSE DISTRESSED IN MIND, BODY, OR ESTATE. Crown 8vo, 3s. 6d.

TOPINARD (DR. PAUL)—

ANTHROPOLOGY. With a Preface by Professor PAUL BROCA. With 49 Illustrations. Demy 8vo, 3s. 6d.

TOVEY (LIEUT.-COL., R.E.)—

MARTIAL LAW AND CUSTOM OF WAR; or, Military Law and Jurisdiction in Troublous Times. Crown 8vo, 6s.

TRAHERNE (MAJOR)—

THE HABITS OF THE SALMON. Crown 8vo, 3s. 6d.

TROLLOPE (ANTHONY)—

THE CHRONICLES OF BARSETSHIRE. A Uniform Edition, in 8 vols., large crown 8vo, handsomely printed, each vol. containing Frontispiece. 6s. each.

THE WARDEN and BARCHESTER TOWERS. 2 vols.
DR. THORNE.
FRAMLEY PARSONAGE.
THE SMALL HOUSE AT ALLINGTON. 2 vols.
LAST CHRONICLE OF BARSET. 2 vols.

LIFE OF CICERO. 2 vols. 8vo. £1 4s.

TROUP (J. ROSE)—

WITH STANLEY'S REAR COLUMN. With Portraits and Illustrations. Second Edition. Demy 8vo, 16s.

UNDERHILL (G. F.)—

IN AND OUT OF THE PIGSKIN. With Illustrations. by WALLIS MACKAY. Crown 8vo. 1s.

VERON (EUGENE)—

ÆSTHETICS. Translated by W. H. ARMSTRONG. Large crown 8vo, 3s. 6d.

WALFORD (MAJOR), R.A.—

PARLIAMENTARY GENERALS OF THE GREAT CIVIL WAR. With Maps. Large crown 8vo, 4s.

WALKER (MRS.)—

UNTRODDEN PATHS IN ROUMANIA. With 77 Illustrations. Demy 8vo, 10s. 6d.

EASTERN LIFE AND SCENERY, with Excursions to Asia Minor, Mitylene, Crete, and Roumania. 2 vols., with Frontispiece to each vol. Crown 8vo, 21s.

WALL (A.)—

A PRINCESS OF CHALCO. With Illustrations. Crown 8vo, 6s.

WARD (JAMES)—

ELEMENTARY PRINCIPLES OF ORNAMENT. With 122 Illustrations in the text. 8vo, 5s.

THE PRINCIPLES OF ORNAMENT. Edited by GEORGE AITCHINSON, A.R.A. Crown 8vo, 7s. 6d.

WATSON (JOHN)—

POACHERS AND POACHING. With Frontispiece. Crown 8vo, 7s. 6d.

SKETCHES OF BRITISH SPORTING FISHES. With Frontispiece. Crown 8vo, 3s. 6d.

WEGG-PROSSER (F. R.)—

GALILEO AND HIS JUDGES. Demy 8vo, 5s.

WELLS (HENRY P.)—

CITY BOYS IN THE WOODS; or, A Trapping Venture in Maine. With upwards of 100 Illustrations. Royal 8vo, 9s.

WHITE (WALTER)—

A MONTH IN YORKSHIRE. With a Map. Fifth Edition. Post 8vo, 4s.

A LONDONER'S WALK TO THE LAND'S END, AND A TRIP TO THE SCILLY ISLES. With 4 Maps. Third Edition. Post 8vo, 4s.

WOODGATE (W. B.)—

A MODERN LAYMAN'S FAITH CONCERNING THE CREED AND THE BREED OF THE "THOROUGHBRED MAN." Demy 8vo, 14s.

WORNUM (R. N.)—

ANALYSIS OF ORNAMENT: THE CHARACTERISTICS OF STYLES. An Introduction to the History of Ornamental Art. With many Illustrations. Ninth Edition. Royal 8vo, cloth, 8s.

WORSAAE (J. J. A.)—

INDUSTRIAL ARTS OF DENMARK, FROM THE EARLIEST TIMES TO THE DANISH CONQUEST OF ENGLAND. With Maps and Woodcuts. Large crown 8vo, 3s. 6d.

WOTTON (MABEL E.)—

A GIRL DIPLOMATIST. Crown 8vo, 3s. 6d.

WRIGHTSON (PROF. JOHN), M.R.A.C., F.C.S., &c; President of the College of Agriculture, Downton.

PRINCIPLES OF AGRICULTURAL PRACTICE AS AN INSTRUCTIONAL SUBJECT. With Geological Map. Second Edition. Crown 8vo, 5s.

FALLOW AND FODDER CROPS. Crown 8vo, 5s.

YOUNG OFFICER'S "DON'T"; or, Hints to Youngsters on Joining. 32mo, 1s.

YOUNGE (C. D.)—

PARALLEL LIVES OF ANCIENT AND MODERN HEROES. New Edition. 12mo, cloth, 4s. 6d.

SOUTH KENSINGTON MUSEUM SCIENCE AND ART HANDBOOKS.

Handsomely printed in large crown 8vo.

Published for the Committee of the Council on Education.

IRONWORK: From the Earliest Times to the End of the Mediæval Period. By J. STARKIE GARDNER. With 57 Illustrations. Crown 8vo, 3s.

MARINE ENGINES AND BOILERS. By GEORGE C. V. HOLMES, Secretary of the Institution of Naval Architects, Whitworth Scholar With Sixty-nine Woodcuts. Large crown 8vo, 3s.

EARLY CHRISTIAN ART IN IRELAND. By MARGARET STOKES. With 106 Woodcuts. Crown 8vo, 4s.
A Library Edition, demy 8vo, 7s. 6d.

FOOD GRAINS OF INDIA. By PROF. A. H. CHURCH, M.A., F.C.S , F.I.C. With Numerous Woodcuts. Small 4to, 6s.

THE ART OF THE SARACENS IN EGYPT. By STANLEY LANE POOLE, B.A., M.A.R S. With 108 Woodcuts. Crown 8vo, 4s.

ENGLISH PORCELAIN: A Handbook to the China made in England during the 18th Century. By PROF. A. H. CHURCH, M.A. With numerous Woodcuts. 3s.

RUSSIAN ART AND ART OBJECTS IN RUSSIA: A Handbook to the reproduction of Goldsmiths' work and other Art Treasures from that country in the South Kensington Museum. By ALFRED MASKELL. With Illustrations. 4s. 6d.

FRENCH POTTERY. By PAUL GASNAULT and EDOUARD GARNIER. With Illustrations and Marks. 3s.

ENGLISH EARTHENWARE: A Handbook to the Wares made in England during the 17th and 18th Centuries. By PROF. A. H. CHURCH, M.A. With numerous Woodcuts. 3s.

INDUSTRIAL ARTS OF DENMARK. From the Earliest Times to the Danish Conquest of England. By J. J. A. WORSAAE, Hon. F.S.A., &c. &c. With Map and Woodcuts. 3s. 6d.

INDUSTRIAL ARTS OF SCANDINAVIA IN THE PAGAN TIME. By HANS HILDEBRAND, Royal Antiquary of Sweden. With numerous Woodcuts. 2s. 6d.

PRECIOUS STONES: Considered in their Scientific and Artistic relations. By PROF. A. H. CHURCH, M.A. With a Coloured Plate and Woodcuts. 2s. 6d.

INDUSTRIAL ARTS OF INDIA. By Sir GEORGE C. M. BIRDWOOD, C.S.I., &c. With Map and Woodcuts. Demy 8vo, 14s.

HANDBOOK TO THE DYCE AND FORSTER COLLECTIONS in the South Kensington Museum. With Portraits and Facsimiles. 2s. 6d.

INDUSTRIAL ARTS IN SPAIN. By JUAN F. RIAÑO. With numerous Woodcuts. 4s.

GLASS. By ALEXANDER NESBITT. With numerous Woodcuts. 2s. 6d.

GOLD AND SILVER SMITHS' WORK. By JOHN HUNGERFORD POLLEN, M.A. With numerous Woodcuts. 2s. 6d.

TAPESTRY. By ALFRED DE CHAMPEAUX. With Woodcuts. 2s. 6d.

BRONZES. By C. DRURY E. FORTNUM, F.S.A. With numerous Woodcuts. 2s. 6d.

SOUTH KENSINGTON MUSEUM SCIENCE & ART HANDBOOKS—*Continued.*

PLAIN WORDS ABOUT WATER. By A. H. CHURCH, M.A. Oxon. With Illustrations. Sewed, 6d.

ANIMAL PRODUCTS: their Preparation, Commercial Uses, and Value. By T. L. SIMMONDS. With Illustrations. 3s. 6d.

FOOD: Some Account of its Sources, Constituents, and Uses. By PROFESSOR A. H. CHURCH, M.A. Oxon. New Edition, enlarged. 3s.

ECONOMIC ENTOMOLOGY. By ANDREW MURRAY, F.L.S. APTERA. With Illustrations. 7s. 6d.

JAPANESE POTTERY. Being a Native Report. With an Introduction and Catalogue by A. W. FRANKS, M.A., F.R.S., F.S.A. With Illustrations and Marks. 2s. 6d.

HANDBOOK TO THE SPECIAL LOAN COLLECTION of Scientific Apparatus. 3s.

INDUSTRIAL ARTS: Historical Sketches. With Numerous Illustrations. 3s.

TEXTILE FABRICS. By the Very Rev. DANIEL ROCK, D.D. With numerous Woodcuts. 2s. 6d.

JONES COLLECTION IN THE SOUTH KENSINGTON MUSEUM. With Portrait and Woodcuts. 2s. 6d.

COLLEGE AND CORPORATION PLATE. A Handbook to the Reproductions of Silver Plate in the South Kensington Museum from Celebrated English Collections. By WILFRED JOSEPH CRIPPS, M.A., F.S.A. With Illustrations. 2s. 6d.

IVORIES: ANCIENT AND MEDIÆVAL. By WILLIAM MASKELL. With numerous Woodcuts. 2s. 6d.

ANCIENT AND MODERN FURNITURE AND WOODWORK. By JOHN HUNGERFORD POLLEN, M.A. With numerous Woodcuts. 2s. 6d.

MAIOLICA. By C. DRURY E. FORTNUM, F.S.A. With numerous Woodcuts. 2s. 6d.

THE CHEMISTRY OF FOODS. With Microscopic Illustrations. By JAMES BELL, Ph.D, &c., Principal of the Somerset House Laboratory.
Part I.—Tea, Coffee, Cocoa, Sugar, &c. 2s. 6d.
Part II.—Milk, Butter, Cheese, Cereals, Prepared Starches, &c. 3s.

MUSICAL INSTRUMENTS. By CARL ENGEL. With numerous Woodcuts. 2s. 6d.

MANUAL OF DESIGN. By RICHARD REDGRAVE, R.A. By GILBERT R. REDGRAVE. With Woodcuts. 2s. 6d.

PERSIAN ART. By MAJOR R. MURDOCK SMITH, R.E. With Map and Woodcuts. Second Edition, enlarged. 2s.

CARLYLE'S (THOMAS) WORKS.

THE ASHBURTON EDITION.

New Edition, handsomely printed, containing all the Portraits and Illustrations, in Twenty Volumes, demy 8vo, 8s. each.

THE FRENCH REVOLUTION AND PAST AND PRESENT. 2 vols.

SARTOR RESARTUS; HEROES AND HERO WORSHIP. 1 vol.

LIFE OF JOHN STERLING—LIFE OF SCHILLER. 1 vol.

LATTER-DAY PAMPHLETS—EARLY KINGS OF NORWAY—ESSAY ON THE PORTRAIT OF JOHN KNOX. 1 vol.

LETTERS AND SPEECHES OF OLIVER CROMWELL. 3 vols.

HISTORY OF FREDERICK THE GREAT. 6 vols.

CRITICAL AND MISCELLANEOUS ESSAYS. 3 vols.

TRANSLATIONS FROM THE GERMAN. 3 vols.

LIBRARY EDITION COMPLETE.

Handsomely printed in 34 vols., demy 8vo, £15 3s.

SARTOR RESARTUS. With a Portrait, 7s. 6d.

THE FRENCH REVOLUTION. A History. 3 vols., each 9s.

LIFE of FREDERICK SCHILLER AND EXAMINATION OF HIS WORKS. With Supplement of 1872. Portrait and Plates, 9s.

CRITICAL AND MISCELLANEOUS ESSAYS. With Portrait. 6 vols., each 9s.

ON HEROES, HERO WORSHIP, AND THE HEROIC IN HISTORY. 7s. 6d.

PAST AND PRESENT. 9s.

OLIVER CROMWELL'S LETTERS AND SPEECHES. With Portraits. 5 vols., each 9s.

LATTER-DAY PAMPHLETS. 9s.

LIFE OF JOHN STERLING. With Portrait, 9s.

HISTORY OF FREDERICK THE SECOND. 10 vols., each 9s.

TRANSLATIONS FROM THE GERMAN. 3 vols., each 9s.

EARLY KINGS OF NORWAY; ESSAY ON THE PORTRAITS OF JOHN KNOX; AND GENERAL INDEX. With Portrait Illustrations. 8vo, cloth, 9s.

PEOPLE'S EDITION.

37 *vols., small crown 8vo,* 37*s.; separate vols.,* 1*s. each.*

SARTOR RESARTUS. With Portrait of Thomas Carlyle.

FRENCH REVOLUTION. A History. 3 vols.

OLIVER CROMWELL'S LETTERS AND SPEECHES. 5 vols. With Portrait of Oliver Cromwell.

ON HEROES AND HERO WORSHIP AND THE HEROIC IN HISTORY.

PAST AND PRESENT.

CRITICAL AND MISCELLANEOUS ESSAYS. 7 vols.

THE LIFE OF SCHILLER, AND EXAMINATION OF HIS WORKS. With Portrait.

LATTER-DAY PAMPHLETS.

WILHELM MEISTER. 3 vols.

LIFE OF JOHN STERLING. With Portrait.

HISTORY OF FREDERICK THE GREAT. 10 vols.

TRANSLATIONS FROM MUSÆUS, TIECK, AND RICHTER. 2 vols.

THE EARLY KINGS OF NORWAY; Essay on the Portraits of Knox.

Or in sets, 37 *vols. in* 18, 37*s.*

CARLYLE'S (THOMAS) WORKS.—*Continued.*

A RE-ISSUE OF THE WORKS OF THOMAS CARLYLE.

Price 2s. 6d. a Volume.

This Edition will include the whole of his writings and translations, together with the Portraits and Maps, and will be

COMPLETED IN 20 CROWN 8vo VOLUMES,

strongly bound in cloth.

SARTOR RESARTUS AND LATTER-DAY PAMPHLETS. With a Portrait of Thomas Carlyle. [*Ready.*

PAST AND PRESENT AND ON HEROES AND HERO WORSHIP. [*Ready.*

LIFE OF JOHN STERLING AND LIFE OF SCHILLER. [*Ready.*

To be followed at intervals by

CRITICAL AND MISCELLANEOUS ESSAYS, EARLY KINGS OF NORWAY AND ESSAY ON THE PORTRAITS OF KNOX. [*In four volumes. Vol. I. ready.*

FRENCH REVOLUTION; a History. [*In two volumes.*

OLIVER CROMWELL'S LETTERS AND SPEECHES, with Portrait of Oliver Cromwell. [*In three volumes.*

HISTORY OF FREDERICK THE GREAT. [*In five volumes.*

WILHELM MEISTER. [*In two volumes.*

TRANSLATIONS FROM MUSÆUS, TIECK AND RICHTER. [*In one volume.*

CHEAP ISSUE.

THE FRENCH REVOLUTION. Complete in 1 vol. With Portrait. Crown 8vo, 2s.

SARTOR RESARTUS, HEROES AND HERO WORSHIP, PAST AND PRESENT, AND CHARTISM. Complete in 1 vol. Crown 8vo, 2s.

OLIVER CROMWELL'S LETTERS AND SPEECHES. Crown 8vo, 2s. 6d.

CRITICAL AND MISCELLANEOUS ESSAYS. 2 vols. 4s.

WILHELM MEISTER. 1 vol., 2s.

LIVES OF SCHILLER AND STERLING. With Portraits. 1 vol. 2s.

SIXPENNY EDITION.

4to, sewed.

SARTOR RESARTUS. Eightieth Thousand.

HEROES AND HERO WORSHIP.

ESSAYS: BURNS, JOHNSON, SCOTT, THE DIAMOND NECKLACE.

The above in 1 vol., cloth, 2s. 6d.

DICKENS'S (CHARLES) WORKS.

ORIGINAL EDITIONS.

In demy 8vo.

THE MYSTERY OF EDWIN DROOD. With Illustrations by S. L. Fildes, and a Portrait engraved by Baker. Cloth, 7s. 6d.

OUR MUTUAL FRIEND. With Forty Illustrations by Marcus Stone. Cloth, £1 1s.

THE PICKWICK PAPERS. With Forty-three Illustrations by Seymour and Phiz. Cloth, £1 1s.

NICHOLAS NICKLEBY. With Forty Illustrations by Phiz. Cloth, £1 1s.

SKETCHES BY "BOZ." With Forty Illustrations by George Cruikshank. Cloth, £1 1s.

MARTIN CHUZZLEWIT. With Forty Illustrations by Phiz. Cloth, £1 1s.

DOMBEY AND SON. With Forty Illustrations by Phiz. Cloth, £1 1s.

DAVID COPPERFIELD. With Forty Illustrations by Phiz. Cloth, £1 1s.

BLEAK HOUSE. With Forty Illustrations by Phiz. Cloth, £1 1s.

LITTLE DORRIT. With Forty Illustrations by Phiz. Cloth, £1 1s.

THE OLD CURIOSITY SHOP. With Seventy-five Illustrations by George Cattermole and H. K. Browne. A New Edition. Uniform with the other volumes, £1 1s.

BARNABY RUDGE: a Tale of the Riots of 'Eighty. With Seventy-eight Illustrations by George Cattermole and H. K. Browne. Uniform with the other volumes, £1 1s.

CHRISTMAS BOOKS: Containing—The Christmas Carol; The Cricket on the Hearth; The Chimes; The Battle of Life; The Haunted House. With all the original Illustrations. Cloth, 12s.

OLIVER TWIST and TALE OF TWO CITIES. In one volume. Cloth, £1 1s.

OLIVER TWIST. Separately. With Twenty-four Illustrations by George Cruikshank. Cloth, 11s.

A TALE OF TWO CITIES. Separately. With Sixteen Illustrations by Phiz. Cloth, 9s.

*** *The remainder of Dickens's Works were not originally printed in demy 8vo.*

DICKENS'S (CHARLES) WORKS.—*Continued.*

LIBRARY EDITION.

In post 8vo. With the Original Illustrations, 30 *vols., cloth,* £12.

				s.	d.
PICKWICK PAPERS	43	Illustrns.,	2 vols.	16	0
NICHOLAS NICKLEBY	39	,,	2 vols.	16	0
MARTIN CHUZZLEWIT	40	,,	2 vols.	16	0
OLD CURIOSITY SHOP & REPRINTED PIECES	36	,,	2 vols.	16	0
BARNABY RUDGE and HARD TIMES	36	,,	2 vols.	16	0
BLEAK HOUSE	40	,,	2 vols.	16	0
LITTLE DORRIT	40	,,	2 vols.	16	0
DOMBEY AND SON	38	,,	2 vols.	16	0
DAVID COPPERFIELD	38	,,	2 vols.	16	0
OUR MUTUAL FRIEND	40	,,	2 vols.	16	0
SKETCHES BY "BOZ"	39	,,	1 vol.	8	0
OLIVER TWIST	24	,,	1 vol.	8	0
CHRISTMAS BOOKS	17	,,	1 vol.	8	0
A TALE OF TWO CITIES	16	,,	1 vol.	8	0
GREAT EXPECTATIONS	8	,,	1 vol.	8	0
PICTURES FROM ITALY & AMERICAN NOTES	8	,,	1 vol.	8	0
UNCOMMERCIAL TRAVELLER	8	,,	1 vol.	8	0
CHILD'S HISTORY OF ENGLAND	8	,,	1 vol.	8	0
EDWIN DROOD and MISCELLANIES	12	,,	1 vol.	8	0
CHRISTMAS STORIES from "Household Words," &c.	14	,,	1 vol.	8	0

Uniform with the above, 10*s.* 6*d.*

THE LIFE OF CHARLES DICKENS. By JOHN FORSTER. With Illustrations.

A NEW EDITION OF ABOVE, WITH THE ORIGINAL ILLUSTRATIONS, IN LARGE CROWN 8vo, 30 VOLS. IN SETS ONLY.

THE "CHARLES DICKENS" EDITION.

In Crown 8vo. In 21 *vols., cloth, with Illustrations,* £3 16*s.*

			s.	d.
PICKWICK PAPERS	8	Illustrations ...	4	0
MARTIN CHUZZLEWIT	8	,, ...	4	0
DOMBEY AND SON	8	,, ...	4	0
NICHOLAS NICKLEBY	8	,, ...	4	0
DAVID COPPERFIELD	8	,, ...	4	0
BLEAK HOUSE	8	,, ...	4	0
LITTLE DORRIT	8	,, ...	4	0
OUR MUTUAL FRIEND...	8	,, ...	4	0
BARNABY RUDGE...	8	,, ...	3	6
OLD CURIOSITY SHOP	8	,, ...	3	6
A CHILD'S HISTORY OF ENGLAND	4	,, ...	3	6
EDWIN DROOD and OTHER STORIES	8	,, ...	3	6
CHRISTMAS STORIES, from "Household Words" ...	8	,, ...	3	6
SKETCHES BY "BOZ"	8	,, ...	3	6
AMERICAN NOTES and REPRINTED PIECES ...	8	,, ...	3	6
CHRISTMAS BOOKS	8	,, ...	3	6
OLIVER TWIST	8	,, ...	3	6
GREAT EXPECTATIONS...	8	,, ...	3	6
TALE OF TWO CITIES	8	,, ...	3	0
HARD TIMES and PICTURES FROM ITALY ...	8	,, ...	3	0
UNCOMMERCIAL TRAVELLER	4	,, ...	3	0

Uniform with the above.

		s.	d.
THE LIFE OF CHARLES DICKENS. Numerous Illustrations.	2 vols.	7	0
THE LETTERS OF CHARLES DICKENS	2 vols.	7	0

DICKENS'S (CHARLES) WORKS.—*Continued.*

THE ILLUSTRATED LIBRARY EDITION.

(WITH LIFE.)

Complete in 32 Volumes. Demy 8vo, 10s. each; or set, £16.

This Edition is printed on a finer paper and in a larger type than has been employed in any previous edition. The type has been cast especially for it, and the page is of a size to admit of the introduction of all the original illustrations.

No such attractive issue has been made of the writings of Mr. Dickens, which, various as have been the forms of publication adapted to the demands of an ever widely-increasing popularity, have never yet been worthily presented in a really handsome library form.

The collection comprises all the minor writings it was Mr. Dickens's wish to preserve.

SKETCHES BY "BOZ." With 40 Illustrations by George Cruikshank.

PICKWICK PAPERS. 2 vols. With 42 Illustrations by Phiz.

OLIVER TWIST. With 24 Illustrations by Cruikshank.

NICHOLAS NICKLEBY. 2 vols. With 40 Illustrations by Phiz.

OLD CURIOSITY SHOP and REPRINTED PIECES. 2 vols. With Illustrations by Cattermole, &c.

BARNABY RUDGE and HARD TIMES. 2 vols. With Illustrations by Cattermole, &c.

MARTIN CHUZZLEWIT. 2 vols. With 40 Illustrations by Phiz.

AMERICAN NOTES and PICTURES FROM ITALY. 1 vol. With 8 Illustrations.

DOMBEY AND SON. 2 vols. With 40 Illustrations by Phiz.

DAVID COPPERFIELD. 2 vols. With 40 Illustrations by Phiz.

BLEAK HOUSE. 2 vols. With 40 Illustrations by Phiz.

LITTLE DORRIT. 2 vols. With 40 Illustrations by Phiz.

A TALE OF TWO CITIES. With 16 Illustrations by Phiz.

THE UNCOMMERCIAL TRAVELLER. With 8 Illustrations by Marcus Stone.

GREAT EXPECTATIONS. With 8 Illustrations by Marcus Stone.

OUR MUTUAL FRIEND. 2 vols. With 40 Illustrations by Marcus Stone.

CHRISTMAS BOOKS. With 17 Illustrations by Sir Edwin Landseer, R.A., Maclise, R.A., &c. &c.

HISTORY OF ENGLAND. With 8 Illustrations by Marcus Stone.

CHRISTMAS STORIES. (From "Household Words" and "All the Year Round.") With 14 Illustrations.

EDWIN DROOD AND OTHER STORIES. With 12 Illustrations by S. L. Fildes.

LIFE OF CHARLES DICKENS. By John Forster. With Portraits. 2 vols. (Not separate.)

DICKENS'S (CHARLES) WORKS.—*Continued.*

HOUSEHOLD EDITION.

(WITH LIFE.)

In 22 Volumes. Crown 4to, cloth, £4 8s. 6d.

MARTIN CHUZZLEWIT, with 59 Illustrations, 5s.
DAVID COPPERFIELD, with 60 Illustrations and a Portrait, 5s.
BLEAK HOUSE, with 61 Illustrations, 5s.
LITTLE DORRIT, with 58 Illustrations, 5s.
PICKWICK PAPERS, with 56 Illustrations, 5s.
OUR MUTUAL FRIEND, with 58 Illustrations, 5s.
NICHOLAS NICKLEBY, with 59 Illustrations, 5s.
DOMBEY AND SON, with 61 Illustrations, 5s.
EDWIN DROOD; REPRINTED PIECES; and other Stories, with 30 Illustrations, 5s.
THE LIFE OF DICKENS. By John Forster. With 40 Illustrations, 5s.
BARNABY RUDGE, with 46 Illustrations, 4s.
OLD CURIOSITY SHOP, with 32 Illustrations, 4s.
CHRISTMAS STORIES, with 23 Illustrations, 4s.
OLIVER TWIST, with 28 Illustrations, 3s.
GREAT EXPECTATIONS, with 26 Illustrations, 3s.
SKETCHES BY "BOZ," with 36 Illustrations, 3s.
UNCOMMERCIAL TRAVELLER, with 26 Illustrations, 3s.
CHRISTMAS BOOKS, with 28 Illustrations, 3s.
THE HISTORY OF ENGLAND, with 15 Illustrations, 3s.
AMERICAN NOTES and PICTURES FROM ITALY, with 18 Illustrations, 3s.
A TALE OF TWO CITIES, with 25 Illustrations, 3s.
HARD TIMES, with 20 Illustrations, 2s. 6d.

CHARLES DICKENS'S CHRISTMAS BOOKS.

REPRINTED FROM THE ORIGINAL PLATES.
Illustrated by John Leech, D. Maclise, R.A., R. Doyle, C. Stanfield, R.A., etc.

Fcap. cloth, 1s. each. Complete in a case, 5s.

A CHRISTMAS CAROL IN PROSE.
THE CHIMES: A Goblin Story.
THE CRICKET ON THE HEARTH: A Fairy Tale of Home.
THE BATTLE OF LIFE. A Love Story.
THE HAUNTED MAN AND THE GHOST'S STORY.

SIXPENNY REPRINTS of DICKENS'S WORKS.

DAVID COPPERFIELD. With numerous Illustrations by Fred Barnard. Medium 8vo.
OLIVER TWIST. With 28 Illustrations by J. Mahoney. Medium 8vo.
READINGS FROM THE WORKS OF CHARLES DICKENS. As selected and read by himself and now published for the first time. Illustrated.
A CHRISTMAS CAROL, AND THE HAUNTED MAN. Illustrated.
THE CHIMES: A Goblin Story, and THE CRICKET ON THE HEARTH. Illustrated.
THE BATTLE OF LIFE: A Love Story, HUNTED DOWN, and A HOLIDAY ROMANCE. Illustrated.

DICKENS'S (CHARLES) WORKS.—*Continued.*

THE CROWN EDITION,

Complete in 17 vols. Containing ALL THE ORIGINAL ILLUSTRATIONS.

And the Letterpress is printed from Type expressly cast for this Edition.

LARGE CROWN OCTAVO.

PRICE FIVE SHILLINGS EACH.

1.—**THE PICKWICK PAPERS.** With Forty-three Illustrations by SEYMOUR and PHIZ.

2.—**NICHOLAS NICKLEBY.** With Forty Illustrations by PHIZ.

3.—**DOMBEY AND SON.** With Forty Illustrations by PHIZ.

4.—**DAVID COPPERFIELD.** With Forty Illustrations by PHIZ.

5.—**SKETCHES BY "BOZ."** With Forty Illustrations by GEO. CRUIKSHANK.

6.—**MARTIN CHUZZLEWIT.** With Forty Illustrations by PHIZ.

7.—**THE OLD CURIOSITY SHOP.** With Seventy-five Illustrations by GEORGE CATTERMOLE and H. K. BROWNE.

8.—**BARNABY RUDGE:** a Tale of the Riots of 'Eighty. With Seventy-eight Illustrations by GEORGE CATTERMOLE and H. K. BROWNE.

9.—**OLIVER TWIST** and **A TALE OF TWO CITIES.** With Twenty-four Illustrations by CRUIKSHANK and Sixteen by *PHIZ.*

10.—**BLEAK HOUSE.** With Forty Illustrations by PHIZ.

11.—**LITTLE DORRIT.** With Forty Illustrations by PHIZ.

12.—**OUR MUTUAL FRIEND.** With Forty Illustrations by MARCUS STONE.

13.—**AMERICAN NOTES; PICTURES FROM ITALY; and A CHILD'S HISTORY OF ENGLAND.** With Sixteen Illustrations by MARCUS STONE.

14.—**CHRISTMAS BOOKS and HARD TIMES.** With Illustrations by LANDSEER, MACLISE, STANFIELD, LEECH, DOYLE, F. WALKER, etc.

15.—**CHRISTMAS STORIES AND OTHER STORIES,** including **HUMPHREY'S CLOCK.** With Illustrations by DALZIEL, CHARLES GREEN, MAHONEY, PHIZ, CATTERMOLE, etc.

16.—**GREAT EXPECTATIONS. UNCOMMERCIAL TRAVELLER.** With Sixteen Illustrations by MARCUS STONE.

17.—**EDWIN DROOD and REPRINTED PIECES.** With Sixteen Illustrations by LUKE FILDES and F. WALKER.

Uniform with the above.

THE LIFE OF CHARLES DICKENS. By JOHN FORSTER. With Portraits and Illustrations. Will be added at the request of numerous Subscribers.

THE DICKENS DICTIONARY. A Key to the Characters and Principal Incidents in the Tales of Charles Dickens. By GILBERT PIERCE, with Additions by WILLIAM A. WHEELER.

THE LAZY TOUR OF TWO IDLE APPRENTICES; NO THOROUGHFARE; THE PERILS OF CERTAIN ENGLISH PRISONERS. By CHARLES DICKENS and WILKIE COLLINS. With Illustrations. Crown 8vo, 5s.

*** These Stories are now reprinted in complete form for the first time.

DICKENS'S (CHARLES) WORKS.—*Continued.*

THE CABINET EDITION.

In 32 vols. small fcap. 8vo, Marble Paper Sides, Cloth Backs, with uncut edges, price Eighteenpence each.

Each Volume contains Eight Illustrations reproduced from the Originals.

In Sets only, bound in blue and red cloth, with cut edges, £2 8s.

CHRISTMAS BOOKS.
MARTIN CHUZZLEWIT, 2 vols.
DAVID COPPERFIELD, 2 vols.
OLIVER TWIST.
GREAT EXPECTATIONS.
NICHOLAS NICKLEBY, 2 vols.
SKETCHES BY "BOZ."
CHRISTMAS STORIES.
THE PICKWICK PAPERS, 2 vols.
BARNABY RUDGE, 2 vols.
BLEAK HOUSE, 2 vols.
AMERICAN NOTES AND PICTURES FROM ITALY.
EDWIN DROOD; AND OTHER STORIES.
THE OLD CURIOSITY SHOP, 2 vols.
A CHILD'S HISTORY OF ENGLAND.
DOMBEY AND SON, 2 vols.
A TALE OF TWO CITIES.
LITTLE DORRIT, 2 vols.
MUTUAL FRIEND, 2 vols.
HARD TIMES.
UNCOMMERCIAL TRAVELLER
REPRINTED PIECES.

A NEW ISSUE OF CHARLES DICKENS'S WORKS.

THE HALF CROWN EDITION.

PRINTED FROM THE EDITION THAT WAS CAREFULLY CORRECTED BY THE AUTHOR IN 1867 AND 1868.

Containing the whole of the original Illustrations. Crown 8vo, 2s. 6d. each.

THE PICKWICK PAPERS. With 43 Illustrations by SEYMOUR and PHIZ.
BARNABY RUDGE: A Tale of the Riots of 'Eighty. With 76 Illustrations by GEORGE CATTERMOLE and H. K. BROWNE.
OLIVER TWIST. With 24 Illustrations by CRUIKSHANK.
THE OLD CURIOSITY SHOP. With 75 Illustrations by GEORGE CATTERMOLE and H. K. BROWNE.
DAVID COPPERFIELD. With 40 Illustrations by PHIZ.
NICHOLAS NICKLEBY. With 40 Illustrations by PHIZ.
MARTIN CHUZZLEWIT. With 40 Illustrations by PHIZ.
DOMBEY AND SON. With 40 Illustrations by PHIZ.
SKETCHES BY "BOZ." With 40 Illustrations by GEORGE CRUIKSHANK.
CHRISTMAS BOOKS. With 64 Illustrations by LANDSEER, DOYLE, MACLISE, LEECH, etc.
BLEAK HOUSE. With 40 Illustrations by PHIZ.
LITTLE DORRIT. With 40 Illustrations by PHIZ.
CHRISTMAS STORIES, from "Household Words" With 14 Illustrations by DALZIEL, GREEN, MAHONEY, etc.
AMERICAN NOTES and REPRINTED PIECES. With 8 Illustrations by MARCUS STONE and F. WALKER.
HARD TIMES and PICTURES FROM ITALY. With 8 Illustrations by F. WALKER and MARCUS STONE.
A CHILD'S HISTORY OF ENGLAND. With 8 Illustrations by MARCUS STONE.
GREAT EXPECTATIONS. With 8 Illustrations by MARCUS STONE.
TALE OF TWO CITIES With 16 Illustrations by PHIZ. [*August.*
UNCOMMERCIAL TRAVELLER. With 8 Illustrations by MARCUS STONE. [*September.*
OUR MUTUAL FRIEND. With 40 Illustrations by MARCUS STONE. [*October.*
EDWIN DROOD and OTHER STORIES. With 12 Illustrations by LUKE FILDES. [*November.*

DICKENS'S (CHARLES) WORKS.—*Continued.*

THE PICTORIAL EDITION,

CONTAINING

UPWARDS OF NINE HUNDRED ENGRAVINGS.

Complete in 17 vols. Royal 8vo, 3s. 6d. each.

DOMBEY AND SON. With 62 Illustrations by F. BARNARD. 3s. 6d.

DAVID COPPERFIELD. With 61 Illustrations by F. BARNARD. 3s. 6d.

NICHOLAS NICKLEBY. With 59 Illustrations by F. BARNARD. 3s. 6d.

BARNABY RUDGE. With 46 Illustrations by F. BARNARD. 3s. 6d.

OLD CURIOSITY SHOP. With 39 Illustrations by CHARLES GREEN. 3s. 6d.

MARTIN CHUZZLEWIT. With 59 Illustrations by F. BARNARD. 3s. 6d.

OLIVER TWIST and A TALE OF TWO CITIES. With 53 Illustrations by J. MAHONEY and F. BARNARD. 3s. 6d.

OUR MUTUAL FRIEND. With 58 Illustrations by J. MAHONEY. 3s. 6d.

BLEAK HOUSE. With 61 Illustrations by F. BARNARD. 3s. 6d.

PICKWICK PAPERS. With 57 Illustrations by PHIZ. 3s. 6d.

LITTLE DORRIT. With 58 Illustrations by J. MAHONEY. 3s. 6d.

GREAT EXPECTATIONS & HARD TIMES. With 50 Illustrations by J. A. FRASER and H. FRENCH. 3s. 6d.

AMERICAN NOTES, PICTURES FROM ITALY, and A CHILD'S HISTORY OF ENGLAND. With 33 Illustrations by FROST, GORDON, THOMSON, and RALSTON. 3s. 6d.

SKETCHES BY "BOZ" and CHRISTMAS BOOKS. With 62 Illustrations by F. BARNARD. 3s. 6d.

CHRISTMAS STORIES and UNCOMMERCIAL TRAVELLER. With 49 Illustrations by E. G. DALZIEL. 3s. 6d.

EDWIN DROOD, REPRINTED PIECES, AND OTHER STORIES. With 30 Illustrations by L FILDES, E. G. DALZIEL, and F. BARNARD. 3s. 6d.

THE LIFE OF CHARLES DICKENS. By JOHN FORSTER. With 40 Illustrations by F. BARNARD and others. 3s. 6d.

THE TWO SHILLING EDITION.

Each Volume contains a Frontispiece. Crown 8vo, *2s.*

The Volumes now ready are—

DOMBEY AND SON.
MARTIN CHUZZLEWIT.
THE PICKWICK PAPERS.
BLEAK HOUSE.
OLD CURIOSITY SHOP.
BARNABY RUDGE.
DAVID COPPERFIELD.
NICHOLAS NICKLEBY.
OUR MUTUAL FRIEND.
CHRISTMAS BOOKS.
OLIVER TWIST.

MR. DICKENS'S READINGS.

Fcap. 8vo, sewed.

CHRISTMAS CAROL IN PROSE. 1s.

CRICKET ON THE HEARTH. 1s.

CHIMES: A GOBLIN STORY. 1s.

STORY OF LITTLE DOMBEY. 1s.

POOR TRAVELLER, BOOTS AT THE HOLLY-TREE INN, and MRS. GAMP. 1s.

Works published at Three Shillings & Sixpence each.

THE VYVYANS; or, The Murder in the Rue Bellechasse. By ANDREE HOPE. Crown 8vo.

IN LOW RELIEF: A Bohemian Transcript. By MORLEY ROBERTS. Crown 8vo.

CROSS CURRENTS: A Novel. By MARY A. DICKENS. New Edition in one volume.

SUSPECTED. By LOUISA STRATENUS.

A GIRL DIPLOMATIST: a Novel. By MABEL E. WOTTON.

AUSTRALIAN LIFE. By FRANCIS ADAMS.

THE SECRET OF THE PRINCESS: a Tale of Country, Camp, Court, Convict, and Cloister Life in Russia. By MRS. SUTHERLAND EDWARDS.

BEYOND THE SEAS; being the surprising Adventures and ingenious Opinions of Ralph, Lord St. Keyne, told by his kinsman, Humphrey St. Keyne. By OSWALD CRAWFURD. Second Edition.

THE STORY OF HELEN DAVENANT. By VIOLET FANE.

A DEPUTY PROVIDENCE. By HENRY MURRAY.

MY "HANSOM" LAYS: Original Verses, Imitations, and Paraphrases. By W. BEATTY-KINGSTON.

SPORT: Fox Hunting, Salmon Fishing, Covert Shooting, Deer Stalking. By the late W. BROMLEY DAVENPORT, M.P. With Illustrations by GENERAL CREALOCK, C.B.

LOG-BOOK OF A FISHERMAN AND ZOOLOGIST. By FRANK BUCKLAND. With Illustrations. Fifth Thousand.

THE HABITS OF THE SALMON. By MAJOR TRAHERNE.

BRITISH SPORTING FISHES. By JOHN WATSON.

ENGLAND: ITS PEOPLE, POLITY, AND PURSUITS. By T. H. S. ESCOTT. New and Revised Edition.

JESUS CHRIST; GOD; AND GOD AND MAN. Conferences delivered at Notre Dame in Paris. By PÈRE LACORDAIRE. Seventh Thousand.

RECOLLECTIONS OF MY YOUTH. By ERNEST RENAN. Translated from the French and revised by Madame RENAN. Second Edition.

HUMAN ORIGINS: EVIDENCE FROM HISTORY AND SCIENCE. By SAMUEL LAING. With Illustrations. Tenth Thousand.

PROBLEMS OF THE FUTURE AND ESSAYS. By SAMUEL LAING. Tenth Thousand.

MODERN SCIENCE AND MODERN THOUGHT. By SAMUEL LAING. Fifteenth Thousand.

A MODERN ZOROASTRIAN. By SAMUEL LAING. Seventh Thousand.

THE SCIENCE OF LANGUAGE: LINGUISTICS, PHILOLOGY, AND ETYMOLOGY. By ABEL HOVELACQUE. With Maps.

SOCIOLOGY. Based upon Ethnology. By DR. CHARLES LETOURNEAU.

BIOLOGY. By DR. CHARLES LETOURNEAU. With 83 Illustrations.

PHILOSOPHY, Historical and Critical. By ANDRÉ LEFÈVRE.

ANTHROPOLOGY. By DR. PAUL TOPINARD. With a Preface by PROFESSOR PAUL BROCA. With 49 Illustrations.

ÆSTHETICS. By EUGENE VERON.

ANIMAL PRODUCTS: Their Preparation, Commercial Uses, and Value. By T. L. SIMMONDS. With Illustrations.

ECONOMIC ENTOMOLOGY. By ANDREW MURRAY, F.L.S. APTERA. With Illustrations.

CHARLES DICKENS AND EVANS, PRINTERS, CRYSTAL PALACE PRESS.

Printed in Dunstable, United Kingdom

91220110R00271